Springer Theses

Recognizing Outstanding Ph.D. Research

Aims and Scope

The series "Springer Theses" brings together a selection of the very best Ph.D. theses from around the world and across the physical sciences. Nominated and endorsed by two recognized specialists, each published volume has been selected for its scientific excellence and the high impact of its contents for the pertinent field of research. For greater accessibility to non-specialists, the published versions include an extended introduction, as well as a foreword by the student's supervisor explaining the special relevance of the work for the field. As a whole, the series will provide a valuable resource both for newcomers to the research fields described, and for other scientists seeking detailed background information on special questions. Finally, it provides an accredited documentation of the valuable contributions made by today's younger generation of scientists.

Theses may be nominated for publication in this series by heads of department at internationally leading universities or institutes and should fulfill all of the following criteria

- They must be written in good English.
- The topic should fall within the confines of Chemistry, Physics, Earth Sciences, Engineering and related interdisciplinary fields such as Materials, Nanoscience, Chemical Engineering, Complex Systems and Biophysics.
- The work reported in the thesis must represent a significant scientific advance.
- If the thesis includes previously published material, permission to reproduce this must be gained from the respective copyright holder (a maximum 30% of the thesis should be a verbatim reproduction from the author's previous publications).
- They must have been examined and passed during the 12 months prior to nomination.
- Each thesis should include a foreword by the supervisor outlining the significance of its content.
- The theses should have a clearly defined structure including an introduction accessible to new PhD students and scientists not expert in the relevant field.

Indexed by zbMATH.

More information about this series at https://link.springer.com/bookseries/8790

Thorben Quast

Beam Test Calorimeter Prototypes for the CMS Calorimeter Endcap Upgrade

Qualification, Performance Validation and Fast Generative Modelling

Doctoral Thesis accepted by
the Rheinisch Westfälische Technische Hochschule,
Aachen, Germany

Springer

Author
Dr. Thorben Quast
CERN
Geneva, Switzerland

Supervisors
Prof. Dr. Martin Erdmann
III Physikalisches Institut A
Rheinisch Westfälische Technische Hochschule
Aachen, Germany

Dr. Eva Sicking
CERN, EP-DT-TP Espl. des Particules 1
Meyrin, Switzerland

Prof. Dr. André David Tinoco Mendes
CERN, EP-CMG-DS Espl. des Particules 1
Meyrin, Switzerland

ISSN 2190-5053 ISSN 2190-5061 (electronic)
Springer Theses
ISBN 978-3-030-90201-8 ISBN 978-3-030-90202-5 (eBook)
https://doi.org/10.1007/978-3-030-90202-5

This Springer imprint is published by the registered company Springer Nature Switzerland AG
The registered company address is: Gewerbestrasse 11, 6330 Cham, Switzerland

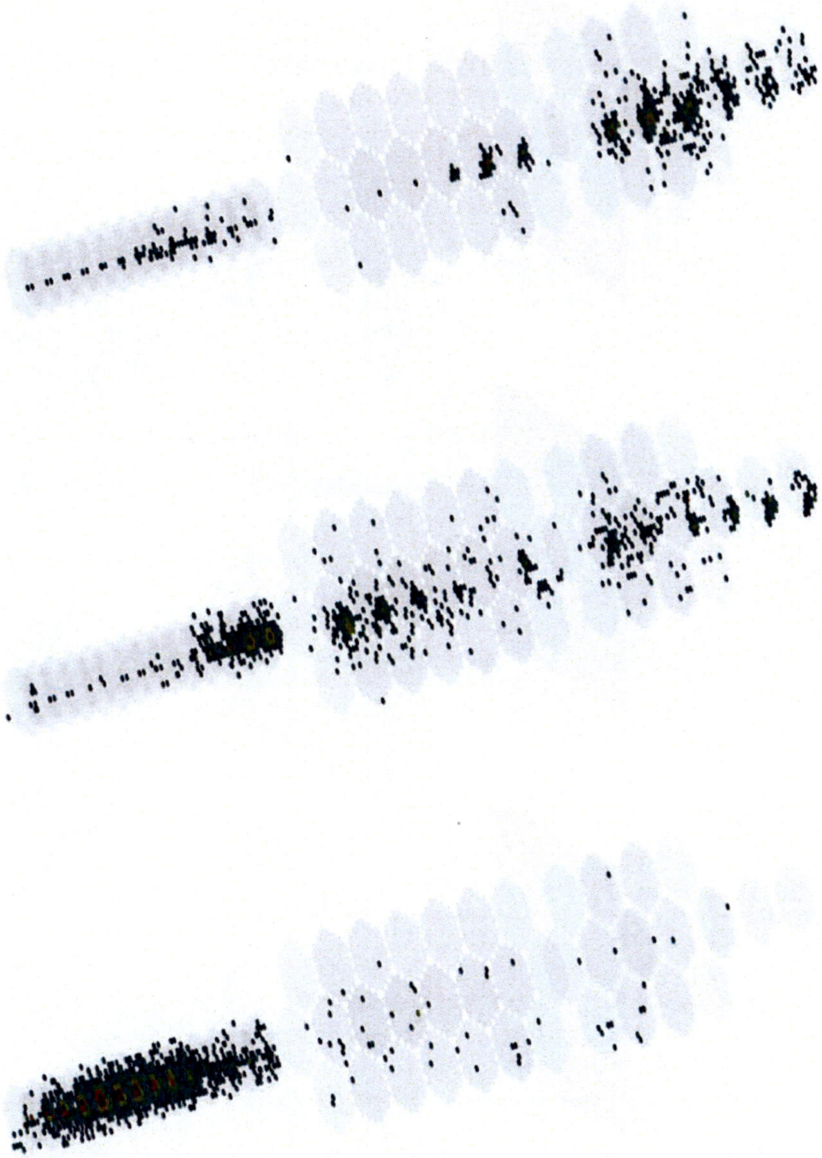

Event displays of the HGCAL prototype calorimeter with 250 GeV/c charged pions.
There is no such thing as a "typical hadronic shower profile".

Supervisors' Foreword

The doctoral thesis of Dr. rer. nat. Thorben Quast contains most relevant results and groundbreaking innovations for detector development and detector simulations. The outstandingly detailed investigations of Dr. Quast concern the construction of a new calorimeter detector to be used in the CMS experiment. The calorimeter is referred to as HGCAL and features a very fine sensor granularity combined with a large target volume as well as precise timing and energy measurements. Dr. Quast carried out extensive research and development work on the suitability of the HGCAL calorimeter concept. His work covered an exceptionally broad spectrum of areas, from testing sensor-related hardware, to hardware setups for test beam experiments, to development of reconstruction and simulation software, to algorithm development and deep learning methods.

In several test beam campaigns Dr. Quast studied all relevant aspects to evaluate prototypes of the HGCAL technology. His results on the final test beam experiment with the large prototype of the calorimeter eventually demonstrate the suitability of the silicon technology targeted. He studied the reconstructed showers of electromagnetic and hadronic particles in the calorimeter, the position and angular resolutions, the energy reconstruction, the depth of first interaction, and the calorimeter ability to discriminate between electrons and pions. All results show the expected properties and performance of this calorimeter concept, demonstrating the suitability of this detector for deployment within the CMS experiment. Beyond this work, Dr. Quast was one of the first particle physicists to have tested the new concept of Generative Adversarial Networks (GAN) for the ultrafast simulation of particle showers. In a GAN, two deep neural networks work against each other during network training, with the so-called generator network simulating energy depositions in the calorimeter's sensors. Dr. Quast has simulated the actual setup of one of the test beam experiments and has shown that particle showers can be generated a thousand times faster with the GAN than with the standard program GEANT4.

Overall, Thorben Quast's dissertation documents these outstanding scientific achievements, for which he received the best possible grade for his doctorate at

RWTH Aachen University. The CMS Collaboration honored his achievements in the form of the 2020 CMS Ph.D. Thesis Award.

July 2021

Martin Erdmann
RWTH Aachen University
Aachen, Germany

Eva Sicking
CERN
Meyrin, Switzerland

André David
CERN
Meyrin, Switzerland

IST Universidade de Lisboa
Lisbon, Portugal

Abstract

In order to cope with the increased radiation level and the challenging pile-up conditions at High Luminosity-LHC, the CMS collaboration will replace its current calorimeter endcaps with the High Granularity Calorimeter (HGCAL) in the mid-2020s. This dissertation addresses two important topics related to the preparation of the HGCAL upgrade: experimental validation of its silicon-based design and fast simulation of its data.

Beam tests at the DESY (Hamburg) and the CERN SPS beam test facilities in 2018 have been the basis for the design validation. The associated experimental infrastructure, the algorithms deployed in the reconstruction of the recorded data, as well as the respective analyses are reported in this thesis: First, core components of the silicon-based prototype modules are characterised and it is demonstrated that the assembled modules are functional. In particular, their efficiency to detect minimum ionising particles (MIPs) traversing the silicon sensors is found to be more than 98% for most of the modules. No indication of charge sharing between the silicon pads is observed. Subsequently, the energy response is calibrated in situ using the beam test data. Equalisation of the different responses among the readout channels is achieved with MIPs hereby deploying the HGCAL prototype as a MIP-tracking device. The relative variation of the inferred calibration constants amounts to 3% for channels on the same readout chip. The calibration of the time-of-arrival information is performed with an external time reference detector. With it, timing resolutions of single cells including the full prototype readout chain around 60 ps in the asymptotic high energy limit are obtained. The calorimetric performance of the HGCAL prototype is validated with particle showers induced by incident positrons and charged pions. For electromagnetic showers, the constant term in the relative energy resolution is measured to be $(0.52 \pm 0.08)\%$, whereas the stochastic term amounts to $(22.2 \pm 0.3)\% \sqrt{\mathrm{GeV}}$. This result is in good agreement with the calorimeter simulation with GEANT4 . The prototype's positioning resolution of the shower axis, after subtracting the contribution from the delay wire chambers in the beam line used as reference, is found to be below 0.4 mm at 300 GeV. At the same energy, the angular resolution in the reconstruction of the electromagnetic shower axis in this prototype is measured to be less than 5mrad. The analysis of the hadronic showers in this thesis makes use state-of-the-art machine learning methods that exploit the calorimeter's granularity. It is indicated that the energy resolution may be improved using software compensation and also that the separation of electromagnetic and charged pion-induced showers in the calorimeter may benefit from such methods. The measurements of the hadronic showers are adequately reproduced by GEANT4-simulation. Altogether, the obtained results from the analysis of the beam test data in this thesis are in agreement with the full functionality of the silicon-based HGCAL design.

The final part of this thesis provides a proof of principle that generative modelling based on deep neural networks in conjunction with the Wasserstein distance is a suitable approach for the fast simulation of HGCAL data: Instead of sequential simulation, a deep neural network-based generative model generates all calorimeter energy depositions simultaneously. This generator network is optimised through an adversarial training process using a critic network guided by the Wasserstein distance. The developed framework in this thesis is applied to both GEANT4 -simulated electromagnetic showers and to positron data from the beam tests. Ultimately, this fast simulation approach is up to four orders of magnitude faster than sequential simulation with GEANT4. It is able to produce realistic calorimeter energy depositions from electromagnetic showers, incorporating their fluctuations and correlations when converted into typical calorimeter observables.

List of Publications Related to this Thesis

(in reverse chronological order)

- **S. Anagul et al.**
  "Response of a CMS HGCAL silicon-pad electromagnetic calorimeter prototype to 20–300 GeV positrons"
  CMS DN-19-019, in preparation for submission to JINST, S. Anagul et al., "Response of a CMS HGCAL silicon-pad electromagnetic calorimeter prototype to 20-300 GeV positrons." In preparation for submission to JINST.
  The author of this thesis has written the text on the overview of the data reconstruction framework. He performed the position resolution study and has written the corresponding section in this publication draft. The text has been revised and copy-edited by the main authors of this publication.

- **B. Acar et al.**
  "Construction, Commissioning and Calibration of CMS CE prototype silicon modules"
  JINST 16 (2021) T04002, B. Acar et al., "Construction and commissioning of CMS CE prototype silicon modules", *JINST* **16** (2021) T04002, doi: 10.1088/1748-0221/16/04/T04002.
  The author of this thesis has written the text on the energy calibration with minimum ionising particles of the 2018 HGCAL beam test prototype. The text has been revised and copy-edited by the main authors of this publication.

- **B. Acar et al.**
  "The DAQ system of the 12,000 Channel CMS High Granularity Calorimeter Prototype"
  JINST 16 (2021) T04001, B. Acar et al., "The DAQ system of the 12,000 Channel CMS High Granularity Calorimeter Prototype", *JINST* **16** (2021) T04001, doi: 10.1088/1748-0221/16/04/T04001.
  The author of this thesis has written the text on the incorporation of the beam characterising detectors and their event synchronisation with the HGCAL beam test prototype. The text has been revised and copy-edited by the main authors of this publication.

- **T. Quast**
  "Qualification, Performance Validation and Fast Generative Modelling of Beam Test Calorimeter Prototypes for the CMS Calorimeter Endcap Upgrade"
  RWTH-2020-06473, T. Quast, *Qualification, Performance Validation and Fast Generative Modelling of Beam Test Calorimeter Prototypes for the CMS Calorimeter Endcap Upgrade.* PhD thesis, RWTH Aachen, Germany, 2020. http://doi.org/10.18154/RWTH-2020-06473.
  This author's doctoral dissertation on which this book is based.

- **P. Ahlburg et al.**
  "EUDAQ - A data acquisition software framework for common beam telescopes"
  JINST 15 (2020) P01038, P. Ahlburg et al., "EUDAQ - A data acquisition software framework for common beam telescopes", *JINST* **15** (2020) P01038, doi: 10.1088/1748-0221/15/01/p01038.
  The author of this thesis has co-written the text on the application of EUDAQ in the context of the HGCAL beam tests. The text has been revised and copy-edited by the main authors of this publication.

- **M. Erdmann, J. Glombitza, and T. Quast**
  "Precise Simulation of Electromagnetic Calorimeter Showers Using a Wasserstein Generative Adversarial Network"
  Comput Softw Big Sci (2019) 3: 4, M. Erdmann, J. Glombitza and T. Quast, "Precise Simulation of Electromagnetic Calorimeter Showers Using a Wasserstein Generative Adversarial Network", *Comput Softw Big Sci* **3** (2019), doi: 10.1007/s41781-018-0019-7.
  The scientific novelty originated from discussions between all three authors. The studies presented in this publication were primarily carried out by the author of this thesis who was therefore the corresponding author with the journal. In particular, the text on the performance benchmarks and on the experimental setup in this publication has been written by the author of this thesis. The full text has been revised and copy-edited by all authors of this publication.

- **N. Akchurin et al.**
  "First beam tests of prototype silicon modules for the CMS High Granularity Endcap Calorimeter"
  JINST 13 (2018) P10023, N. Akchurin et al., "First beam tests of prototype silicon modules for the CMS High Granularity Endcap Calorimeter", *JINST* **13** (2018) P10023, doi: 10.1088/1748-0221/13/10/p10023.
  The author of this thesis has performed the studies on the position resolution and has written the corresponding section in this publication. The text has been revised and copy-edited by the corresponding author of this publication.

- **CMS Collaboration**
  "The Phase-2 Upgrade of the CMS Endcap Calorimeter"
  CERN-LHCC-2017-023, CMS Collaboration, *The Phase-2 Upgrade of the CMS Endcap Calorimeter*. CERN-LHCC-2017-023 (CMS-TDR-019), 2017. https://cds.cern.ch/record/2293646.
  The author of this thesis has performed the electrical sensor characterisation measurements and the analysis of the data leading to Figure 7.14 of this report. Furthermore, he performed the study on the position resolution of the 2016 HGCAL prototype whose result is shown in Figure 5.21 (right) therein. The text was written by the main editors of this report.

- **Thorben Quast**
  "Construction and beam-tests of silicon-tungsten prototype modules for the CMS High Granularity Calorimeter for HL-LHC"
  JINST 13 (February 2018) C02044, T. Quast, "Construction and beam-tests of silicon-tungsten prototype modules for the CMS High Granularity Calorimeter for HL-LHC", *JINST* **13** (2018) C02044 , doi: 10.1088/1748-0221/13/02/c02044.
  The author of this thesis is the main author of these proceedings. The text was revised by the CMS collaboration.

Acknowledgments

At this point, I would like to extend my gratitude to all the people who have contributed to the realisation of this dissertation.

Foremost, I would like to thank Prof. Dr. Martin Erdmann for the supervision of this work. I am appreciative of your competency, your friendly nature as well as your trust in me, and have considered your support in the last year a privilege. The discussions and suggestions in the regular working group meetings have had significant impact on the successful completion of this work.

Similarly, I would like to thank Dr. Eva Sicking for an excellent local supervision at the European Organization for Nuclear Research CERN. You always had an open door to listen and address my concerns, be it in technical issues, organisatorial matters or career perspectives. Without you, this dissertation would not have been possible.

Secondly, I would like to thank Prof. Dr. André David Tinoco Mendes for the cooperation in the conduction of the HGCAL beam tests and for accepting to review this thesis as a referee.

This work was sponsored by the Wolfgang Gentner Programme of the German Federal Ministry of Education and Research. I am grateful to Dr. Lucie Linssen for having pointed me to this opportunity in 2016.

A large fraction of the results in this work are based on measurements which would not have been possible without the effort of many people:

I am thanking the DESY test beam coordinators as well as the CALICE AHCAL team at DESY for their helpful support during our test in March 2018. The HGCAL measurement programme there has received funding from the AIDA-2020 project.

I am grateful to Dr. Nikolaos Charitonidis and Dr. Alexander Gerbershagen for the configuration of the H2 test beam line at CERN which has yielded particle beams of good quality. Without this, the beam test measurements with the HGCAL prototype would not have been possible. I am also indebted to Dr. Dragoslav Lazic who was exposed to many of my requests concerning infrastructural aspects of the delay wire chambers at the H2 beam line.

Furthermore, I would like to thank Dr. Arnaud Steen for his supporting work in the context of the HGCAL beam tests and Dr. Shilpi Jain for taking care of the corresponding simulations.

For the numerous discussions of the analysis results, I would like to thank the entire EP-LCD HGCAL group at CERN, led by Dr. Eva Sicking, as well the CMS HGCAL System Test group, led by Dr. Dave Barney, Dr. Artur Lobanov and Dr. Rong-Shyang Lu.

I am indebted to Jonas Glombitza who gave me the decisive hint about the applicability of the Wasserstein metric for the generative modelling of calorimeter data. Furthermore, your detailed knowledge of and tips on deep learning tools have been very insightful in the following work-intensive weeks.

I would like to thank the entire VISPA working group at RWTH Aachen University, especially Benjamin Fischer, Dennis Noll and Yannik Rath, for the provision of computing resources and for the technical support during the conduction of the WGAN studies.

Prior to its original submission, this dissertation has been proofread entirely by Dr. Eva Sicking and Prof. Dr. Martin Erdmann for which I am particularly thankful.

Furthermore, I am very appreciative of the following people who have given valuable comments to various parts of the thesis draft:

Florian Pitters has reviewed the draft on the electrical sensor characterisation in Sect. 6.1. Ryan Quinn has provided comments on the description of the HGCAL beam test DAQ hardware in Sect. 6.2.1 and Dr. Arnaud Steen on the application of EUDAQ in Sect. 6.2.2. Dr. Nikolaos Charitonidis has given advice on drafting the text on the SPS Beam Test Facility in Sect. 6.4.1. The correctness of the description of the HGCAL beam test setups at DESY in Sect. 6.3 and at CERN in Sects. 6.4.2–6.4.4 was confirmed by Dr. Artur Lobanov. Dr. Arabella Martelli has provided instructive feedback on the calibration of the HGCAL prototype's timing information and for the assessment of its resolution in Sects. 7.2.5, 7.5 and 9.3. As for the results on the performance validation, I am thanking Matteo Bonanomi and Dr. Catherine Adloff for their remarks to Sect. 10.2 and Dr. Katja Krüger for hers on Sect. 10.3. Jonas Glombitza has given additional tips to the unpublished parts of Chap. 11.

Also in the context of drafting this thesis, I am thanking the world-class physicist Dr. Marcel Rieger for proof-reading individual passages. Your remarks to particle physics at the LHC in Chap. 2, to the documentation of the data reconstruction workflow in Sect. 7.1, and to the application of machine learning for the classification of electron- and pion-induced showers with the HGCAL prototype in Sect. 10.4 have been more than helpful.

I am thanking Dominik Arominski for a pleasant atmosphere in our shared offices throughout the last three years and for the occasional exchange of culinary specialties from our travels.

Soprattutto, ringrazio Emilia, la mia meravigliosa fidanzata e futura moglie, per aver accettato il nostro ridotto tempo prezioso insieme mentre scrivevo questa tesi. Mi hai sostenuto per tutto quel tempo di lunghe serate davanti alla scrivania e mi hai anche aiutato a ricordare le finezze dello Standard Model. Senza di te, gli ultimi mesi sarebbero stati un "disastro".

Meiner Familie, insbesondere meinem zuletzt verstorbenen Opa, danke ich für die Unterstützung in allen Belangen und für ihr unbegrenztes Vertrauen in mich und meinen Entscheidungen. I am thanking Donald W. Schnitker for keeping me down to earth. Letzter Dank gilt natürlich meinen Freunden, die stets für Rückhalt und Lebensfreude sorgen.

Shortly before the start of my physics studies in 2011, Dirk Nowitzki has achieved his life goal following many years of steady effort. I have taken his determination as an example ever since.

All dreams are crazy. Until they come true.

June 2020 Dr. Thorben Quast

Contents

Chapter 1
Introduction

Science is the investigation of relations in nature and the extension of our understanding of them. While ancient scientists tried to explain natural phenomena as a whole, the typical approach in modern science is to reduce complex questions into smaller and more elementary problems. In this sense, particle physics may be considered the most elementary field of science. It studies the smallest known structures, treated as point-like particles, that constitute all observable matter and give rise to the forces between them.

The widely accepted theory in particle physics is the Standard Model. Although it is the most successful theory in this field, it is known to be incomplete. It lacks explanation for decisive phenomena on cosmological scales, for instance the origin of dark matter or the overabundance of matter over anti-matter. Most strikingly, it fails to explain gravity. Furthermore, the Standard Model is arguably rather descriptive as it incorporates numerous empiric parameters. Only experimental testing of this theory may open the door towards more fundamental insights into the constituents from which the universe is built.

One way of doing so is via collisions of particles at high energies. Today, the Large Hadron Collider (*LHC*) [1] at CERN in Switzerland is the most powerful machine for this purpose. In the LHC, protons are accelerated to energies up to 7TeV and are eventually brought to collision at four interaction points. In those collisions, the proton constituents may interact with each other, hereby producing new particles that emerge from the interaction. The outgoing particles are then detected in sophisticated apparatuses, such as the Compact Muon Solenoid (*CMS*) [2] at LHC Point 5, which in turn enables inference of the underlying interactions.

Like many other branches of science, accelerator-driven high-energy particle physics relies on a large variety of expertise that includes theory predictions, accelerator physics, detector physics, computer science, engineering and statistics. A combination of all this expertise at the LHC and its experiments has led to the discovery of the Higgs boson in 2012 [3, 4]. With it, the Standard Model of Particle Physics

T. Quast, *Beam Test Calorimeter Prototypes for the CMS Calorimeter Endcap Upgrade*, Springer Theses, https://doi.org/10.1007/978-3-030-90202-5_1

could finally be completed. Unfortunately, there have been so far no clear indications of conceivable physics beyond it.

To improve the sensitivity of their measurements, particle physicists are continuously aiming at enhancing their experimental setups. One of the upcoming major improvements is the approved upgrade of the LHC towards higher luminosity (*HL-LHC*) in the 2020s. At HL-LHC, the provided collision rates will be increased significantly with respect to its current values. By this means, the statistical uncertainties of the LHC measurements will be reduced and the sensitivity to detect rare phenomena in the collisions will be enhanced. Chapter 2 gives a brief introduction to the theoretical foundations and the experimental concepts of particle physics at the LHC, focussing on the CMS detector.

Together with the HL-LHC accelerator upgrades, the LHC experiments need to be upgraded as well to cope with the associated challenging collision environment. This thesis deals with the upgrade of one specific detector element of CMS, that belongs to the class of calorimeters. In particle physics, calorimeters are devices that measure the energy of incident particles. Albeit the detection principle is simple, there are different concepts in its implementation. Chapter 3 provides an overview of calorimetry.

The detector upgrade investigated in this thesis is the so-called CMS highly granular calorimeter (*HGCAL*) [5] to replace the current endcap calorimeters for HL-LHC. In essence, HGCAL is a sampling calorimeter that is largely based on silicon wafers as sensitive material with 0.5-1 cm^2 sized pads. With its high granularity, the operation of HGCAL will ultimately commence a "new era in calorimetry" [5]. Until then, the realisation of HGCAL remains an ambitious detector project not only for its requirements but also for its tight timeline. The planned material costs amount to more than 60 MCHF alone, driven by the cost of the silicon wafers. The motivation for the choice of HGCAL, details on its design and on the intended performance are discussed in Chap. 4.

In this context, the exact design choices are based on calorimeter simulation and on assumptions on the electronic performances, and experimental validation of the HGCAL concept is imperative. Hence, prototypes of the silicon-based modules were assembled and were arranged between passive absorber plates to form prototypes of the HGCAL. Then, those prototype modules and calorimeters were exposed to particle beam in order to test their response to radiation and to assess their functionality under realistic experimental conditions.

In this thesis, essential contributions to the conduction of the HGCAL beam tests, to the experimental qualification of the prototype modules, to the validation of the calorimetric performance and to the fast simulation of its data are documented. The overall contribution strategy is defined in Chap. 5, whereby the stated criteria therein are tracked throughout the next chapters.

In general, conduction of beam test experiments requires a broad spectrum of expertise. Beam tests nowadays, in fact, may easily "reach a level of complexity that is comparable to fixed target experiments in the past," says E. Elsen, the CERN Director for Research and Computing. The work for this thesis has established parts of the experimental infrastructure, described in Chap. 6, that was an integral ingredient in the fruitful conduction of the HGCAL beam test experiments in 2018. The hereby (co-)developed algorithms for the reconstruction of the HGCAL prototype data and for those of the beam characterising detectors are elaborated in Chap. 7.

The analysis of the recorded data is subject to Chaps. 8–10. Chapter 8 characterises individual module components and assesses the proper functionality of the assembled modules in particle beam. The in situ calibration of the modules is discussed in Chap. 9. The energy scale for each sensitive readout cell is calibrated with minimum ionising particles and the intercalibration of the different electronic gains is endorsed. Moreover, the timing capabilities of the prototype readout chip to detect particle showers is studied for single channels. The evaluation of the HGCAL prototype's calorimetric performance with incident positrons and charged pions, with finally more than 90 functional modules, is presented in Chap. 10. State-of-the-art analysis techniques are applied in this thesis to make full use of the prototype calorimeter's granularity to improve its performance. In addition, the results on the calorimetric performance are complemented by their comparison to sequential calorimeter simulation with GEANT4 when possible.

While doing so, one quickly realises that sequential calorimeter simulation is a computing intensive task. At HL-LHC, the computing resources for the latter are anticipated to be limited, rendering fast simulation techniques an attractive alternative for the future. Traditionally, fast simulation approaches are engineered by experts in the field and are highly specific to the underlying detector geometry. Over the last years, Generative Adversarial Networks (*GANs*) [6] have become an established tool that could provide a more generic solution to this task. In Chap. 11, a variant of GANs, the Wasserstein GAN [7, 8], is applied for the generation of electromagnetic showers in the HGCAL topology.

Last, all the findings of this work are summarised, and an outlook as well as the conclusions are presented in Chap. 12.

Unit Convention

Except for Sect. 2.1, the International System of Units is used throughout this thesis. In particular, the speed of light in vacuum is denoted as $c = 299792458\,\mathrm{ms}^{-1}$.

References

1. Evans L, Bryant P (2008) LHC machine. JINST 3:S08001. http://dx.doi.org/10.1088/1748-0221/3/08/s08001
2. CMS Collaboration (2008) The CMS experiment at the CERN LHC. JINST 3:S08004. http://dx.doi.org/10.1088/1748-0221/3/08/S08004
3. ATLAS Collaboration (2012) Observation of a new particle in the search for the Standard Model Higgs boson with the ATLAS detector at the LHC. Phys Lett B 716:1–29. http://dx.doi.org/10.1016/j.physletb.2012.08.020
4. CMS Collaboration (2012) Observation of a new boson at a mass of 125 GeV with the CMS experiment at the LHC. Phys Lett B 716:30–61. http://dx.doi.org/10.1016/j.physletb.2012.08.021
5. CMS Collaboration (2017) The phase-2 upgrade of the CMS endcap calorimeter. CERN-LHCC-2017-023 (CMS-TDR-019). https://cds.cern.ch/record/2293646
6. Goodfellow IJ, et al. (2014) Generative adversarial networks. http://arxiv.org/abs/1406.2661 [stat.ML]
7. Arjovsky M, Chintala S, Bottou L (2017) Wasserstein GAN. http://arxiv.org/abs/1701.07875 [stat.ML]
8. Gulrajani I, et al. (2017) Improved training of wasserstein GANs. http://arxiv.org/abs/1704.00028 [cs.LG]

Chapter 2
Particle Physics at the Large Hadron Collider

The identification of the most fundamental rules that govern the evolution of the universe, including all its matter and forces, is a long ongoing quest of mankind. For illustration of this task, one may consider the two-dimensional model universe in Fig. 2.1. After its initial state is randomly generated, its evolution is defined by three simple rules, in this example taken from Conway's Game of Life, cf. Chap. 8 in Ref. [1]. Eventually, the model universe converges into a state with reproducing or constant structures, analogous to life on Earth. Hereby, the exact evolution appears rather untraceable at the large scale and it requires investigation of small scale structures to infer the three underlying rules that determine this model universe's dynamic.

Physicists attempt to find the corresponding rules for our universe. Due to the well-established principles of quantum mechanics, those rules are not deterministic on microscopic scales but rather allow for probabilistic predictions. Furthermore, they must obey the principle of relativity in which space and time are treated equally. Currently, the widely accepted theory that is able to predict most of the observable phenomenology at the microscopic level is the Standard Model of Particle Physics (*SM*). In this theory, particles are described as excited states of abstract quantum fields. They mediate the forces and the matter that form the atoms, the building blocks of life. Section 2.1 explains the core concepts of the Standard Model of Particle Physics.

However, it is well known that the Standard Model of Particle Physics is incomplete or at most only an approximation of something more fundamental. Consequently, extensive experimental testing is essential. For this purpose, the theory can be used to predict likelihoods for certain interaction processes of colliding particles. Having the interaction likelihood of a certain process predicted, it can be verified experimentally by repeated observation of the interaction that start from the same initial conditions. Practically, this is done for example through repeated collisions of particles at a known energy. Those collisions ideally occur at high rates and high centre-of-mass energies because potential new phenomena in most theories beyond the SM are expected to be both very rare and tend to happen at high energies. The Large Hadron Collider (*LHC*) is the machine that provides collisions of protons

T. Quast, *Beam Test Calorimeter Prototypes for the CMS Calorimeter Endcap Upgrade*, Springer Theses, https://doi.org/10.1007/978-3-030-90202-5_2

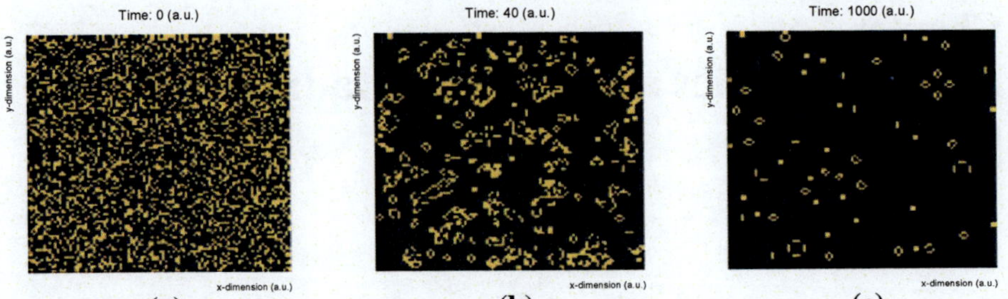

(a) **(b)** **(c)**

Fig. 2.1 The evolution of a model universe following the rules of Conway's Game of Life

at unprecedented man-made centre-of-mass energies and rates. An overview of the Large Hadron Collider is given in Sect. 2.2.

The measurement of the final states from the interactions is the task of dedicated particle detectors. Their objective is the identification and characterisation of all particles emerging from a collision. Section 2.3 presents the Compact Muon Solenoid (*CMS*) experiment which is one of the major experiments located at one of the four interaction points of the LHC.

The LHC will be upgraded in the 2020's to provide collisions at even higher rates. By this means, it will enable access to probe even rarer conceivable phenomena which are predicted by theories that reach beyond the scope of the SM. This running phase of the LHC is referred to as the High-Luminosity LHC. Its features and the necessary accelerator upgrades are outlined in Sect. 2.4.

2.1 Standard Model of Particle Physics

The Standard Model of Particle Physics (*SM*) is the theoretical framework that describes the fundamental components of matter and the three forces, that determine all its phenomenology at the microscopic level: electromagnetism, the strong and the weak forces. Gravity, the dominant force at macroscopic scales, is not part of it. The Standard Model is a quantum field theory based on the combination of special relativity [2], quantum mechanics and classical field theory. The fundamental objects in this theory are quantum fields that span the entire universe. The physical manifestation of those fields are their quantised, excited states, which can also be interpreted as point-like, elementary particles. Particles are both the building blocks of all observable matter as well as the mediators of the forces that act between them.

Giving a complete introduction to the Standard Model is far beyond the scope of this thesis. In Sect. 2.1.1, an overview of the particles and their interactions in the SM is given. Section 2.1.2 summarises the theoretical foundation of the SM. Finally, a few of its known limitations are mentioned in Sect. 2.1.3.

The structure of this section is inspired by the CERN 2015 Summer Student lectures of Prof. Michael Krämer (RWTH Aachen). The used equations are quoted from Refs. [3, 4]. Natural units, i.e. $c = \hbar = k_B = \epsilon_0 := 1$, as well as Einstein's convention of summing over repeated indexes are used in this section to facilitate the notation.

2.1.1 Matter and Forces

The Standard Model is a Yang–Mills theory [5] that encapsulates the known fundamental symmetries. It is based on the gauge symmetry groups $SU(3)_C \otimes SU(2)_L \otimes U(1)_Y$, where the subscripts indicate the quantum numbers associated to the specific symmetries: C for the colour charge, L for the weak isospin and Y for the hypercharge. As a direct consequence of Noether's theorem [6], quantum numbers are conserved quantities. In the SM, each particle is assigned a set of quantum numbers. Also, each particle has an associated partner with equal mass but otherwise inverted quantum numbers. Those partners are called *anti-particles*.

All observable matter is made of so-called fermion particles whereas the interactions are mediated by gauge boson particles. Fermions and bosons differ in view of their spin: fermions have a half-integer spin, and the spin of bosons takes integer values. Fermions can be further categorised into three generations which is motivated by similarities in their quantum numbers. Furthermore, fermions are organised in two classes: leptons and quarks. Leptons and quarks are subject to the electroweak forces and quarks interact also through the strong force.

Quarks carry elementary charge, a weak isospin and a colour charge. They are *confined*, meaning they can only exist in bound states (*hadrons*). Those can be created from two (*mesons*), of three (*baryons*), of four and even of five valence quarks, see e.g. the recent measurement [7] by the LHCb experiment. The proton (up, up, down quarks) is the only bound quark state that is stable with a mean life of more than $2.1 \cdot 10^{29}$ yr [8]. Together with neutrons (up, down, down), they constitute the building blocks of atomic nuclei, so-called *nucleons*, in which also the neutron remains stable. Other than that, most bound quark states disintegrate in less than a fraction of a microsecond in their respective rest frame. Noteworthy are the charged pions π^+, consisting of an up and an anti-down quark, that decay on average after $\sim$26 ns, and the charged kaons K^+, consisting of an up and an anti-strange quark, that decay on average after $\sim$12 ns.

By contrast, leptons can exist freely. They do not carry any colour but are assigned a weak isospin. Electrons, muons and taus also carry electric charge, while the neutrinos are electrically neutral. Among the electrically charged ones, the electron is the only stable one. Together with the nucleons, electrons form atoms. The other two electrically charged leptons, i.e. the muon and the tau, have a mean lifetime of 2.2 µs and 0.3 ps, respectively. The mass range of the charged leptons spans four orders of magnitude whereas the SM neutrinos are massless. However, neutrino oscillation experiments indicate that the neutrinos actually have non-zero mass, see e.g. Ref. [9].

Table 2.1 List of fermions in the SM, categorised into three generations (Gen.). Anti-particles are not listed. Mass values are taken from Ref. [8]. The listed interactions will be described next. EM symbolises the electromagnetic force. The uncertainties on the charged lepton masses are less than the respective least significant digits

	Gen.	Name	Mass	Electric	Forces			
				Charge [e]	EM	Weak	Strong	BEH
Quark	I	up (u)	$2.2^{+0.5}_{-0.4}$ MeV	2/3	✓	✓	✓	✓
		down (d)	$4.7^{+0.5}_{-0.3}$ MeV	−1/3	✓	✓	✓	✓
	II	charm (c)	$1.275^{+0.025}_{-0.035}$ GeV	2/3	✓	✓	✓	✓
		strange (s)	95^{+9}_{-3} MeV	−1/3	✓	✓	✓	✓
	III	top (t)	173.0 ± 0.4 GeV	2/3	✓	✓	✓	✓
		bottom (b)	$4.18^{+0.04}_{-0.03}$ GeV	−1/3	✓	✓	✓	✓
Lepton	I	electron (e)	511 keV	−1	✓	✓		✓
		e neutrino (ν_e)	< 2 eV	0		✓		
	II	muon (μ)	105.658 MeV	−1	✓	✓		✓
		μ neutrino (ν_μ)	< 0.19 MeV	0		✓		
	III	tau (τ)	1777 MeV	−1	✓	✓		✓
		τ neutrino (ν_τ)	< 18.2 MeV	0		✓		

Table 2.2 SM interactions and the corresponding gauge bosons. Mass values are taken from Ref. [8]

Mediates	Gauge Boson	Mass [GeV]	El. Charge [e]	Subject to
EM	Photon	$< 1 \cdot 10^{-18}$ eV	0	EM
Strong	8 gluons	0	0	Strong
Weak	$W^{\pm}$	80.385 ± 0.015	± 1	BEH, EM, Weak
	Z	91.1876 ± 0.0021	0	BEH, Weak
BEH	Higgs	125.09 ± 0.24	0	BEH, Weak

Table 2.1 gives an overview of the matter particles of the SM and the possible interactions which are discussed next.

Gauge bosons are the mediators of the three fundamental forces in the SM. Those forces are the electromagnetic force, whose classical analogue is described by Maxwell's equation, the weak (nuclear) force, that e.g. is responsible for the β-decay of atoms, and finally the strong (nuclear) force, that holds the nucleons together. The fourth interaction does not correspond to a force but yields the origin of mass of the particles. The corresponding mechanism (*BEH*) is mediated by the so-called Higgs boson. Table 2.2 gives an overview of the gauge bosons and the associated SM interactions.

The photon is the carrier of the electromagnetic force and couples to all fermions and bosons that have electric charge. Since the photon is massless and does not interact with itself, the reach of the electromagnetic force is infinite. In daily life, electromagnetism (*EM*) is the most encountered force among the ones described in the SM. It was historically the first one that could be described in a consistent relativistic and quantum mechanical framework called *Quantum Electrodynamics* (*QED*) [10–13].

The strong force is mediated through massless gluons which couple to all particles with colour charge, namely to quarks and to themselves. As the name suggests, this force is the strongest of all known forces. However, it is effective only on small distances, namely at the order of the extension of atomic nuclei (~1 fm). The theory of the strong force is referred to as *Quantum Chromodynamics (QCD)* [14–16]. Due to the self-interaction of the gluons, QCD incorporates the concept of a *running coupling constant*, which states that the coupling strength between coloured particles decreases at high momentum transfers (small distances) and increases at low-momentum transfers (large distances). The property of the strong force to become weaker at decreasing distances is also called *asymptotic freedom* and justifies the validity of QCD perturbation theory for particle collisions at high energies. At lower energies, QCD perturbation theory fails and phenomenological theories such as lattice QCD need to be applied. In this particular theory, the result of the running strong coupling is a phenomenological "string potential" between coloured quarks at distance r, cf. Eq. 2.1 [17].

$$V_{q\bar{q}}(r) = -\frac{4}{3} \cdot \frac{\alpha_{\text{str}}}{r} + \sigma \cdot r \ , \quad \alpha_{\text{str}} \text{ and } \sigma \text{ are constants} \tag{2.1}$$

Its linear term makes it energetically preferable to create additional quarks from the vacuum to form bound quark states. For this particular reason, quarks are confined and only exist in bound states.

The W and Z bosons are the mediators of the weak force. The reach of this force is also limited to the extent of atomic nuclei. It is about five orders of magnitude weaker than the strong force, hence explaining its name. As discussed in the next section, the weak and the electromagnetic force can be described in the unified *theory of electroweak interactions* [18, 19]. In this theory, only fermions from the same generation are allowed to undergo weak interactions with each other. In addition, only fermions of the same chirality, i.e. a quantity that is closely related to the particle's spin, may interact with each other. According to this electroweak theory alone, neither the involved fermions nor the W and Z bosons would have a mass. In contrast to gluons and photons however, the W and Z bosons are, in fact, quite massive. Most fermions also have sizeable masses.

A theoretical framework to address this shortcoming was proposed in 1964 by Englert, Brout and Higgs [20, 21], which is commonly referred to as the *Brout–Englert–Higgs mechanism* (*BEH*). The core prediction of this mechanism is the existence of a scalar quantum field whose excitation is the so-called Higgs boson. This scalar field does not correspond to a fundamental force but renders mass to the W and Z bosons, and to all fermions, except the neutrinos.

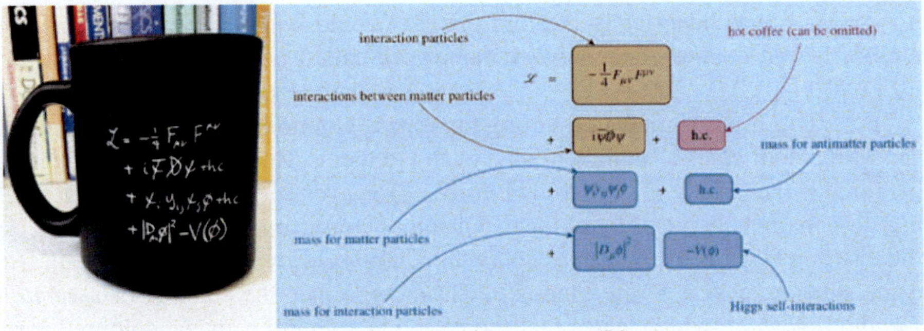

(a) (b)

Fig. 2.2 **a** Compact version of the SM Lagrangian on a coffee mug. **b** The terms coloured in brown are governed by the electromagnetic, weak, and strong forces, while those that are coloured blue are governed by interactions with the Brout–Englert–Higgs field. Figures and captions taken from Ref. [22]

2.1.2 Architecture of the Lagrangian

The dynamic of the quantum fields, from which the particles emerge, is implicitly described by the Lagrange density $\mathcal{L}$. In the context of the SM, $\mathcal{L}$ is simply referred to as *Lagrangian* and it is constructed from axiomatically imposed symmetries. In particular, the *action*, defined as the space-time integral over the SM Lagrangian, should remain unchanged when certain group transformations are applied to the field representations.

The most compact form of the SM Lagrangian is depicted on the popular black coffee mugs [22] that can be acquired at the CERN gift shop, cf. Fig. 2.2a. It consists of distinct terms, cf. Fig. 2.2b, which shall be motivated in the following.

The Gauge Sector: *Interaction Particles*, *Interactions Between Matter Particles*
To start the construction of the SM Lagrangian, an essential requirement is that it must be invariant under Lorentz transformation to obey the laws of special relativity. This condition is met by the Lagrangian for a free Dirac field ψ, given by Eq. 2.2.

$$\mathcal{L}_{\text{Dirac}} = \bar{\psi}\left(i \cdot \gamma^{\mu}\partial_{\mu} - m\right)\psi \tag{2.2}$$

This Lagrangian can be used to derive the *Dirac equation* [23], i.e. a relativistic version of the Schrödinger equation from non-relativistic quantum mechanics. In Eq. 2.2, ψ is a spinor that describes particles and antiparticles with spin $\pm\frac{1}{2}$, i.e. fermions, in the absence of any interaction. Partial derivatives ∂_{μ} are applied with respect to the four-dimensional space-time coordinates x_{μ}. γ_{μ} are the Dirac matrices and $\bar{\psi} := \psi^{\dagger}\gamma_0$.

Then, the SM interactions arise from the convention that the SM Lagrangian must be invariant under local n-dimensional gauge transformations $U(\vec{\theta})$. *Local* means that the transformation parameters $\vec{\theta}$ depend on the space-time coordinates. In general, $\mathcal{L}_{\text{Dirac}}$ is not invariant under local gauge transformations because $\partial_\mu (U\Phi) \neq U(\partial_\mu \Phi)$.

An instructive example to understand how gauge bosons arise from an imposed local gauge invariance is demonstrated by applying a $U(1)$ transformation to the free Dirac field Lagrangian in Eq. 2.2. The $U(1)$ transformation is equivalent to the multiplication of the spinor ψ with a complex number, cf. Eq. 2.3.

$$\psi \rightarrow \psi' = U(\theta(x))\psi \overset{U(1)}{\widehat{=}} \exp(-i\theta(x))\psi \tag{2.3}$$

Note that $\mathcal{L}_{\text{Dirac}}(\psi')$ is not equal to $\mathcal{L}_{\text{Dirac}}(\psi)$ as it is explicitly demonstrated in Eq. 2.4

$$\mathcal{L}_{\text{Dirac}}(\psi') = \mathcal{L}_{\text{Dirac}}(\psi) + \bar{\psi}\gamma^\mu \left(\partial_\mu \theta(x)\right)\psi \tag{2.4}$$

In order to achieve invariance under $U(1)$ transformation, a gauge field A_μ needs to be introduced. The resulting Lagrangian $\mathcal{L}_{\text{QED}}$ is precisely the one describing QED. A_μ is interpreted as the photon field. It transforms as $A'_\mu \rightarrow A_\mu - \partial_\mu \theta(x)$, whose classical analogue is the arbitrary gauge of the electromagnetic potentials in Maxwell's equations [24]. Finally, the Dirac Lagrangian in Eq. 2.2 is extended to yield the $U(1)$ invariant QED Lagrangian according to Eq. 2.5.

$$\begin{aligned} \partial_\mu \rightarrow D_\mu &:= \partial_\mu + i \cdot eA_\mu Q\,, \quad F_{\mu\nu} := \partial_\mu A_\nu - \partial_\nu A_\mu \\ \mathcal{L}_{\text{Dirac}} \rightarrow \mathcal{L}_{\text{QED}} &= -\frac{1}{4}F^{\mu\nu}F_{\mu\nu} + \bar{\psi}\left(i \cdot \gamma^\mu D_\mu - m\right)\psi \end{aligned} \tag{2.5}$$

Therein, Q represents the electric charge. D_μ is referred to as the *covariant derivative*. $F^{\mu\nu}$ is the *field strength tensor* and describes the dynamics of the associated gauge boson field.

In this instructive example, $U(1)$ is an abelian gauge group, such that the associated group generators commute. The concept of enforcing local gauge invariance has been extended to non-abelian gauge $SU(n)$ groups by Yang and Mills [5]. The generic $SU(n)$ invariant Yang–Mills Lagrangian $\mathcal{L}_{\text{YM}}$ has a similar form as $\mathcal{L}_{\text{QED}}$ with generalised expressions for the field strength tensor and the covariant derivative, cf. Eq. 2.6.

$$\begin{aligned} \mathcal{L}_{\text{YM}} &= -\frac{1}{4}\sum_{a=1}^{n^2-1} F^a_{\mu\nu}F^{a\mu\nu} + \bar{\psi}\left(i\gamma^\mu D_\mu - m\right)\psi \\ F^a_{\mu\nu} &= \partial_\mu A^a_\nu - \partial_\nu A^a_\mu - g f_{abc} A^b_\mu A^c_\nu\,, \quad D_\mu = \left(\partial_\mu + igT_a A^a_\mu\right) \end{aligned} \tag{2.6}$$

In $\mathcal{L}_{\text{YM}}$, T_a are the generators of the respective $SU(n)$ group, f_{abc} are its structure constants, A^a_μ are new gauge boson particle fields ensuring the invariance, and g is an arbitrary coupling. The generalised strength tensor $F^a_{\mu\nu}F^{a\mu\nu}$ contains terms which are cubic or quartic in the gauge particle fields. This implies self-interactions of the associated bosons. Both QCD and the theory of electroweak interactions are special cases of a Yang–Mills theory:

- **QCD** is founded on enforcing symmetry under $SU(3)$ transformation. It acts on quark fields q_j with corresponding masses m_j. Each quark field is organised in $SU(3)_C$ triplets, where each component is a colour charge (red, green, blue). Gluons are combinations from eight gluon fields g^A_μ. They are the mediators of the strong force and also carry colour charge. The corresponding QCD Lagrangian $\mathcal{L}_{\text{QCD}}$ is given by Eq. 2.7.

$$\begin{aligned}\mathcal{L}_{\text{QCD}} &= -\frac{1}{4}\sum_{a=1}^{8} F^{a\mu\nu}F^a_{\mu\nu} + \sum_{j=1}^{n_f=6} \bar{q}_j \left(i\cdot D_\mu\gamma^\mu - m_j\right) q_j \\ D_\mu &= \partial_\mu + i\cdot g_s\sum_{A=1}^{8} t^A g^A_\mu\,,\quad F^a_{\mu\nu} = \partial_\mu g^a_\nu - \partial_\nu g^a_\mu - g_s C_{abc} g^b_\mu g^c_\nu\end{aligned} \tag{2.7}$$

  g_s is the coupling constant of the strong force and t^A are the $SU(3)$ generators. The generators do not commute but fulfill the relation $[t^a, t^b] = iC_{abc}t^c$, where C_{abc} stands for the $SU(3)$ structure constants. The non-vanishing commutation allows for interactions among three and four gluons.

- The **electroweak** (*EW*) interaction arises from enforcing symmetry under $SU(2)_L \otimes U(1)_Y$ transformations. The EW interaction acts on fermions denoted as Ψ. Fermion fields are arranged into $SU(2)_L$ doublets or singlets depending on the chirality states. For instance, a lepton l with its neutrino ν_l from the same generation form a left-handed doublet symbolised as $\Psi = (l,\ \nu_l)^T$. Since the SM neutrinos are assumed to be massless, no right-handed state is foreseen. Thus, the right-handed singlet has the shape $\Psi = l$. One defines $\Psi_{L/R} := \frac{1}{2}(1 \mp \gamma_5)\Psi$ as the left/right-handed chirality states. The corresponding Yang–Mills Lagrangian for electroweak interactions $\mathcal{L}_{\text{EW}}$ in the limit of negligible fermion masses is given by Eq. 2.8.

$$\begin{aligned}\mathcal{L}_{\text{EW}} &= -\frac{1}{4}\sum_{A=1}^{3} F^{A\mu\nu}F^A_{\mu\nu} - \frac{1}{4}B^{\mu\nu}B_{\mu\nu} + \bar{\psi}_L i\gamma^\mu D_\mu \psi_L + \bar{\psi}_R i\gamma^\mu D_\mu \psi_R \\ F^A_{\mu\nu} &= \partial_\mu W^A_\nu - \partial_\nu W^A_\mu - g\epsilon_{ABC}W^B_\mu W^C_\nu\,,\quad B_{\mu\nu} = \partial_\mu B_\nu - \partial_\nu B_\mu \\ D_\mu &= \partial_\mu + ig\sum_{A=1}^{3} t^A W^A_\mu + ig'\frac{1}{2}YB_\mu\end{aligned} \tag{2.8}$$

  t^A and $Y/2$ are the $SU(2)$ and $U(1)$ generators in the respective representations $\psi_{L,R}$ with $t^A\psi_R = 0$. ϵ_{ABC} denotes the Levi-Civita tensor, and g as well as g' are two coupling constants. Since the generators t^A do not commute, cubic and quadratic terms in the gauge fields are present implying self-interactions of the associated bosons. Linear combinations of the mathematical fields W^A_μ and B_μ

yield the physical W and Z bosons as carriers of the weak interaction, as well as the aforementioned photon.

The EW Symmetry Breaking Sector: ***Mass for Interaction Particles, Higgs Self-Interaction***

None of the Yang–Mills Lagrangians contain quadratic terms in the gauge bosons that are interpretable as their mass terms. To include such terms, the SM Lagrangian is extended by $\mathcal{L}_{\text{Higgs}}$, cf. Eq. 2.9, which is equivalent to the terms labelled "mass for interaction particles" and "Higgs self-interaction" in Fig. 2.2b.

$$\mathcal{L}_{\text{Higgs}} = \left(D_\mu \Phi\right)^\dagger \left(D^\mu \Phi\right) - V\left(\Phi^\dagger \Phi\right) \tag{2.9}$$

In $\mathcal{L}_{\text{Higgs}}$, Φ is defined as a scalar field that transforms as a doublet under a $SU(2)$ transformation. D_μ is the electroweak covariant derivative as defined in Eq. 2.8. V is a potential with up to quadratic terms in Φ, cf. Eq. 2.10.

$$V(\Phi^\dagger \Phi) = -\mu^2 \Phi^\dagger \Phi + \frac{\lambda}{2}\left(\Phi^\dagger \Phi\right)^2 \tag{2.10}$$

Both parameters μ^2 and λ in that potential are required to be greater than zero. The consequence is a ground state Φ_0 of Φ that fulfills $|\Phi_0| = \nu := \sqrt{\mu^2/\lambda}$. The choice of the ground state is arbitrary due to the potential's rotationally symmetric minimum. This circumstance is referred to as *spontaneous electroweak symmetry breaking*. The Higgs particle H is defined as an excitation of the Φ field from its ground state. It is convenient to choose a specific gauge and to write the scalar field Φ according to Eq. 2.11.

$$\Phi = \begin{pmatrix} 0 \\ \nu + \frac{H}{\sqrt{2}} \end{pmatrix} \tag{2.11}$$

Then, this form may be inserted into Eq. 2.9. Combining $\mathcal{L}_{\text{Higgs}}$ and $\mathcal{L}_{\text{EW}}$, the effect of the symmetry breaking results in massive W and Z boson while leaving the photon massless. The potential V may be rewritten for excited values around its ground state as done in Eq. 2.12.

$$V = -\frac{\mu^2 \nu^2}{2} + \mu^2 H^2 + \frac{\mu^2}{\sqrt{2}\nu} H^3 + \frac{\mu^2}{8\nu^2} H^4 \tag{2.12}$$

From this, one infers that the theoretically predicted Higgs boson has a mass that is proportional to the parameter μ. The discovery of a Higgs boson-like particle with a mass consistent with the experimental SM constraints was announced by the ATLAS and CMS collaborations in 2012 [25, 26], 48 years after its existence had been postulated. In addition, Eq. 2.12 implies that the Higgs boson must interact with itself, which still remains to be established experimentally.

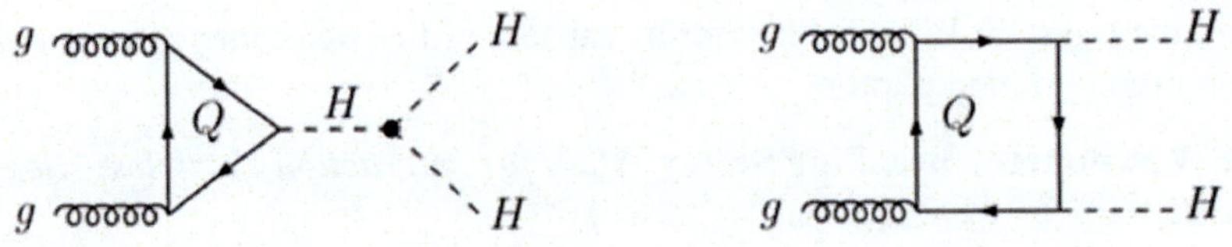

Fig. 2.3 Example Feynman diagrams, here for the production of a pair of Higgs bosons via fusion of two gluons. Diagrams taken from Ref. [28]

The Flavour Sector: ***Mass for Matter Particles***
A standard mass term for free fermions of the form $m\bar{\psi}\psi$ would contain mixtures of left- and right-handed chirality states which are forbidden in the electroweak Lagrangian. Again, the Higgs mechanism can be applied to explain the origin of the empirically found fermion masses. For this purpose, a Yukawa term $\mathcal{L}_{\text{Yukawa}}$ is added to the Lagrangian, cf. Eq. 2.13.

$$\mathcal{L}_{\text{Yukawa}} = -\lambda_f \left(\bar{\psi}_L \Phi \psi_R + \bar{\psi}_R \Phi \psi_L \right) \tag{2.13}$$

Insertion of the field Φ around its ground state with the Higgs boson as its excitations, the fermion masses m_f can be identified to be proportional to the product $\lambda_f \cdot v$ of the Yukawa coupling strength λ_f and the ground potential value v. Moreover, the Higgs boson now also couples to the fermions, whereby the coupling strength is proportional to the quotient m_f/v. In a more generalised formulation of $\mathcal{L}_{\text{Yukawa}}$ that includes all three generations of fermions, the coupling strengths are replaced by a flavour-mixing matrix. This mixing is given by the so-called *CKM matrix* [8] and it implies flavour-changing weak interactions.

The Full SM Lagrangian
Summing up all the discussed terms, the full SM Lagrangian $\mathcal{L}_{\text{SM}}$ is obtained as in Eq. 2.14.

$$\mathcal{L}_{\text{SM}} = \mathcal{L}_{\text{QCD}} + \mathcal{L}_{\text{EW}} + \mathcal{L}_{\text{Higgs}} + \mathcal{L}_{\text{Yukawa}} \tag{2.14}$$

Given the full SM Lagrangian, the Euler–Lagrange equation, cf. Eq. 2.15, can in principle be used to derive the equations of motions of the fields ϕ, analogous to classical field theory.

$$\partial_\mu \frac{\partial \mathcal{L}}{\partial(\partial_\mu \phi)} = \frac{\partial \mathcal{L}}{\partial \phi} \tag{2.15}$$

Yet, it is practically impossible to precisely solve the resulting differential equations. Instead, this may be done up to limited orders in perturbation theory. In this context, "Feynman diagrams were invented in 1948 to help physicists find their way out of a morass of calculations" [27]. Figure 2.3 shows two example Feynman diagrams [13] for the production of a pair of Higgs bosons via fusion of two gluons in leading order of perturbation theory.

Feynman diagrams are not only graphical illustrations of the possible interactions of elementary particles but, more importantly, they represent recipes that dictate the computation of matrix elements. Matrix elements are essential ingredients to predictions of interaction cross sections, of particle decay rates and of other quantities that are experimentally verifiable.

2.1.3 Limitations

The Standard Model is an impressive achievement in theoretical physics and has withstood all its experimental tests so far. However, there are crucial phenomena that the SM yet fails to describe. Most prominently, cosmological problems such as the nature of dark matter [29] and dark energy [30], as well as the origin of the matter-over-anti-matter asymmetry cannot be explained with the SM. Furthermore, the unification with gravity, i.e. the establishment of a verifiable theory that unifies the SM with General Relativity [31] in a consistent mathematical framework, is an outstanding major issue in theoretical physics. Another criticism is targeted at the overall more descriptive rather than explanatory nature of the SM. The hierarchy of the fermion masses and the reason for the at least 26 free parameters (neutrino masses and mixings included) of the SM are not yet derived from first principles.

Therefore, it is of utmost importance to experimentally test the Standard Model of Particle Physics under all possible aspects. The observation of significant deviations from the corresponding SM predictions would lead to important hints at where to modify, extend or fully revise this theory.

2.2 Large Hadron Collider

Controlled collisions of elementary particles at unprecedented centre-of-mass energies are provided by the Large Hadron Collider (*LHC*). The LHC is the largest and most powerful synchrotron ever built. It is primarily designed to accelerate, store and collide counter-rotating bunches of protons[1] with energies up to 7TeV. More than ten thousand researchers and support staff were involved in the planning and construction of this machine. After years of construction, the first bunch of protons circulated in September 2008. Ever since, the LHC has produced collisions that have enabled thorough testing of the SM and searches for physics beyond the SM. This section gives a brief overview on the LHC, on its pre-accelerator chain, on its operation parameters and explains the implications to the physics measurements. More details are available in Ref. [32] and in the LHC's technical design reports in Refs. [33–35].

The LHC is a ring accelerator that consists of eight straight sections and eight arcs. In total, the LHC is 26.7 km long and is located in a tunnel 45–170 m underground which previously hosted the Large Electron-Positron Collider [36], LHC's predecessor. An ultrahigh vacuum of around 10^{-13} bar in the beam pipes is main-

[1] The LHC also runs with heavy ions, which is not subject in this thesis.

tained throughout the accelerator. The bending of the particle trajectories is done by 1232 15 m long dipole magnets with a magnetic field strength of up to 8.3 T. This extraordinary strength is realised by electric currents of 11.8 kA in the magnet coils. In order to achieve such high currents, the coils are made from superconducting NbTi and are operated at temperatures down to 1.9 K. As the inner tunnel is only 3.7 m wide, a twin-bore magnet scheme with two sets of coils and beam pipes in one mechanical structure was adopted [37], instead of two separate ones.

Prior to their injection into the LHC, the protons undergo a series of pre-accelerations. The full acceleration complex at CERN is depicted in Fig. 2.4. At the beginning of the injection chain, a duoplasmatron creates the proton beam from a hydrogen source. The protons are subsequently accelerated from 90 keV to a kinetic energy of 50 MeV in the LINAC2. Thereafter, they reach the Proton Synchrotron Booster (*PSB*) where the proton bunches are formed. Protons in the PSB are further accelerated to relativistic energies and are transferred into the Proton Synchrotron (*PS*) with a momentum of 1.4 GeV/c. Next, the PS accelerates the bunches further to 25 GeV/c and ejects them into the Super Proton Synchrotron (*SPS*) [38]. After acceleration in the SPS, the protons have reached momenta of 400 or 450 GeV/c, depending on its cycle. From there, they are transferred to the LHC or are delivered

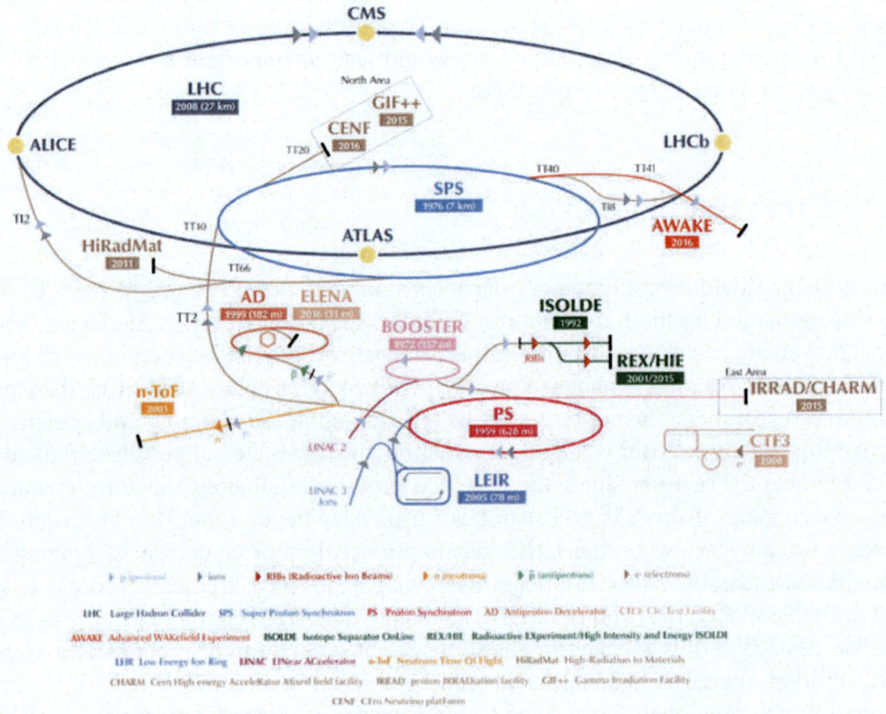

Fig. 2.4 CERN accelerator complex and the experiments as of 2017. Original figure taken from Ref. [39]. Point 5 and the North Area are highlighted

to fixed target experiments in the SPS North Area, cf. Sect. 6.4.1. The minimum spacing between proton bunches in the LHC is 25 ns, or equivalently 7.5 m, corresponding to a maximum of 2808 bunches per beam. In 2018, filling the LHC took about 15 min.

The acceleration of the protons in LHC is done via electromagnetic waves with a radiofrequency of $f_{\text{RF}} = 400.789\,\text{MHz}$. The waves are generated in eight radiofrequency cavities per beam. At each of the eight cavities, a single proton gains approximately 2 MeV in kinetic energy. By this means, the theoretical acceleration time from the injection to the LHC design energy would amount to 36 s.

Electromagnetic repulsion of the protons in the bunches and the fact that not all protons are equally affected by the acceleration in the cavities require permanent focussing of the proton beam. 392 quadrupole magnets are operated to tighten the beam in the transverse direction. Corrections of the longitudinal momentum spread and other higher-order corrections are realised by sextupole, octupole and decapole magnets. Due to the gradual ramping up of the LHC magnets, the practically achieved acceleration time is rather 20 min.

After full acceleration, the counter-rotating proton bunches are brought to collision at four locations within dedicated detector apparatuses [40–43]. In order to maximise the collision rate, the proton bunches are squeezed to transverse cross-section diameters close to 10 µm. The average expected number of interactions dN_P that corresponds to a process P per time dt can be calculated using Eq. 2.16.

$$\frac{dN_P}{dt} = \sigma_P \cdot L \tag{2.16}$$

σ_P stands for the cross section of the interaction process which generally depends on the centre-of-mass energy. It can be computed using Feynman diagrams together with the consideration of the parton-distribution functions of the process initiating quarks or gluons in the protons. L symbolises the instantaneous luminosity that is solely determined by accelerator parameters as stated by Eq. 2.17.

$$L \propto \frac{N_b^2 \cdot n_b \cdot F}{\epsilon_n \cdot \beta^*} \tag{2.17}$$

In this equation, N_b is the number of particles per bunch, the number of bunches per beam is denoted as n_b, and F symbolises a geometric reduction factor due to the angle of the colliding beams. The normalised transverse beam emittance ϵ_n is a quantity that relates to the compactness of the beam, both in spatial as well as in momentum space, and reflects the quality of the bunch preparation in the pre-acceleration chain. β^* is the so-called beta function which describes the beam's transverse spread at the interaction point and is determined by the beam optics.

In 2018, LHC operated at a centre-of-mass energy of 13 TeV. Its peak luminosity (at point 5) was $21.4\,\text{nb}^{-1}\,\text{Hz}$, the average integrated luminosity per day was $903.3\,\text{pb}^{-1}$ and $67.86\,\text{fb}^{-1}$ of integrated luminosity was provided in total [44].

Assuming a 27.78 fb [45] cross section for the inclusive production of a pair of Higgs bosons, roughly 1900 pairs of Higgs bosons were produced in the collisions in 2018.

The measurement and reconstruction of all outgoing, stable particles from the interaction is done by sophisticated detectors. The Compact Muon Solenoid (*CMS*) experiment is the one that is located 100 m underground at LHC point 5 and is described in the next section.

2.3 Compact Muon Solenoid Experiment

The Compact Muon Solenoid (*CMS*) is a multipurpose detector [42]. It can be understood as a high-speed camera with, by now [46], more than $100 \cdot 10^6$ pixels that takes three-dimensional photographs of the collisions at a rate of 40 MHz. CMS is designed to observe a broad phase space of new physics phenomena that might occur in the proton-proton collisions. The detector is operated and its data are analysed by a collaboration of more than 4000 physicists, engineers, computer scientists and technicians from approximately 200 institutes in more than 40 nations.

The detector's main purpose is the measurement of all stable particles that emerge from the primary interaction process. Among those particles are protons, neutrons, kaons, pions, electrons, photons and muons, whereas the presence of neutrinos can only be inferred indirectly. For each particle, one is interested in its vectorial momentum $\mathbf{p}$, energy, charge sign and its origin. The latter is important because of *pile-up*, i.e. additional inelastic proton-proton collisions besides the one of interest. In general, pile-up causes substantial activity in the detector and needs to be treated by rejection of signals from particles that do not originate from the primary interaction vertex. Assuming an inelastic proton-proton cross section of 68.1 mb [47] in the fiducial region and taking the LHC peak luminosity in 2018 as a reference, cf. Sect. 2.2, the average number of collisions per bunch crossing at LHC may easily amount to values at the order of 40.

Section 2.3.1 describes the CMS detector design and highlights the current subdetector systems. For interpretation of the recorded detector data, the CMS collaboration follows the so-called Particle Flow reconstruction approach [48] which is outlined in Sect. 2.3.2.

2.3.1 CMS Detector Design

The CMS detector consists of specialised particle sub-detectors that are arranged cylindrically around the beam pipe. The entire detector is 15 m in diameter, 28.7 m long and weighs about 14 kt. Its name is derived from its core features: Compared to ATLAS [41], i.e. the other multipurpose detector at LHC, the CMS detector is relatively compact (C), muons can be measured accurately with it (M), and it contains a powerful solenoid magnet (S).

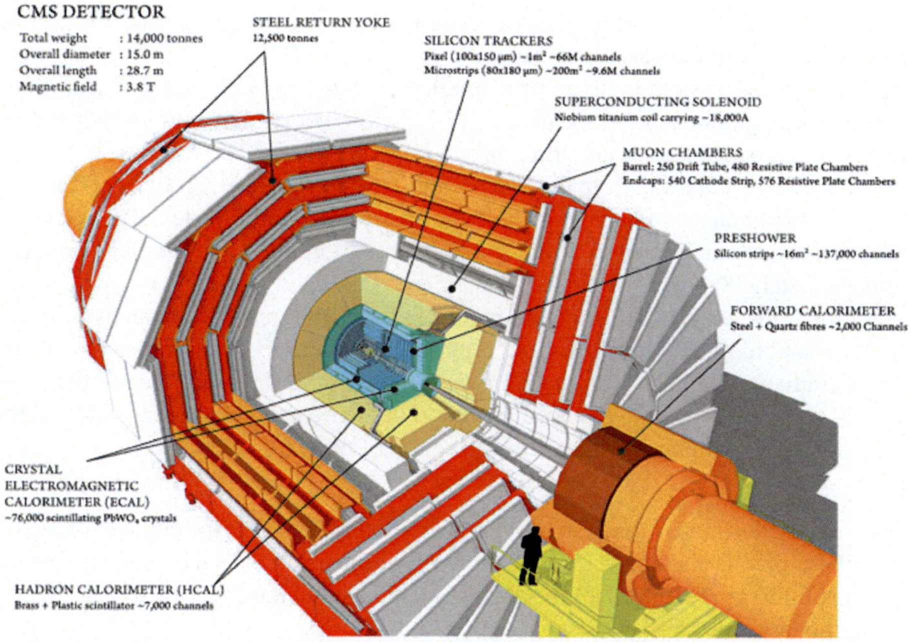

Fig. 2.5 Overview of the CMS detector and all its sub-detectors. Figure taken from Ref. [49]

Figure 2.5 shows an overview of the CMS detector. By convention, a right-handed coordinate system is defined such that the z-axis coincides with the beam axis, the x-axis points to the centre of the LHC and the y-axis is directed upwards. It is beneficial to describe the direction with respect to the beam axis using the so-called pseudorapidity η, cf. Eq. 2.18, because differences in η are approximately Lorentz invariant at relativistic particle energies.

$$\eta := -\ln \tan \frac{\theta}{2}\,, \quad \text{with } \tan\theta := \frac{\sqrt{p_x^2 + p_y^2}}{p_z} \tag{2.18}$$

In the following, the detector components are described, starting with the innermost toward the outside: The identification of particle trajectories (*tracks*) is performed by the tracking system, the calorimeters measure the particle energy, the tracks of charged particles are bent by the magnetic field and the outermost sub-detectors detect high-energetic muons. Since storing the full detector data from all collisions is impossible, CMS incorporates a two-staged triggering system [50] that reduces the stored event rate from the 40 MHz collision rate to $\mathcal{O}(1\,\text{kHz})$.

Tracking System

The CMS tracking system [51] is made of concentric layers of silicon-pixel and silicon-strip detectors located around the interaction point. Charged particles traversing the sensitive silicon diodes generate measurable signals, cf. Sect. 4.3, and create tracker hits. Using advanced pattern recognition techniques [52], the hits can be combined to form the particle tracks.

The tracking system consists of a smaller, highly-pixelated vertex detector and a larger, less granular tracker. Since 2017 [53], the vertex detector comprises four layers in the barrel region and three layers in each endcap. The first vertex layer is located only ~4 cm away from the LHC beam. With a pixel size of 100 μm × 150 μm, it allows for precise reconstruction of the primary interaction vertex. The tracker is further outside and consists of silicon strips. It is more than 5 m long and contains approximately 200 m^2 of silicon.

In total, the entire CMS tracking includes 75 million readout channels. It needs to withstand high particle fluxes and high radiation levels. To minimise the effects due to radiation damages, the tracking system is operated at -10°C. Further challenges are imposed by the required precision of the detector alignment [54] and the combinatorial complexity in the reconstruction of the particle tracks.

Calorimeter System

In general, calorimeters are destructive devices that fully absorb the incident particles, except for muons and neutrinos, for inference of the energy. As it is essential for the results of this work, Chap. 3 of this thesis is dedicated to the concepts of calorimetry in high-energy particle physics.

Apart from the forward calorimeters, the components of the CMS calorimeter system are installed within the solenoid volume in two regions: the barrel and the endcaps. The system can be divided into two components with different functionality. The first one is the electromagnetic calorimeter (ECAL) that measures the energy of electromagnetically induced showers and the second one is the hadronic calorimeter (HCAL) for characterisation of hadronically induced showers.

The ECAL [55] is a uniform calorimeter whose sensitive material are scintillating lead-tungstate ($PbWO_4$) crystals. The amount of produced light in the crystal is proportional to the deposited energy of the traversing particle shower and is read out by photodetectors. Each of the 61200 $PbWO_4$ crystals in the barrel region has a squared cross section of 2.2 cm × 2.2 cm and is 23 cm long (equivalent to ~26 radiation lengths). The two endcaps cover the range $1.479 < |\eta| < 3.0$ and are located 314 cm away from the interaction point. The 7324 crystals therein are slightly larger, namely 2.86 cm × 2.86 cm, and are 22 cm long.

The ECAL performance for single particles has been first evaluated in beam test campaigns. The electromagnetic energy resolution σ_e/μ_e as a function of the particle energy E can be parameterised using Eq. 2.19 [42].

$$\left(\frac{\sigma_e}{\mu_e}\right)^2 = \left(\frac{0.12}{E/\mathrm{GeV}}\right)^2 + \left(\frac{2.8\%}{\sqrt{E/\mathrm{GeV}}}\right)^2 + (0.3\%)^2 \, . \tag{2.19}$$

Using Z boson decays, the performance has been re-evaluated with LHC collisions [56]. It is found that the Z-mass resolution amounts to 2% when the decay products are measured in the barrel, while it is 2–5% in the endcaps. In order to maintain this resolution, the loss of crystal transparency due to radiation effects [57] during the lifetime of the LHC needs to be corrected for. This is achieved through continuous calibration with a dedicated light monitoring system [58, 59].

Hadronic showers are not stopped in the ECAL but reach the HCAL [60, 61]. The HCAL is a sampling calorimeters that uses 3.7 mm thick plastic scintillators as active material and brass as passive absorbers. Similar to the ECAL crystals, the scintillators are affected by radiation-induced degradation. The created light in the scintillators is guided through wavelength shifting fibres to silicon photomultipliers (*SiPM*). Due to space limitations, the barrel is made of 15 sampling layers, interspersed with 5 cm thick absorbers. It is about 79 cm thick, equivalent to approximately 5 nuclear interaction lengths. The endcaps comprise one initial sampling layer, followed by 18 layers grouped together into a single tower. The absorber thickness in the endcaps is 8 cm such that the total thickness in the endcaps is equivalent to 10.5 nuclear interaction lengths. The HCAL has a rather non-granular structure with 9072 readout channels in total of which 2592 are in the barrel region and 2592 are in the endcaps ($1.3 < |\eta| < 3.0$). The remaining 3888 channels are in the HCAL's forward endcaps ($3.0 < |\eta| < 5.0$) that extend CMS' hermicity, as well as in an area outside the solenoid volume that is instrumented with scintillator tiles.

The energy resolution for hadronically induced showers has been assessed separately for the HCAL barrel, endcap and forward regions [62–65]. The result for the endcap ($\sigma_{\text{h, HE}}/\mu_{\text{h, HE}}$) is quoted in Eq. 2.20 [64].

$$\left(\frac{\sigma_{\text{h, HE}}}{\mu_{\text{h, HE}}}\right)^2 = \left(\frac{106.1\%}{\sqrt{E/\text{GeV}}}\right)^2 + (4.0\%)^2 \tag{2.20}$$

In front of the ECAL and the HCAL in the endcap regions, the preshower calorimeter is located. It consists of two layers of 10^5 individually readout silicon strips interspersed with lead. Its granularity aids in improving the spatial resolution of the energy measurement, especially for identifying decays of neutral pions into collimated photons.

Solenoid Magnet

The solenoid magnet is a 12.5 m long cylinder with a diameter of 6.3 m. It weighs more than 12 kt, and hence it is the heaviest part of the entire CMS detector. Like the LHC dipole magnets, the coil is made of superconducting NbTi cables that are cooled down to 4.5 K. By this means, it can carry an electric current of up to 18 kA, which then creates a uniform, 3.8 T strong magnetic field inside its volume. The field is returned by an outer return yoke that is split into three layers and interspersed with the muon detectors. More than 2 GJ of energy is stored in the magnetic field during operation. A dedicated quench protection system prevents negative effects from an abrupt breakdown of the magnetic field, e.g. due to loss of superconductivity in the coils.

Muon System

The outermost sub-detector system includes the muon detectors [66]. Those are gaseous detectors interleaved by the iron return yoke and are only traversed by muons and by undetectable neutrinos. A traversing muon produces a detectable

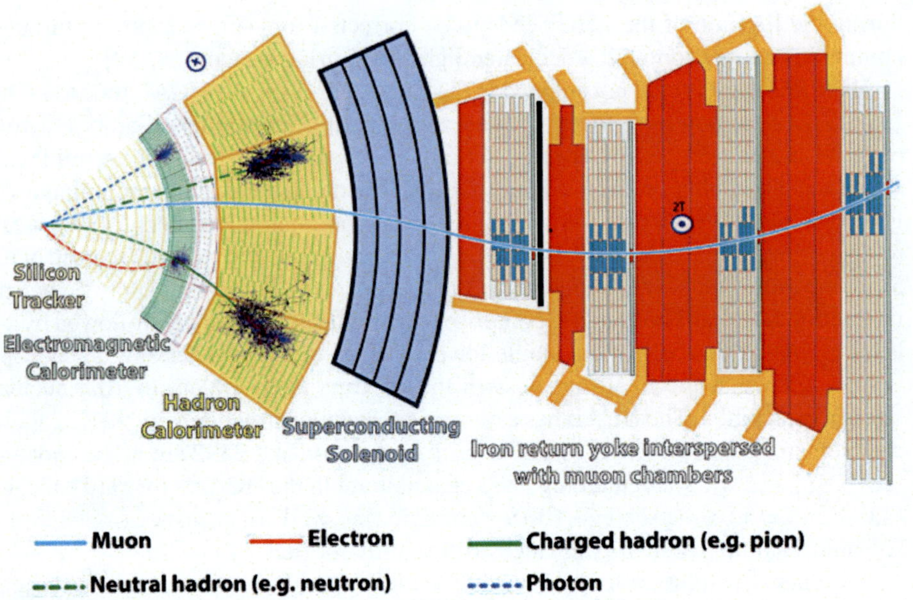

Fig. 2.6 Cross section view of the CMS detector and the typical signatures of different particle types therein. The Particle Flow algorithm exploits the measurements from all sub-detectors for the interpretation of the data as physics objects. Figure taken from Ref. [67]

electron avalanche caused by ionisation of the contained gas. The barrel layers are instrumented with 250 drift tube chambers which have a drift time of 380 ns. 3–4 layers, depending on the region, of 468 cathode strip detectors are located in the endcap region. Both the endcap and the barrel region are additionally complemented with fast resistive plate chambers that are used for triggering.

2.3.2 *Particle Flow Reconstruction*

The measured data in the CMS detector correspond to hits in the tracking and muon systems as well as energy depositions in the calorimeters. Given this data, the goal of the reconstruction is the identification and kinematic characterisation of the particles that traverse the detector. Furthermore, the reconstruction aims at the identification of all interaction vertexes, especially the primary ones with which detector activity due to pile-up can be mitigated.

For this purpose, CMS uses the Particle Flow algorithm [48]. This algorithm makes use of the measurements from all sub-detectors and considers the typical signatures of the different particles therein, cf. Fig. 2.6. Its output are physics objects that are interpreted as muons, photons, electrons, neutral and charged hadrons. These can be used further to derive composite objects, such as:

- Particle jets, that are interpreted as the product of hadronisation and fragmentation of a quark or a gluon from the interaction process.
- Missing transverse energy, which is attributed to the presence of neutrinos (or other new particles).

The physics objects provide a full description of the collision and represent the starting point for many physics analyses with the CMS detector.

Particle Flow proceeds in three successive steps:

1. In the first step, the data of each sub-detector is treated independently. Hits in the tracking and in the muon systems are combined to particle tracks, and energy depositions in the calorimeters are combined to energy clusters.
2. Next, the locally reconstructed tracks and clusters are linked to each other. The deposited energy in calorimeter clusters to which multiple particle tracks can be assigned is split up and linked to the respective tracks.
3. The linking is evaluated globally and possible ambiguities are resolved. Among all matching hypotheses, only the best ones are kept. Finally, the derived objects (e.g. jets, missing transverse energy) are computed.

A successful application of Particle Flow requires a precise track resolution in the tracking system and sufficient granularity of the calorimeter system. The latter is important both for the spatial resolution of the energy clusters and for the separation of collimated particles. Provided sufficient granularity, the ambiguities in the assignment of particle tracks to energy clusters, when multiple particles impinge, are minimised. Moreover, the identification of energy depositions from neutral particles, that do not leave hits in the tracking system, is enhanced with granular calorimeters. This is particularly crucial for the optimisation of the jet energy resolution using Particle Flow.

2.4 High Luminosity LHC Upgrade

LHC will remain the most powerful particle accelerator for another two decades. In 2006, the European Strategy for particle physics recommended to "fully exploit the physics potential of the LHC", and has reconfirmed this recommendation in 2013. Reducing the statistical precision of physics measurements at LHC by half after 2020 would however take more than ten years if its current instantaneous luminosity was kept. Hence, a major increase of the accelerator's instantaneous luminosity is foreseen in the mid 2020s. The design centre-of-mass energy of 14 TeV remains unchanged. This luminosity increase requires major upgrades of the LHC acceleration complex. The upgraded LHC is referred to as High Luminosity-LHC, or *HL-LHC*. The HL-LHC upgrade plan, features and operation parameters discussed in this section are quoted from its technical design report [68].

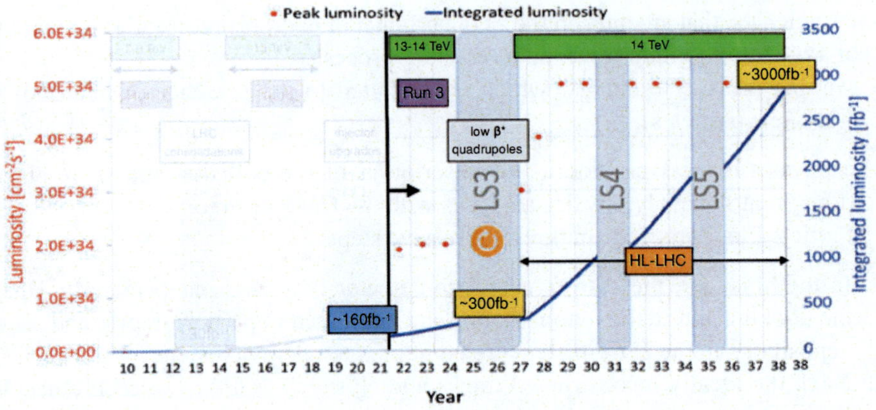

Fig. 2.7 Nominal LHC and HL-LHC luminosity scenario. Figure taken and modified from Ref. [68]

At the HL-LHC a nominal instantaneous luminosity of $L = 5 \cdot 10^{34}\,\mathrm{cm}^{-2}\mathrm{s}^{-1}$ is foreseen that corresponds to five times the LHC design value. With 160 days of physics operation per year and physics efficiencies of 50%, $250\,\mathrm{fb}^{-1}$ of integrated luminosity can be collected each year, and $3000\,\mathrm{fb}^{-1}$ in total before 2040.

For this, HL-LHC relies on novel magnet and accelerator technology such as several 11-12 T Nb_3Sn-based magnets, superconducting cavities for beam rotation (crab cavities) [69], new beam collimation techniques and high power superconducting links. These technologies will allow to increase the number of protons per bunch from LHC's design value of $1.15 \cdot 10^{11}$ to $2.2 \cdot 10^{11}$, to lower the normalised emittance from 3.75 µm to 2.50 µm, and to reduce the minimum β^* from 0.55 m to 0.2 m. In order to fully profit from the lowered β^*, the crab cavities, for which the civil engineering has already started in 2019, are essential for restoring the geometric reduction factor to the LHC value. In addition, they will be used for levelling the instantaneous luminosity at a constant value.

The higher instantaneous luminosity leads to higher radiation levels. The bottlenecks for operating the current LHC in the anticipated radiation environment are sudden electric breakdowns of the beam focussing magnets at the interaction points, cryogenics, collimators and the radiation hardness of the electronics. Besides, renovations are being considered for the quench protection system, for the machine protection systems, to e.g. deal with a breakdown of the kicker magnets, and for the system for remote exchange or repair of beam instrumentation.

Altogether, HL-LHC is a major and challenging upgrade. The HL-LHC design study was initiated in 2010 [70] and its proposal [71] was finally approved in June 2016. Besides the aforementioned civil engineering for the crab cavities, the upgrade of its injectors [72] is a related, ongoing activity during the current long shutdown of the LHC. The total material cost for the HL-LHC upgrade is estimated to be around 950 MCHF.

HL-LHC's nominal luminosity scenario is shown in Fig. 2.7. In this scenario, many decisive operation parameters are considered with a 50% margin. Therefore, it is possible that the instantaneous luminosity may even reach $L = 7.5 \cdot 10^{34}\ \mathrm{cm^{-2}s^{-1}}$. This corresponds to an average pile-up of 200. It is not straightforward to predict if the experiments will be able to make full use of the collisions in such a busy environment. Already in the nominal scenario, the average pile-up is anticipated to be around 140 and upgrades to the LHC experiments are necessary. The replacement of the CMS endcap calorimeter, introduced in Chap. 4, is one of the approved upgrade projects that will aim at successful detector operation and measurements at HL-LHC.

References

1. Hawking S, Mlodinow L (2010) The grand design. Bantam Books, New York
2. Einstein A (1905) On the electrodynamics of moving bodies. http://dx.doi.org/10.1002/andp.200590006
3. Griffiths D (2008) Introduction to elementary particles, 2nd edn. WILEY-VCH Verlag GmbH & Co., Weinheim
4. Altarelli G (2013) Collider physics within the standard model: a Primer. http://arxiv.org/abs/1303.2842 [hep-ph]
5. Yang CN, Mills RL (1954) Conservation of isotopic spin and isotopic gauge invariance. Phys Rev 96:191–195. http://dx.doi.org/10.1103/PhysRev.96.191
6. Noether E (1918) Invarianten beliebiger Differentialausdrücke. Nachrichten von der Gesellschaft der Wissenschaften zu Göttingen, Mathematisch-Physikalische Klasse, pp 37–44. http://eudml.org/doc/59011
7. Aaij R, et al. (2019) Observation of a narrow pentaquark state, $P_c(4312)^+$, and of the two-peak structure of the $P_c(4450)^+$. Phys Rev Lett 122:222001. http://dx.doi.org/10.1103/physrevlett.122.222001
8. Patrignani C, et al. (Particle Data Group) The review of particle physics. Chin Phys C 40: 100001. http://dx.doi.org/10.1088/1674-1137/40/10/100001
9. Super-Kamiokande Collaboration (1998) Evidence for oscillation of atmospheric neutrinos. Phys Rev Lett81:1562. http://dx.doi.org/10.1103/physrevlett.81.1562
10. Schwinger J (1948) Quantum electrodynamics. I. A covariant formulation. Phys Rev 74:1439–1461. http://dx.doi.org/10.1103/PhysRev.74.1439
11. Schwinger J (1949) Quantum electrodynamics. II. Vacuum polarization and self-energy. Phys Rev 75:651–679. http://dx.doi.org/10.1103/PhysRev.75.651
12. Schwinger J (1949) Quantum electrodynamics. III. The electromagnetic properties of the electron-radiative corrections to scattering. Phys Rev 76:790–817. http://dx.doi.org/10.1103/PhysRev.76.790
13. Feynman RP (1949) Space-time approach to quantum electrodynamics. Phys Rev 76. http://dx.doi.org/10.1103/PhysRev.76.769
14. Weinberg S (1973) Non-abelian gauge theories of the strong interactions. Phys Rev Lett 31:494–497. http://dx.doi.org/10.1103/PhysRevLett.31.494
15. Gross DJ, Wilczek F (1973) Ultraviolet behavior of non-abelian gauge theories. Phys Rev Lett 30:1343–1346. http://dx.doi.org/10.1103/PhysRevLett.30.1343
16. Fritzsch H, Gell-Mann M, Leutwyler H (1973) Advantages of the color octet gluon picture. Phys Lett B 47:365–368. http://dx.doi.org/10.1016/0370-2693(73)90625-4
17. Kaczmarek O, Zantow F (2005) Static quark-antiquark interactions in zero and finite temperature QCD: I. Heavy quark free energies, running coupling, and quarkonium binding. Phys Rev D 71:114510. http://dx.doi.org/10.1103/PhysRevD.71.114510

18. Glashow SL (1961) Partial-symmetries of weak interactions. Nucl Phys 22:579–588. http://dx.doi.org/10.1016/0029-5582(61)90469-2
19. Weinberg S (1967) A model of leptons. Phys Rev Lett 19:1264–1266. http://dx.doi.org/10.1103/PhysRevLett.19.1264
20. Higgs PW (1964) Broken symmetries, massless particles and gauge fields. Phys Lett 12:132–133. http://dx.doi.org/10.1016/0031-9163(64)91136-9
21. Englert F, Brout R (1964) Broken symmetry and the mass of gauge vector mesons. Phys Rev Lett 13:321–323. http://dx.doi.org/10.1103/PhysRevLett.13.321
22. Woithe J, Wiener GJ, Van der Veken F (2017) Let's have a coffee with the standard model of particle physics! Phys Edu 52:034001. http://dx.doi.org/10.1088/1361-6552/aa5b25
23. Dirac PAM (1928) The quantum theory of the electron. Proc R Soc Lond A 117:610–624. http://dx.doi.org/10.1098/rspa.1928.0023
24. Fließbach T (2008) Elektrodynamik - Lehrbuch zur Theoretischen Physik II. Spektrum Akademischer Verlag, Heidelberg
25. ATLAS Collaboration (2012) Observation of a new particle in the search for the standard model Higgs boson with the ATLAS detector at the LHC. Phys Lett B 716:1–29. http://dx.doi.org/10.1016/j.physletb.2012.08.020
26. CMS Collaboration (2012) Observation of a new boson at a mass of 125 GeV with the CMS experiment at the LHC. Phys Lett B 716:30–61. http://dx.doi.org/10.1016/j.physletb.2012.08.021
27. Kaiser D (2005) Physics and Feynman's diagrams. Am Sci 93:156–165. https://web.mit.edu/dikaiser/www/FdsAmSci.pdf
28. Baglio J, et al. (2013) The measurement of the Higgs self-coupling at the LHC: theoretical status. JHEP 2013:151. http://dx.doi.org/10.1007/JHEP04(2013)151
29. Bertone G, Hooper D, Silk J (2005) Particle dark matter: evidence, candidates and constraints. Phys Rep 405:279–390. http://dx.doi.org/10.1016/j.physrep.2004.08.031
30. Peebles PJE, Ratra B (2003) The cosmological constant and dark energy. Rev Mod Phys 75:559–606. http://dx.doi.org/10.1103/RevModPhys.75.559
31. Einstein A (1916) Die Grundlage der allgemeinen Relativitätstheorie. Annalen der Physik 49. http://dx.doi.org/10.1002/andp.19163540702
32. Evans L, Bryant P (2008) LHC machine. JINST 3:S08001. http://dx.doi.org/10.1088/1748-0221/3/08/s08001
33. Brüning O, et al. (2004) LHC design report volume 1: LHC main ring. CERN yellow reports: monographs. http://dx.doi.org/10.5170/CERN-2004-003-V-1
34. Brüning O, et al. (2004) LHC design report volume 2: LHC infrastructure and general services. CERN yellow reports: monographs. http://dx.doi.org/10.5170/CERN-2004-003-V-2
35. Benedikt M, et al. (2004) LHC design report volume 3: LHC injector chain. CERN yellow reports: monographs. http://dx.doi.org/10.5170/CERN-2004-003-V-3
36. Assmann RW, Lamont M, Myers S (2002) A brief history of the LEP collider. Nucl Phys B, Proc Suppl 109:17–31. http://dx.doi.org/10.1016/S0920-5632(02)90005-8
37. Wyss C (200) The LHC magnet programme: from accelerator physics requirements to production in industry. https://cds.cern.ch/record/466530
38. CERN (2019) The super proton synchrotron. https://home.cern/science/accelerators/super-proton-synchrotron. Accessed 20 July 2019
39. Mobs E (2016) The CERN accelerator complex. Complexe des accélérateurs du CERN.. https://cds.cern.ch/record/2197559
40. ALICE Collaboration (2008) The ALICE experiment at the CERN LHC. A large ion collider experiment. JINST 3:S08002. http://dx.doi.org/10.1088/1748-0221/3/08/S08002
41. ATLAS Collaboration (2008) The ATLAS experiment at the CERN large hadron collider. JINST 3:S08003. http://dx.doi.org/10.1088/1748-0221/3/08/S08003
42. CMS Collaboration (2008) The CMS experiment at the CERN LHC. JINST 3:S08004. http://dx.doi.org/10.1088/1748-0221/3/08/S08004
43. LHCb Collaboration (2008) The LHCb detector at the LHC. JINST 3:S08005. http://dx.doi.org/10.1088/1748-0221/3/08/S08005

44. CMS Collaboration (2019) CMS luminosity - public results. https://twiki.cern.ch/twiki/bin/view/CMSPublic/LumiPublicResults#2018_proton_proton_collisions. Accessed 28 Nov 2019
45. Grazzini M, et al. (2018) Higgs boson pair production at NNLO with top quark mass effects. JHEP 59. https://doi.org/10.1007/JHEP05(2018)059
46. Sonneveld J (2018) Commissioning and first results from the CMS phase 1 upgrade pixel detector. arXiv:1807.08987 physics.ins-det
47. Aaboud M, et al. (2016) Measurement of the inelastic proton-proton cross section at $\sqrt{s} =$ 13 TeV with the ATLAS detector. Phys Rev Lett 117:182002. http://dx.doi.org/10.1103/PhysRevLett.117.182002
48. CMS Collaboration (2017) Particle-flow reconstruction and global event description with the CMS detector. JINST 12. http://dx.doi.org/10.1088/1748-0221/12/10/p10003
49. About CMS. https://cms.cern/detector. Accessed 09 Oct 2019
50. CMS Collaboration (2017) The CMS trigger system. JINST 12:P01020. http://dx.doi.org/10.1088/1748-0221/12/01/P01020
51. CMS Collaboration (1997) The CMS tracker system project: technical design report. CERN-LHCC-98-006. http://cds.cern.ch/record/368412
52. CMS Collaboration (2014) Description and performance of track and primary-vertex reconstruction with the CMS tracker. JINST 9:P10009. http://dx.doi.org/10.1088/1748-0221/9/10/p10009
53. Dominguez A, et al. (2012) CMS technical design report for the pixel detector upgrade. CERN-LHCC-2012-016 (CMS-TDR-11). https://cds.cern.ch/record/1481838
54. Schleper P, et al. (2008) Alignment of the CMS silicon tracker using Millepede II. J Phys: Confer Ser 119. http://dx.doi.org/10.1088/1742-6596/119/3/032040
55. CMS Collaboration (1997) The CMS electromagnetic calorimeter project: technical design report. CERN-LHCC-97-033. https://cds.cern.ch/record/349375
56. CMS Collaboration (2013) Energy calibration and resolution of the CMS electromagnetic calorimeter in pp collisions at $\sqrt{s} = 7$ TeV. JINST 8. http://dx.doi.org/10.1088/1748-0221/8/09/p09009
57. Adzic P, et al. (2010) Radiation hardness qualification of $PbWO_4$ scintillation crystals for the CMS electromagnetic calorimeter. JINST 5:P03010. http://dx.doi.org/10.1088/1748-0221/5/03/P03010
58. Zhang L, et al. (2005) Performance of the monitoring light source for the CMS lead tungstate crystal calorimeter. IEEE Trans Nucl Sci 52:1123–1130. http://dx.doi.org/10.1109/TNS.2005.852661
59. Anfreville M, et al. (2008) Laser monitoring system for the CMS lead tungstate crystal calorimeter. Nucl Instrum Meth A 594:292–320. http://dx.doi.org/10.1016/j.nima.2008.01.104
60. CMS Collaboration (1997) The CMS hadron calorimeter project: technical design report. CERN-LHCC-97-031. http://cds.cern.ch/record/357153
61. CMS Collaboration (2012) CMS technical design report for the phase 1 upgrade of the hadron calorimeter. CERN-LHCC-2012-015. https://cds.cern.ch/record/1481837
62. Chatrchyan S, et al. (2010) Performance of the CMS hadron calorimeter with cosmic ray muons and LHC beam data. JINST 5:T03012. http://dx.doi.org/10.1088/1748-0221/5/03/T03012
63. Abdullin S et al (2008) Design, performance, and calibration of CMS hadron-barrel calorimeter wedges. Eur Phys J C 55:159–171. https://doi.org/10.1140/epjc/s10052-008-0573-y
64. Baiatian G, et al. (2008) Design, performance, and calibration of cms hadron endcap calorimeters. http://cds.cern.ch/record/1103003
65. Abdullin S (2008) Design, performance, and calibration of CMS forward calorimeter wedges. Eur Phys J C 53:139–166. http://dx.doi.org/10.1140/epjc/s10052-007-0459-4
66. CMS Collaboration (1997) The CMS muon project: technical design report. CERN-LHCC-97-032. https://cds.cern.ch/record/343814
67. Barney D (2016) CMS detector slice. http://cds.cern.ch/record/2120661. CMS-PHO-GEN-2016-001
68. Apollinari G, et al. (2017) High-luminosity large hadron collider (HL-LHC): technical design report V. 0.1. CERN yellow reports: monographs. CERN. https://cds.cern.ch/record/2284929

69. CERN (2018) Crab cavities animation. http://dx.doi.org/10.17181/cds.2621681
70. Rossi L (2011) LHC upgrade plans: options and strategy. https://cds.cern.ch/record/1407911
71. CERN Finance Committee (2016) The high-luminosity LHC project. 298th meeting of scientific policy committee. CERN/SPC/1068. https://cds.cern.ch/record/2199189
72. Garoby R, et al. (2013) Status and plans for the upgrade of the LHC injectors. CERN-ATS-2013-059. https://cds.cern.ch/record/1566636

Chapter 3
Shower Physics and Calorimetry

Particle energy measurements are an integral part in many particle physics experiments nowadays. This task is performed by calorimeters. Besides the completion of the four-vector of isolated, charged particles, the relevance of calorimeters has increased ever since the discovery of the W boson [1] because of their central role in the reconstruction of the energy flow in complex event signatures (jets and missing transverse energy).

Calorimeters are destructive devices. An incident particle enters the calorimeter, interacts with its material and produces a *shower* of secondary particles of lower energy.

The secondary particles produce the detectable signals whereby their energies may differ from the energy of the incident particle easily by multiple orders of magnitude. Although the calorimeter operation principle is simple, calorimeters are non-trivial detectors due to the wealth of distinct interaction processes of the shower particles with the material. Section 3.1 summarises the relevant interactions in this context.

The evolution of the particle cascades substantially depends on the type of the incident particle. In general, one distinguishes showers that are initiated by electromagnetically interacting particles and showers due to hadron incidences. The characteristics of the respective shower developments are described in Sect. 3.2. The shown equations in the first two sections of this chapter are taken from Refs. [2, 3].

The inference of the energy of the incident particle is based on the generated signals in the sensitive elements located in the calorimeter. Silicon and scintillators are chosen as sensitive material for HGCAL. The usage of silicon as sensors is discussed later in Sect. 4.3. Before, Sect. 3.3 explains how the full incidence energy can be derived and presents fundamental concepts related to the calorimetric performance. Thereafter, the two most commonly applied calorimeter designs, including their advantages and disadvantages, are outlined in Sect. 3.4.

Finally, simulation of the calorimeter performance is important in design studies as well as in the interpretation of the recorded, experimental data. Albeit it is usually

T. Quast, *Beam Test Calorimeter Prototypes for the CMS Calorimeter Endcap Upgrade*, Springer Theses, https://doi.org/10.1007/978-3-030-90202-5_3

computing intensive, accurate simulation of particle showers and their interaction with the calorimeter material are typically done sequentially. Section 3.5 elaborates the simulation paradigm of GEANT4, which is the (sequential) simulation framework deployed in this thesis.

3.1 Particle Interactions with Matter

The most abundant secondary particles in a typical calorimeter used in high-energy particle physics are electrons, positrons, photons, as well as stable charged and neutral hadrons. Those shower particles interact with the calorimeter material in various ways. In this section, the underlying interaction processes are summarised.

Charged Particles
Charged particles interact with the electromagnetic fields of the atoms in the material. For a relativistic particle, whose momentum in units of its rest mass ($= \beta \cdot \gamma$) is between 0.1 and 1000, the dominating energy loss is via the interaction with the atomic shell electrons. Hereby, sufficient energy is transferred to bring electrons into unbound states. This process is called ionisation. The mean energy loss is often indicated per distance and per density (X) of the material. It is also referred to as *mass stopping power* and is denoted as $\langle dE/dX \rangle$. For ionisation, it can be computed using the "Bethe equation", cf. Eq. 3.1 [2]. It is accurate within a few percent for heavy charged particles, i.e. electrons and positrons excluded, and for intermediate-Z materials.

$$\langle -\frac{dE}{dX}|_{\text{ionisation}} \rangle = K \cdot z^2 \cdot \frac{Z}{A} \cdot \frac{1}{\beta^2} \cdot \left(\frac{1}{2} \cdot \ln \frac{2 \cdot m_e \cdot c^2 \cdot \beta^2 \cdot \gamma^2 \cdot W_{\max}}{I^2} - \beta^2 - \frac{\delta(\beta \cdot \gamma)}{2} \right) \quad (3.1)$$

In this equation, z is the electric charge number of the incident particle and m_e is the electron mass. Z and A are the atomic and mass numbers of the material, $W_{\max}$ is the maximum transferable kinetic energy of a free electron colliding with an atom and I is the material-specific mean excitation energy. K is a material-independent proportionality constant. $\delta(\beta \cdot \gamma)$ is the so-called "density effect" correction that accounts for the polarisation of the material, or more precisely, for the shielding of the electric field far from the particle's path. An example of the mass stopping power for positive muons in copper as a function of their momentum is depicted as the red curve in Fig. 3.1. In this example, the mean energy loss reaches a minimum for $\beta \cdot \gamma \sim 3.5$. Particles with momenta corresponding to this minimum are referred to as *minimum ionising particles*.

Due to their low mass, their spin and the identity to the target electrons, the ionisation losses for electrons and positrons are inherently different from Eq. 3.1. Incident electrons undergo Møller scattering with the atomic electrons (taken as free electrons), whereas positrons are affected by Bhabha scattering. In addition to the nominal ionisation, both processes may lead to large energy transfers to atomic electrons. The resulting, fairly complicated, parameterisation of the mean energy

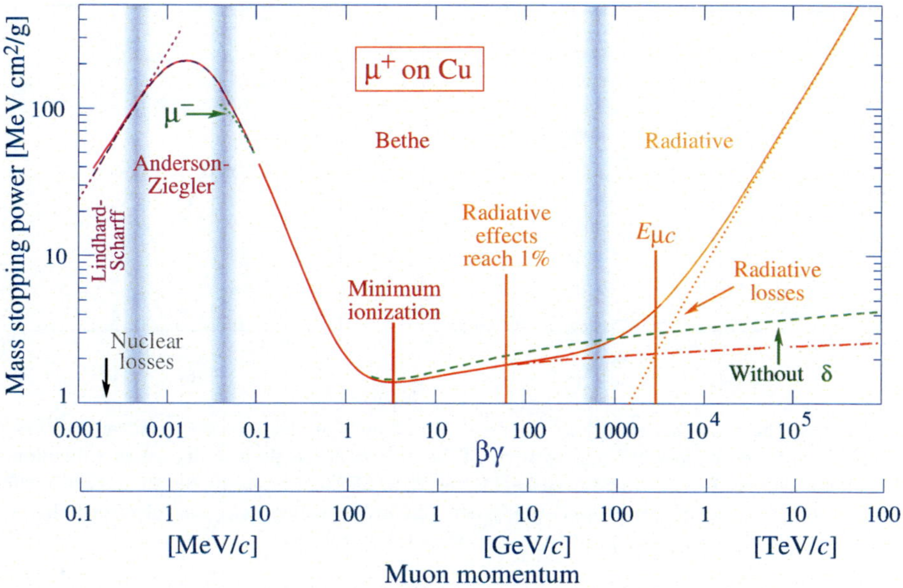

Fig. 3.1 The mean energy loss (mass stopping power) of positive muons in copper as a function of their momentum. The solid curve represents the total energy loss. The prediction by the "Bethe equation", cf. Eq. 3.1, is drawn in red. Figure taken from Ref. [2]

loss for electrons (positrons) are stated in Eq. 33.24 (33.25) in Ref. [2]. In essence, the difference between electrons and positrons is only effective at lower energies and practically irrelevant for energies above $\mathcal{O}\,(10\,\text{MeV})$.

Bremsstrahlung is the effect of photon emission by a charged particle in the presence of an electromagnetic field, in this case of the atoms. For increasing momentum of the incident particle, energy losses due to bremsstrahlung become more and more dominant. The *critical energy* ($\epsilon_{\text{crit.}}$) of a material is defined as the energy at which the average losses due to bremsstrahlung equals the average losses due to ionisation, cf. Eq. 3.2.[1]

$$\epsilon_{\text{crit.}} := E, \text{ when } \langle \frac{dE}{dX}|_{\text{ionisation}} \rangle = \langle \frac{dE}{dX}|_{\text{brems}} \rangle \tag{3.2}$$

For practical purposes, $\epsilon_{\text{crit.}}$ of a material is typically quoted for electrons or for positrons. However, it must, in principle, be defined for each particle type individually. Namely, bremsstrahlung losses are inversely proportional to the squared mass of the incident particle. Hence, lighter charged particles are affected more severely than heavier ones. For instance, muons in the GeV energy regime lose energy exclusively via ionisation with the material. Consequently, their critical energy exceeds the one for electrons by multiple orders of magnitude. At energies higher than $\epsilon_{\text{crit.}}$, the bremsstrahlung losses are approximately proportional to the particle's energy, cf.

[1] The difference to alternative definitions, e.g. by Rossi, are irrelevant to this discussion.

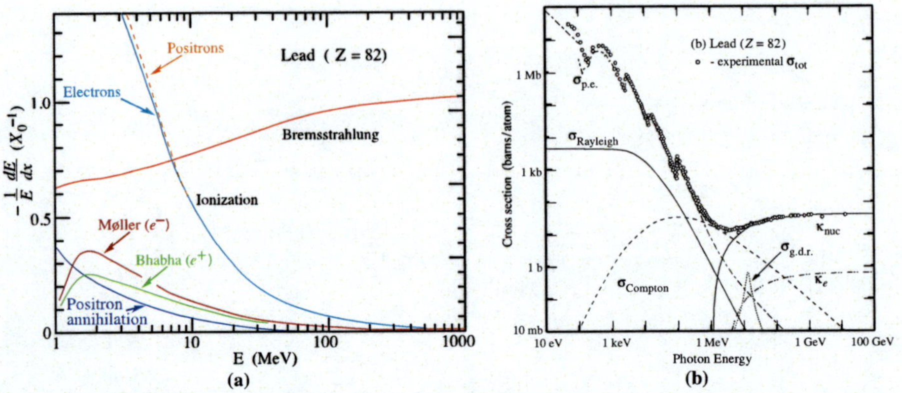

Fig. 3.2 **a** The fractional energy loss per radiation length of electrons and positrons in lead as a function of their energy. **b** Cross sections of the photon interactions with lead as a function of the photon energy. The shown processes are Rayleigh scattering (σ_{Rayleigh}), the photoelectric effect ($\sigma_{\text{p.e.}}$), Compton scattering (σ_{Compton}), pair production by the nucleus (κ_{nuc}) and electron fields (κ_{e}), as well as photo-nuclear absorption ($\sigma_{\text{g.d.r}}$). Figures taken from Ref. [2]

Eq. 3.3.

$$-\langle \frac{dE}{dX}|_{\text{brems}} \rangle \approx \frac{E}{X_0'} \,, \quad E \gg \epsilon_{\text{crit.}} \tag{3.3}$$

In this expression, X_0' is a fixed constant that generally depends on the particle type and on the material. X_0' will be specified further in Sect. 3.2.1.

The previously discussed processes that contribute to the energy loss of electrons and positrons in lead are presented in Fig. 3.2a. As can be seen, annihilation of positrons with atomic electrons is another sizeable contribution at lower energies. The result of this annihilation is the creation of a pair of photons. The different interactions of photons with matter are explained next.

Photons

The cross sections of the subsequently elaborated processes, of how photons interact with matter, as a function of the photon energy are shown for lead in Fig. 3.2b.

- Photons may scatter elastically with the atomic shell electrons in a process called Rayleigh scattering. Since no energy is transferred in this interaction, Rayleigh scattering does not contribute to the shower energy.
- The photoelectric effect [4] describes the interaction of photons with the atom in which shell electrons are expelled and leave the atom with some kinetic energy. It is the most dominant energy loss for low energetic photons.
- Photons may also scatter directly with shell electrons in a process called Compton scattering [5]. Substantial energy is transferred to the scattered electrons such that they leave the atom afterwards. Compton scattering is the most dominant energy loss at intermediate photon energies.

- If the energy of the photon is sufficiently high, it may convert into an electron-positron pair. This process requires the presence of an electric field, e.g. of the electron cloud or the atomic nucleus. This process is referred to as pair production and is dominant at high photon energies.
- For energies around $\mathcal{O}\,(10\,\text{MeV})$, photons may also excite the atomic nuclei into resonant states, most notably at the Giant Dipole Resonance [6]. After excitation, the target nucleus is broken up and emits nucleons as well as other high-energetic photons. However, the cross section for nuclear photon interaction is small compared to pair production. Interactions with the atomic nuclei predominantly occur through incident hadrons which are discussed in the following.

Hadrons

Like any charged particle, charged hadrons lose their energy via ionisation and bremsstrahlung, whereby the latter is only relevant for very high particle energies and is practically negligible in many applications. Besides, they can interact with the atomic nuclei via the strong force.

In this regard, nuclear spallation is the most common effect. Spallation is a two-stage process. It is initiated by an intranuclear cascade in which the incident hadron interacts quasi-freely with the atomic nucleons. In such a cascade, some of the nucleons may escape the nucleus. If the energy suffices, the entire nucleus may even break up (nuclear fission). The second stage of nuclear spallation is the so-called evaporation stage. There, the excited nucleus is de-excited via isotropic emission, cf. Fig. 2.26 in Ref. [3], of free nucleons and photons with energies at the order of 1 MeV. Whereas the cross section of the aforementioned electromagnetic interactions for charged particles and photons are well predictable using Quantum Electrodynamics, the prediction of spallation cross sections is based on phenomenological models and empirical measurements. The total cross section for fixed-target proton-proton interactions at 100 GeV amounts to $\sim$38 mb [3]. In general, it is found to scale with the projectile's size. For example, the pion-proton cross section at the same energy is only $\sim$24 mb.

Furthermore, the decay of neutral hadrons is noteworthy. Neutral pions practically decay instantaneously as their mean lifetime at rest is only $\sim 10^{-16}$ s. In 98.823% [2] of all cases, they decay into a pair of photons, which then undergo electromagnetic interactions. Slow neutrons may also decay due to their finite lifetime of around 880 s. However, slow neutrons are mostly captured by nuclei in the matter they traverse, whereby additional photons are emitted.

Altogether, hadron interactions with the material result in additional energy transfers either to overcome the binding energy in evaporation processes, to the recoil of the target nuclei or to the creation of non-interacting neutrinos. This energy is not detectable in the calorimeter and hence is regarded as *invisible*.

Table 3.1 Properties of selected materials characterising the energy loss and evolution of particle showers therein. Z and A are identical to their definitions in Eq. 3.1. ρ is the mass density and $dE/dx|_{\min}$ is the minimum energy loss per distance dx due to ionisation. X_0 is the radiation length and R_M denotes the Molière radius, which are both introduced in Sect. 3.2.1. $\lambda_{n(\pi)}$ is the nuclear (pion) interaction length, as defined in Sect. 3.2.2. All values are taken from Ref. [7]

Material	Z	A	ρ [g/cm^3]	$dE/dx\|_{\min}$ [MeV/cm]	$\epsilon_{\text{crit.}}$ (e^-) [MeV]	X_0 [cm]	R_M [cm]	λ_n [cm]	$\frac{\lambda_\pi}{\lambda_n}$
Carbon (C)	6	12.01	2.27	3.952	81.67	18.85	4.894	37.89	1.37
Silicon (Si)	14	28.09	2.33	3.876	40.19	9.370	4.944	46.52	1.27
Iron (Fe)	26	55.85	7.87	11.43	21.68	1.757	1.719	16.77	1.22
Copper (Cu)	29	63.55	8.96	12.57	19.42	1.436	1.568	15.32	1.21
Tungsten (W)	74	183.84	19.3	22.10	7.97	0.3504	0.9327	9.946	1.14
Lead (Pb)	82	207.2	11.4	12.74	7.43	0.5612	1.602	17.59	1.13

3.2 Development of Particle Showers

Instead of tracing all involved interactions at the microscopic level, a macroscopic description of the longitudinal and transverse shower development is expedient in most applications in calorimetry. For this purpose, material-dependent scaling quantities are introduced. They are listed in Table 3.1 for a few selected materials that were integrated in the HGCAL prototype.

The definitions of these scaling quantities will be provided in the following sections in which a phenomenological overview on electromagnetic and hadronic showers is given.

Electromagnetic showers are induced by incidences of photons, electrons or positrons, whereas hadronic showers are caused by incidences of hadrons. Due to the distinct interactions of the respective shower constituents with the material, the evolution of both shower types is fundamentally different.

3.2.1 Electromagnetic Showers

The main constituents of electromagnetic showers are electrons, positrons and photons. Qualitatively, photon emission due to bremsstrahlung and photon decays into electron-positron pairs cause successive multiplication of the shower particles. Figure 3.3a illustrates a representative cascade induced by a photon. With progressing ageing of the shower, the average energy per shower particle reduces, until no secondary particles can be produced any further.

The longitudinal shower energy density per depth segment can be modelled with a gamma distribution. In this context, the characteristic scale is the so-called *radiation length*, symbolised as X_0. It is defined as X_0' in Eq. 3.3 specified for electrons.

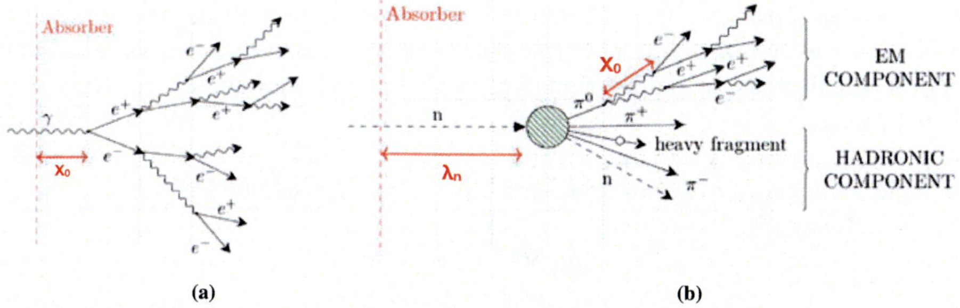

Fig. 3.3 Illustration of the development of **a** electromagnetically induced showers and **b** hadronically induced showers. Figures taken from Ref. [8]. (Courtesy Dr. Yasmine Israeli.)

Equivalently, X_0 corresponds to the average length after which a high-energetic electron has lost $1 - e^{-1} \approx 63.2\%$ of its original energy due to bremsstrahlung. A related concept is the concept of the photon interaction length. It is denoted as λ_{pair} and is defined as the mean free path of a photon prior to its decay into an electron-positron pair. As stated in Eq. 3.4, the photon interaction length of a material is proportional to the radiation length in the limit of high photon energies [9].

$$\lambda_{\text{pair}} \overbrace{\rightarrow}^{E\rightarrow\infty} \frac{9}{7} \cdot X_0 \tag{3.4}$$

The steady multiplication of shower particles stops when the average energy of the shower particles reaches the critical energy. Beyond this point, the average electrons and positrons lose their energy mainly via ionisation. Bremsstrahlung, and with it creation of further shower-intensifying photons, is more and more suppressed such that the shower slowly diminishes. From such considerations, the shower maximum $t_{\max}$, defined as the depth of the maximum shower energy density, can be computed from the incidence energy E using Eq. 3.5.

$$\frac{t_{\max}}{X_0} = \ln \frac{E}{\epsilon_{\text{crit.}}} + c_{e/\gamma} \tag{3.5}$$

Due to the numerical difference in X_0 and λ_{pair}, the coefficients $c_{e/\gamma}$ take distinct values for showers that are induced by electrons/positrons and for those that are caused by photons. Empirical studies [10] have found that it requires a depth D_{EM} corresponding to more than three times the maximum shower depth D_{EM} in order to contain 98% of the shower energy, cf. Eq. 3.6.

$$D_{\text{EM}}(98\%) \;=\; 3 \cdot (t_{\max} + 1.5X_0) \tag{3.6}$$

The shower development transverse to the main propagation axis is caused by multiple scattering of low-energy electrons in the Coulomb fields of the atoms. In addition, back-scattered particles from Compton scattering represent a sizeable contribution to the transverse shower spread. The associated scaling quantity is the Molière radius R_M. It is defined as the radius of an hypothetical cylinder around the shower's main axis in which 90% of the shower energy is deposited. Like X_0, R_M is a material-dependent quantity. From empirical studies [10], one has found that it requires a cylinder radius R_{EM} of at least twice the Molière radius to contain 95% of the shower energy, cf. Eq. 3.7.

$$R_{\mathrm{EM}}(95\%) = 2 \cdot R_M \tag{3.7}$$

3.2.2 *Hadronic Showers*

The showering cascade due to an incident hadron is initiated by a nuclear interaction of the projectile particle with an atomic nucleus. The probability dn for this primary interaction to occur in a thickness element $d\lambda$ at a depth λ can be computed from the total interaction cross section σ and the number density of the target nuclei in the material N_n, cf. Eq. 3.8.

$$\begin{aligned} dn(\lambda) &= \mathrm{P}(\text{interaction in } d\lambda \text{ at depth } \lambda) \cdot \mathrm{P}(\text{no interaction until depth } \lambda) \\ &= (\sigma \cdot N_n \cdot d\lambda) \cdot \lim_{dx \to 0} (1 - \sigma \cdot N_n \cdot dx)^{\frac{\lambda}{dx}} \\ &= \frac{d\lambda}{\lambda_0} \cdot \lim_{h \to \infty} (1 - \frac{\lambda}{\lambda_0 \cdot h})^h , \quad \text{with } \lambda_0 := \frac{1}{\sigma \cdot N_n} \\ &= \frac{d\lambda}{\lambda_0} \cdot \exp\left(-\frac{\lambda}{\lambda_0}\right) \\ \leftrightarrow \frac{dn}{d\lambda}(\lambda) &\propto \exp\left(-\frac{\lambda}{\lambda_0}\right) \end{aligned} \tag{3.8}$$

In this equation, λ_0 is the so-called *interaction length*. It is equivalent to the mean free path of a hadron before it undergoes a nuclear interaction, and it defines the scale that characterises the evolution of hadronic showers in a given material. As the involved total cross section depends on the projectile particle, the interaction length for nucleons, i.e. the nuclear interaction length λ_n, differs from the pion interaction length λ_π, i.e. for incident pions.

Secondary hadrons initiate further nuclear interactions that cause sub-showers at various locations in the material rendering hadronically induced showers more sparse, wider and deeper than their electromagnetic counterpart. The hadronic shower shape is driven by shower-by-shower fluctuations. For this reason, *there is no such thing as a "typical hadronic shower profile"* [11]. However, scaling laws, analogous to electromagnetic showers, that relate the shower containment to the material depth in units of the nuclear interaction length can be determined empirically, e.g. see Eq. 29 in Ref. [10].

Following the primary interaction, the subsequent cascade of shower particles consists of an electromagnetically and a hadronically interacting component, cf. Fig. 3.3b. The latter component includes undetectable constituents contributing to the invisible fractions of the total shower energy. In general, the average shower energy carried by the electromagnetic component f_{em} tends to increase with the incident particle energy E. This increase can qualitatively be explained by the fact that the production of a π_0 is an irreversible process because of its most likely decay into (electromagnetically interacting) photons. An example parameterisation of f_{em} for proton-induced showers in tin is given by Eq. 3.9 [12].

$$\langle f_{\text{em}} \rangle = 1 - \left(\frac{E}{E_0}\right)^{k-1} \quad , \quad k \sim 0.8 \, , \quad E_0 \sim 3\,\text{GeV} \tag{3.9}$$

Similar to the shower shape, the shower-by-shower fluctuations in the composition of hadronic showers are considerable. The implications on calorimetric measurements are discussed as part of the next section.

3.3 Basic Concepts in Calorimetry

The development of particle showers and the interaction of the constituents with the material has been discussed in the previous two sections. Now, the fundamental concepts related to the measurement of the incident particle energy E in a calorimeter are introduced. In fact, the term calorimetry is somewhat misleading as it traditionally refers to measurements of heat transfers. The expected temperature increase of a 300 GeV electromagnetic shower stopped in 10 kg of lead is only $\mathcal{O}$ (10 nK) and thus it is not a suitable, observable quantity. In particle-physics applications, a calorimeter is built such that the shower particles traverse some material that is sensitive to the shower particles. The shower particles then produce a measurable signal therein, e.g. via generated charge due to ionisation or via light either due to scintillation, transition- or due to Cherenkov radiation. Without saturation effects, the signal size is proportional to the shower energy that is deposited in the sensitive material. Minimum ionising particles can be used for defining the scale of the deposited energy, cf. Sect. 9.1.

The amount of detected shower energy in the sensors is called *visible energy*, denoted as E_{vis}. Depending on the calorimeter type, the visible energy may be anywhere between a tiny fraction to the full amount of the true incidence energy E. *Linearity* of a calorimeter implies proportionality between the mean visible energy and the corresponding incident particle energy, cf. Eq. 3.10.

$$\langle E_{\text{vis}}(E) \rangle \propto E \tag{3.10}$$

In general, any calorimeter should be designed to be linear, at least in the energy range for which it is intended. Otherwise, systematic biases may be propagated to the energy measurement of decaying particles, hereby impairing the calorimetric performance. For instance, one may consider a 300 GeV neutral pion which decays into a pair of photons with 150 GeV each. In a calorimeter with a slight nonlinearity, here assumed to be $E = 100\langle E_{\text{vis}}\rangle + 10\,\text{GeV}$, the pion energy would be underestimated systematically to be 290 GeV.

Besides the linearity, the second important figure of merit of a calorimeter is its *energy resolution*. As the evolution of the shower and the interactions of the particles with the sensitive material are stochastic processes, the visible energy exhibits a spread ΔE_{vis}. Linearity provided, this spread is in one-to-one correspondence with the energy resolution ΔE and it scales with the incident particle energy. The relative energy resolution can be parameterised as a quadratic sum of up to three terms, cf. Eq. 3.11.

$$\left(\frac{\Delta E_{\text{vis}}}{E_{\text{vis}}}\right)^2 \overset{\text{Equation 3.10}}{\widehat{=}} \left(\frac{\Delta E}{E}\right)^2 = \left(\frac{S}{\sqrt{E}}\right)^2 + C^2 + \left(\frac{N}{E}\right)^2 \tag{3.11}$$

The constants S, C and N are unique for each calorimeter. S is the coefficient of the so-called *stochastic term*. The stochastic term originates from the fact that the visible energy is proportional to the number of signal-generating shower particles in the sensitive material, which in turn is described by Poisson statistics. The *constant term* C incorporates the effects due to detector miscalibration, material inhomogeneity and other nonlinearities in the measurement of the deposited energies. In some devices, electronic noise in the instruments may also require the inclusion of a *noise term* $N \cdot E^{-1}$. Two explicit examples of the relative energy resolution, namely for the current CMS ECAL and HCAL, are given by Eqs. 2.19 and 2.20.

The calorimeter *response* is another core concept in calorimetry. It is defined as the measurable fraction of the shower energy. In a mathematical sense, it is a stochastic number with sizeable, intrinsic variation. The response is usually attributed to a calorimeter as a global characteristic. Yet, its definition may also be applied to the individual sensitive entities. Since the multitude of possible interactions with the calorimeter material is far greater for hadronic showers, the variation in the hadronic response is typically larger, and the resolution is worse, than for electron, positron or photon-induced showers. In the context of calorimetry in this thesis, e symbolises the mean response to electromagnetic showers and h to the hadronic component of hadronically induced showers. Then, the mean response to pion-induced showers can be written as in Eq. 3.12.

$$\pi = e \cdot [f_{\text{em}} + (1 - f_{\text{em}}) \cdot h/e] \tag{3.12}$$

A calorimeter is characterised to be *compensating*, if $h/e = 1$. This can be achieved intrinsically, i.e. on hardware level, by properly tuning the choice and arrangement of the calorimeter materials. A famous example that followed this approach is the

ZEUS calorimeter [13]. Non-compensation, i.e. $h/e \neq 1$, has two generally undesired consequences:

- As mentioned at the end of Sect. 3.2, fluctuations in the fraction of the electromagnetic shower component are sizeable. Those fluctuations represent an additional variation in the response to hadronic showers, cf. Eq. 3.12, and with it worsen the associated energy resolution. Hence, the hadronic shower energy resolution of non-compensating calorimeters tends to be worse than for compensating ones.
- Owing to the slight energy dependence of f_{em}, cf. Eq. 3.9, the response to hadronically induced showers in a non-compensating calorimeter is often nonlinear.

The impact of $h/e \neq 1$ must be studied specifically in each case and a generic quantification shall not be attempted here.

With additional efforts in the reconstruction of the incidence energy beyond Eq. 3.10, the positive effects of compensation may also be obtained in an intrinsically non-compensating calorimeter:

- In dual-readout calorimetry [14], one measures the shower energy using two processes with different h/e-ratios. This enables an event-by-event determination of f_{em} and with it a more accurate reconstruction of the incident hadron energy.
- Another approach is software compensation [15, 16]. In this procedure, the deposited energies are weighted in the sum of the visible energy according to their likelihood to originate from the electromagnetic or hadronic shower component. That degree can be quantified using the measured shower energy density. For this purpose, readout granularity of the calorimeter, cf. Sect. 3.4, is essential.

Apart from the energy measurement, a non-exhaustive list of possible calorimeter tasks includes: characterisation of the shower profiles and their start, inference of the shower impact position and direction of its axis, particle identification. While those tasks are not crucial for the energy measurement of single particles, they are indispensable for experiments using Particle Flow reconstruction like CMS, cf. Sect. 2.3.2. The exact application specifications and conceivable engineering limitations often influence the choice of a calorimeter design. The next section will present the two most common types.

3.4 Calorimeter Types

Calorimeters can be classified into two types, namely into homogeneous and sampling calorimeters. They are based on different design paradigms and have advantages and disadvantages over one another.

- In a ***homogeneous calorimeter***, the full detector volume is sensitive to the shower particles. For the application in high-energy particle physics, the energy measurement hereby is often based on light production in this material, e.g. in dense scintillating crystals. A currently existing example is the CMS ECAL, cf. Sect. 2.3.1.

Since the full shower is contained in the sensitive material, the number of signal-creating particles is maximised leading to overall good energy resolutions with homogeneous calorimeters. A possible disadvantage is the limited knowledge on the location of the energy depositions in the calorimeter.

- A ***sampling calorimeter*** consists of alternating layers of passive and sensitive material. The passive layers generate the particle showers. They are usually made of materials that are dense with respect to the targeted shower type. This can be lead for the measurement of electromagnetic showers or more cost-effective iron for hadronic showers. The sensitive material, also referred to as active material, samples the shower energy density in between the passive layers. Plastic scintillators, silicon or liquified noble gases are common choices for active materials. Currently existing examples of sampling calorimeters are the CMS HCAL, cf. Sect. 2.3.1, and the ATLAS liquid-argon calorimeter [17]. Also the upgraded CMS calorimeter endcap will belong to this class of calorimeters. By design, a sampling calorimeter can enable characterisation of the longitudinal shower profile. However, as the fraction of shower particles in the sensitive volumes, that contribute to E_{vis}, is smaller than for homogeneous calorimeters, the energy resolution, dominated by the stochastic term, tends to be worse in sampling calorimeters than for homogeneous calorimeters.

The feature of ***granularity*** can in principle be attributed to both calorimeter types. It is implied if the readout of the energy depositions is performed in a grid transverse to the principal shower direction. Although the degree of granularity is not well-defined, it is practical to consider a calorimeter to be granular if the size of the readout channels is smaller than the associated Molière radius. By this means, the transverse extent of electromagnetic showers is sampled with multiple sensitive entities and the distinguishability of closely separated incident particles in the calorimeter is enhanced.

3.5 Simulation with GEANT4

GEANT4 [18–20] simulates the propagation and interaction of particles in and with matter using the Monte Carlo method [21]. Nowadays, GEANT4 is the mostly utilised toolkit for the simulation of particle showers and calorimeter data. Since 2004, GEANT4 is used by the ATLAS, the CMS and the LHCb collaborations for the simulation of their detector responses. In order to estimate the effects of particle showers in a given material, the user has to build his or her own GEANT4 application. In such an application, the exact geometry needs to be implemented, all the physics processes must be defined, initial conditions need to be set, and the relevant output information must be computed. Hence, GEANT4 applications are typically highly specific to their use cases.

The shower simulation itself proceeds sequentially in the sense that each particle including its interactions with the material is tracked individually. In this context,

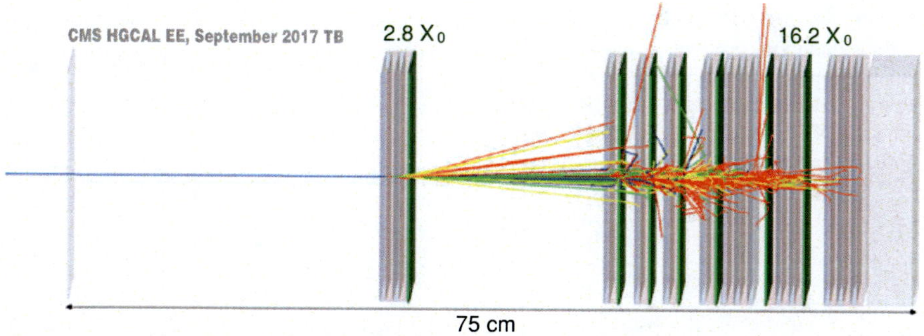

Fig. 3.4 GEANT4 simulation of a 90 GeV electron-induced particle shower. The tracks are colour-coded according to the momentum of the assigned shower particle (blue: highest momentum, red: lowest). Only constituents with momenta above 50 MeV/c are depicted. The shown geometry is inspired by a 7-layer CMS HGCAL beam test prototype that was tested in September 2017

a *track* represents the current instant in the simulation and can be understood as a snapshot of a particle. A track is updated in a series of *steps*. The steps are governed by the involved interaction processes. They occur both along the step, after the step and before the step, the latter only if the associated particle is at rest. Effectively, interactions change the dynamic physics quantities assigned to the track, they may generate secondary tracks (secondary particles) or they may stop the progressing of the track (particle decay or absorption).

Inside a volume defined as *sensitive*, the steps are accessible to obtain various delta information. Among other things, those include information on the energy loss of the associated particle in that volume. Lost energies, in turn, may be used to compute sums of deposited energies in the sensitive volumes, so-called simulated *hits*.

The sequential processing and updating of the tracks is referred to as an *event*. The set of tracks is stored in a stack. Tracks are popped from the stack and the stepping is performed. Conceivable secondary tracks from the interaction processes are pushed back. Tracks disappear if they leave the outermost volume (*world*), if they decay or undergo an inelastic collision, if their kinetic energy reaches zero and no rest interaction is defined, or if the user terminates it. The event simulation proceeds until no tracks are left in the stack to be processed further. An example from the simulation of a 90 GeV electron-induced particle shower in a simplified sampling geometry is shown in Fig. 3.4.

There is a large variety of physics processes [22] implemented in GEANT4. The available processes can be categorised into electromagnetic and hadronic interactions, decays, transportations as well as parameterised models. For the appropriate inclusion of all processes, GEANT4 follows an atomistic rather than an integral approach. In particular, the user is offered the flexibility to specify the desired processes for each involved shower particle individually. Furthermore, one can define cut-off ranges to set the applicability of certain processes, e.g. to optimise the simulation speed.

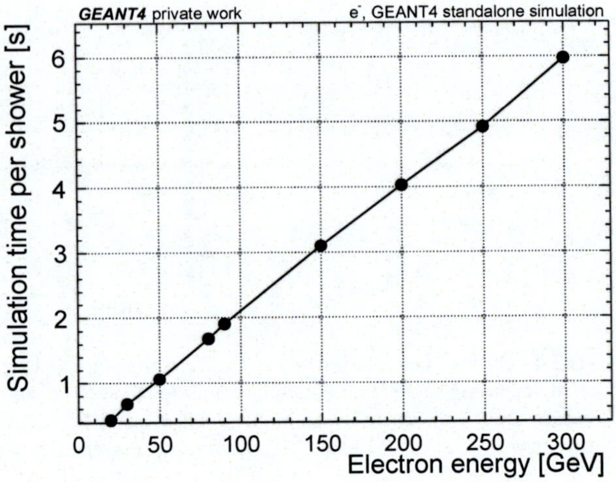

Fig. 3.5 Average CPU time for the simulation (without visualisation) of electron-induced showers with GEANT4 in the geometry depicted in Fig. 3.4. CPU: Intel© Xeon© CPU E5-1620

The physics of electromagnetic showers is relatively well modelled due to the availability of analytic QED calculations. They consist of a complete set of electromagnetic interactions which are valid from 1 keV up to the order of PeV. Furthermore, sophisticated approximations are deployed for low-energy photons, electrons/positrons and charged hadrons. By comparison, phenomenological models are largely used for the inelastic collisions between hadrons because respective analytic calculations are not always possible. Among those phenomenological models, there is the Bertini cascade [23] that describes low-energy hadron collisions with the nucleons treated as a compound. At high energies, the quark and gluon contents of the nucleons become more relevant for which Quark-Gluon [24] as well as the Fritiof [25] string models are more appropriate.

In general, this atomistic approach of specifying the interaction processes allows for tuning the simulation accuracy and the required computing resources. As different simulation results would otherwise be hard to compare, GEANT4 also provides common *physics lists* with predefined interaction definitions and particle-specific settings. The recommended physics list in high-energy particle physics experiments is *FTFP_BERT*. It incorporates a standard set of electromagnetic physics, the Bertini cascade for hadronic interactions below 5 GeV and the Fritiof model for hadronic interactions above 4 GeV. FTFP_BERT and one of its extensions that includes a CMS-specific tune of electromagnetic interactions (*FTFP_BERT_EMM*) are used throughout this thesis.

The drawback of the accurate, sequential simulation with GEANT4 is the associated computational cost. In good approximation, the computational time for the simulation of one shower increases linearly with the incident particle energy, cf. Fig. 3.5. This observation is consistent with the intuitive proportionality of the shower energy to the number of shower particles, or equivalently to the number of tracks, to be traced. As an example, the calorimeter simulation of the CMS experiment with GEANT4 may amount to the order of a few minutes for complex event signatures in the proton-proton collisions. Substantial improvements in the simulation time are not expected since GEANT4 is already highly optimised and "no hot spots of computing performance" are present [26]. In the mid- to long-term perspective, the R&D efforts conducted in the GeantV project [27] aim at the exploitation of vectorisation techniques and fine-grained parallelism. Computing gains ranging from half to one order of magnitude might be achieved hereby.

References

1. UA1 Collaboration (1983) Experimental observation of isolated large transverse energy electrons with associated missing energy at $\sqrt{s}$=540GeV. Phys Lett B122:103–116. https://doi.org/10.1016/0370-2693(83)91177-2
2. Patrignani C, et al. (Particle Data Group) The review of particle physics. Chin Phys C 40:100001. https://doi.org/10.1088/1674-1137/40/10/100001
3. Wigmans R (2017) Calorimetry - energy measurements in particle physics, No. 107, 2nd edn. Oxford Science Publications, Oxford
4. Einstein A (1905) Über einen die Erzeugung und Verwandlung des Lichtes betreffenden heuristischen Gesichtspunkt. Annalen der Physik 322:132–148. https://doi.org/10.1002/andp.19053220607
5. Compton AH (1923) A quantum theory of the scattering of X-rays by light elements. Phys Rev 21:483–502. https://doi.org/10.1103/PhysRev.21.483
6. Berman BL, Fultz SC (1975) Measurements of the giant dipole resonance with monoenergetic photons. Rev Mod Phys 47:713–761. https://doi.org/10.1103/RevModPhys.47.713
7. Particle Data Group (2012) Atomic and nuclear properties of materials for more than 300 materials. http://pdg.lbl.gov/2012/AtomicNuclearProperties/. Accessed 07 Sep 2019
8. Israeli Y (2018) Energy reconstruction in highly granular calorimeters for future electron-positron colliders. PhD thesis, Technische Universität München, Germany. http://mediatum.ub.tum.de/?id=1459326
9. Ott K (1953) Die Einzelprozesse der Elektronen und Lichtquanten. Springer, Berlin, pp 320–329. https://doi.org/10.1007/978-3-642-87230-3_26
10. Almadi U (1981) Fluctuations in calorimetry measurements. Physica Scripta 23:409–424. https://doi.org/10.1088/0031-8949/23/4a/012
11. Livan M, Wigmans R (2013) Misconceptions about calorimetry. arXiv:1704.00661 [physics.ins-det]
12. Gabriel TA, et al. (1994) Energy dependence of hadronic activity. Nucl Instrum Meth A 338: 336–347. https://doi.org/10.1016/0168-9002(94)91317-X
13. Bernstein A, et al. (1993) Beam tests of the ZEUS barrel calorimeter. Nucl Instrum Meth A 336:23–52. https://doi.org/10.1016/0168-9002(93)91078-2
14. Lee S, Livan M, Wigmans R (2018) Dual-readout calorimetry. Rev Mod Phys 90:025002. https://doi.org/10.1103/RevModPhys.90.025002
15. Adloff C, et al. (2012) Hadronic energy resolution of a highly granular scintillator-steel hadron calorimeter using software compensation techniques. JINST 7:P09017. https://doi.org/10.1088/1748-0221/7/09/p09017
16. Repond J, et al. (2018) Hadronic energy resolution of a combined high granularity scintillator calorimeter system. JINST 13:P12022. https://doi.org/10.1088/1748-0221/13/12/p12022
17. ATLAS Collaboration (1996) ATLAS liquid-argon calorimeter: technical design report. CERN-LHCC-96-041. https://cds.cern.ch/record/331061
18. Agostinelli S, et al. (2003) GEANT4: a simulation toolkit. Nucl Instrum Meth 506 no. 250–303 (2003). https://doi.org/10.1016/S0168-9002(03)01368-8

19. Allison J, et al. (2006) Geant4 developments and applications. IEEE Trans Nucl Sci 53:270–278. https://doi.org/10.1109/TNS.2006.869826
20. Allison J, et al. (2016) Recent developments in Geant4. Nucl Instrum Meth A 835:186. https://doi.org/10.1016/j.nima.2016.06.125
21. Metropolis N, Ulam S (1949) The Monte Carlo method. J Am Stat Assoc 44:335–341. https://doi.org/10.2307/2280232
22. Geant4 Collaboration (2019) Physics reference manual. http://geant4-userdoc.web.cern.ch/geant4-userdoc/UsersGuides/PhysicsReferenceManual/fo/PhysicsReferenceManual.pdf. Accessed 05 Nov 2019
23. Wright DH, Kelsey MH (2015) The Geant4 bertini cascade. Nucl Instrum Meth A 804:175–188. https://doi.org/10.1016/j.nima.2015.09.058
24. Folger G, Wellisch JP (2003) String parton models in Geant4. arXiv:0306007 [nucl-th]
25. Andersson B, Gustafson G, Nilsson-Almqvist B (1987) A model for low-p_T hadronic reactions with generalizations to hadron-nucleus and nucleus-nucleus collisions. Nucl Phys B 281:289–309. https://doi.org/10.1016/0550-3213(87)90257-4
26. Banerjee S, et al. (2019) CMS simulation in the HL-LHC Era. https://hepsoftwarefoundation.org/cwp/hsf-cwp-011-CMS-Simu-CWP-SDiego.pdf. Accessed 01 Nov 2019
27. Amadio G, et al. (2015) The GeantV project: preparing the future of simulation. J Phys: Confer Ser 664:072006. https://doi.org/10.1088/1742-6596/664/7/072006

Chapter 4
CMS Calorimeter Endcap Upgrade (HGCAL)

As stated in Sect. 2.4, the High Luminosity LHC (*HL-LHC*) will provide more than five times the design instantaneous luminosity of the LHC. Despite this being indispensable to measurements with rare topologies that rely on a large set of collisions to analyse, this increase in luminosity would pose significant challenges to the detection of particles with the current CMS detector. Therefore, the CMS collaboration is planning upgrades and even entire replacements of some of its subdetectors, cf. Fig. 4.1. The largely upgraded CMS detector is also referred to as the *Phase 2* CMS detector. Only by this means, physicists may effectively profit from the enormous collision datasets to be accumulated at the HL-LHC.

Among these upgrades is the replacement of both calorimeter endcaps. The necessity of this particular upgrade is explained in Sect. 4.1. Ever since its technical proposal [2] in 2015, the calorimeter upgrade proceeds along an ambitious timeline with hundreds of physicists, engineers, and supporting staff involved.

Given the engineering challenges, it is not surprising that in the course of this project's evolution so far, design choices have been modified constantly. Even some naming conventions have changed during the time of working for this thesis. Originally, the upgraded calorimeter endcap was referred to as *HGCAL* which stands for *High Granularity Calorimeter* and with it contains the core characteristic of its design. By now, the usage of the term *Calorimeter Endcap (CE)* is preferred in official reports. Nonetheless, this thesis mostly makes use of the initial abbreviation *HGCAL* for reference to this calorimeter.

Section 4.2 summarises the design of the upgraded calorimeter endcap. The explanation of all aspects related to HGCAL are beyond the scope of this thesis. Implementation details are elaborated in the Technical Design Report [3]. Rather, two topics that are of relevance for this thesis are briefly discussed here. Section 4.3 explains the use of silicon as sensitive material for large fractions of HGCAL. Afterwards, Sect. 4.4 explains how the event reconstruction in the CMS detector will profit from HGCAL.

T. Quast, *Beam Test Calorimeter Prototypes for the CMS Calorimeter Endcap Upgrade*, Springer Theses, https://doi.org/10.1007/978-3-030-90202-5_4

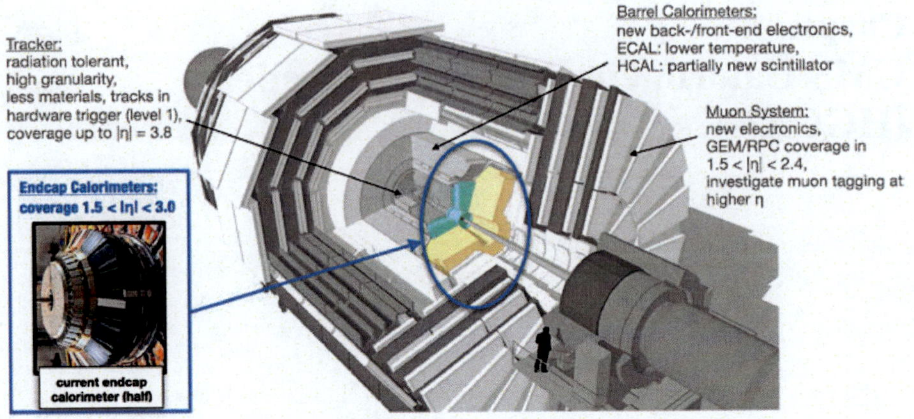

Fig. 4.1 Highlighting of selected Phase 2 upgrades to the CMS detector foreseen for the HL-LHC. Figure taken and extended from Ref. [1]

Especially the silicon sensors are a driving factor to the total cost of the final calorimeter, cf. Table 6.1 in Ref. [3]. A general experimental proof of concept of the proposed silicon-based design is thus essential. For this purpose, HGCAL prototypes have been assembled and tested under realistic conditions with particle beam since 2016. The so far largest beam test with more than 90 prototype modules equipped with a preliminary readout chip took place at the end of 2018. Section 4.5 describes the assembly of the beam test prototype modules and of the HGCAL prototype calorimeters which they formed.

4.1 Motivation and Requirements

The energy measurement and particle identification with the current CMS calorimeter endcaps at HL-LHC would be drastically impaired because of two intertwined reasons.

Increased Confusion due to Pile-Up Events
In 2018, the mean number of inelastic proton-proton collisions occurring at the same bunch crossing, i.e. *pile-up*, in the CMS detector amounted to $\sim$32[1] [4]. Figure 4.2a visualises the recorded activity in the CMS detector for a bunch crossing with 78 pile-up events recorded in 2012. What was considered exceptional back then will only be a bit more than half of the level of pile-up at HL-LHC. There, the mean pile-up is expected to be $\sim$128 and 16% of all bunch crossings will even have more than 140 inelastic collisions. Due to this increase, the assignment of measured energy depositions in the calorimeters to the particles emerging from the hard collision of

[1] $\sim$37 if one assumes the theoretically predicted cross section for inelastic proton-proton collisions.

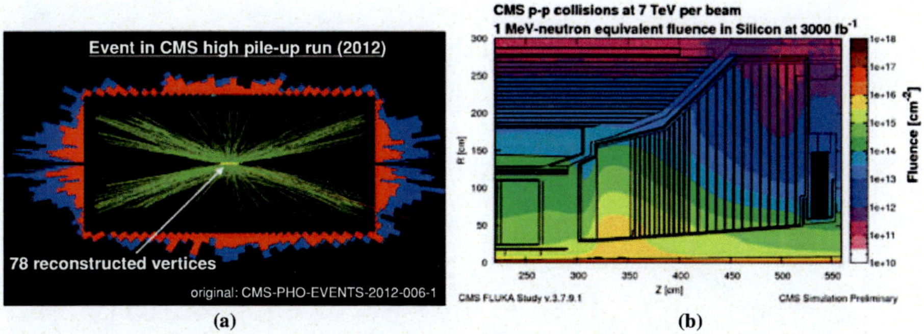

Fig. 4.2 **a** Event display of a collision with 78 proton-proton collisions in the same bunch crossing (pile-up) recorded with the CMS detector in 2012. Figure taken and modified from Ref. [5]. **b** Expected fluences in the calorimeter endcap at the end of the HL-LHC's operational lifetime. Figure taken from Ref. [3]

interest becomes more difficult. Energy depositions from pile-up would tend to be confused as energy from the showering particles of interest.

Degrading Sensors due to Radiation Damage
Even if pile-up was not an issue, the current ECAL's sensitive material could not cope with the expected radiation levels during the HL-LHC operation. Fluences up to 10^{16} 1 MeV-neutron-equivalents per cm^2 are anticipated in the forward regions of the endcap calorimeters, cf. Fig. 4.2b. These fluences exceed the design value of the installed $PbWO_4$ crystals in the CMS ECAL by more than one order of magnitude. The light output of the crystals is reduced, like it is shown in Fig. 4.3a, and consequently so is their energy response. A full compensation of this degradation using the available light monitoring system [6, 7] will not be feasible. The implication on the energy resolution of isolated electromagnetic showers after exposure to a fluence corresponding to the HL-LHC's operational lifetime, or equivalently 3000 fb^{-1}, is exemplarily shown in Fig. 4.3b. In this example, the stochastic term reaches 8–9% which is a fourfold [2] of its value after the current ECAL's nominal lifetime of 500 fb^{-1}.

Solution: All-New Calorimeter
Combining both the decreasing crystal performance and the increasing ambiguity in the association of energy depositions to particles, Fig. 4.3c demonstrates the worsened resolution in the detection of photons from the decay of a Higgs boson to two photons. Regular exchange of degraded $PbWO_4$ crystals of the current calorimeter endcap is not an option at HL-LHC. Therefore, it is clear that a new calorimeter must be installed replacing the current calorimeter endcaps. As the previous argumentation can be extended to the hadronic calorimeter as well, the latter is also subject to this upgrade.

The sensitive material of the upgraded calorimeter must be radiation tolerant for regions with especially harsh radiation levels. Pile-up can be treated in the offline event reconstruction adequately if the measurement of energy depositions

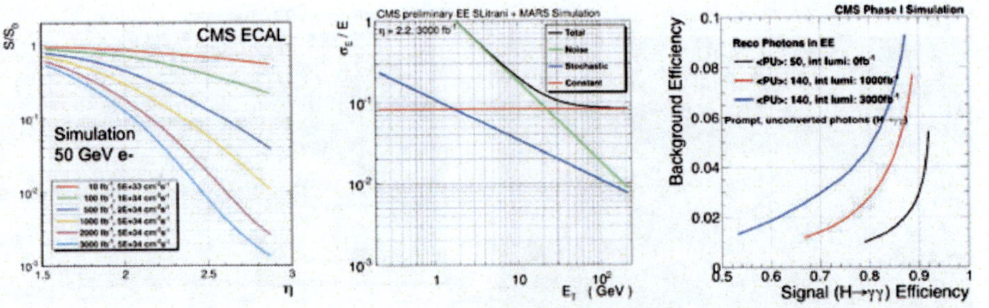

(a) **(b)** **(c)**

Fig. 4.3 **a** Expectation of the relative light output S/S_0 of the ECAL crystals for electron showers of 50 GeV as a function of the pseudorapidity η for various ageing conditions, corresponding to different integrated luminosities. **b** The energy resolution including ageing effects at $|\eta| = 2.2$ after 3000 fb^{-1} versus the transverse energy E_T. **c** Selection efficiency for photons from the Higgs boson decay to two photons versus the background efficiency estimated with the jet in γ+jet events for the ECAL Endcap for three MC samples which simulate: unaged detector at the end of Phase-I pile-up conditions, conditions early in HL-LHC operation, end of Phase-II conditions. Figures and their captions are taken from Ref. [2]

is sufficiently granular. Furthermore, analogous to other studies on future calorimeters, e.g. in Ref. [8], the additional recording of the time of each energy deposition, with $\mathcal{O}(50\,\text{ps})$ precision for the case of HL-LHC collision, can be helpful in the rejection of pile-up.

It should be annotated that the HGCAL design is aimed at offering "robustness and good performance through the full HL-LHC operational lifetime" [9]. Ultimately, HGCAL's calorimetric performance should be comparable to, and not be necessarily better than, the current CMS endcap calorimeter. However, it should do so in a much more busy environment. Hence, HGCAL represents a true improvement to the current CMS calorimeter. Its foreseen design is presented in the next section.

4.2 HGCAL Design Overview

Figure 4.4 gives an overview of the Calorimeter Endcap design. The CE will contain both an electromagnetic (*CE-E*) and a hadronic compartment (*CE-H*). The CE-E will be made of 26 sensitive layers interspersed with Cu, CuW and Pb absorber plates, equivalent to a total depth of 27.7 radiation lengths (X_0). By comparison, the sampling in the CE-H is less frequent. It will comprise 21 sensitive layers with steel as passive material in between. Dominated by the hadronic compartment, the total depth of HGCAL will be approximately 8.5 nuclear interaction lengths (λ_n).

Silicon will be used as active sensor material throughout the CE-E and otherwise in the detector regions that are exposed to high levels of radiation. Not only are the radiation-induced damages in silicon acceptable and to some extent treatable,

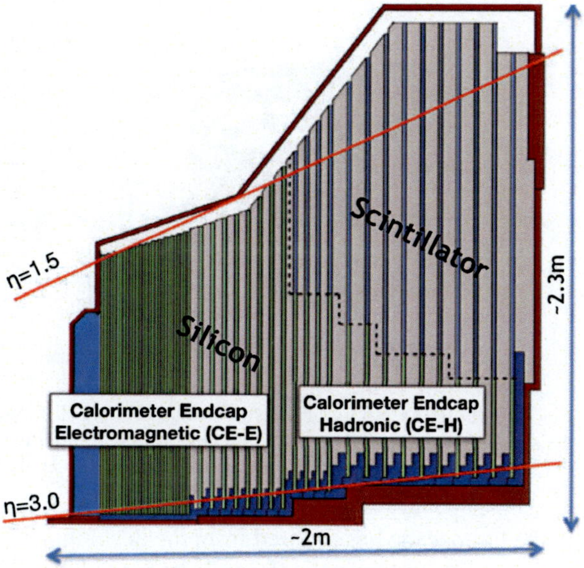

Fig. 4.4 Overview of the 2019 CMS calorimeter endcap upgrade design. Note that in 2021, the design was slightly modified: The sampling layers in the CE-E were reduced from 28 to 26, and in the CE-H from 22 to 21, respectively. Figure taken and modified from Ref. [10]

but silicon also provides fast signals allowing for time measurements of energy depositions with the required precision to effectively reject pile-up. The usage of silicon as active material is described in Sect. 4.3. Here, it is remarked that the total area of silicon in HGCAL will be more than $600\,m^2$. In order to cover this area in the most cost-effective manner, a hexagonal geometry for the silicon sensors was decided on. To reduce the number of modules, the baseline of the sensor size has been adjusted from the initially foreseen 6″ design to 8″. The basic sensitive units are $\sim$0.5–1 cm^2 large diodes on the silicon sensors. Each of the approximately six million diodes will be readout individually.

Ultimately, 30,000 silicon-based modules will be made from a sandwich structure of a printed circuit board housing the front-end electronics, the silicon sensor, gold-plated Kapton© as a biasing layer and a baseplate mainly for mechanical rigidity and cooling, cf. left side of Fig. 4.5. In the CE-E, the silicon modules will be used to create self-supporting sandwich structures which are commonly referred to as *cassettes*. For this, the modules will be installed on both sides of a copper cooling plate and will be closed with lead plates.

In regions with relatively low levels of radiation in the hadronic compartment, squared scintillating tiles with SiPM readout will be used as sensitive material. Their size will start at $4\,cm^2$ and will increase with increasing detector radius up to $30\,cm^2$. The total number of associated readout channels is 400,000, i.e. more than one order of magnitude less than the silicon-based region. Full lateral coverage in the transition region between the silicon- and the scintillator tiles coupled to SiPMs, cf. right side of Fig. 4.5, requires the incorporation of numerous variants of the hexagonal silicon modules. Still, scintillator plus SiPM sensors are compact and cost-effective and thus they are a good choice to equip the entire calorimeter at a reasonable cost. Their R&D

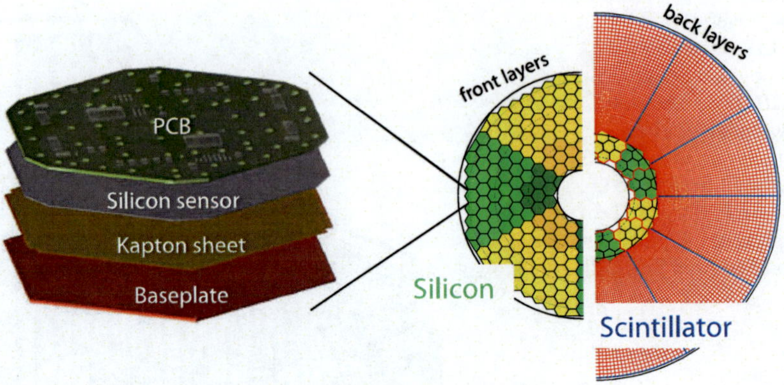

Fig. 4.5 (Left) Assembly of the silicon-based HGCAL modules and (right) their arrangement in the all-silicon and mixed layers. Graphics taken and modified from Ref. [3]

in the last decade has profited from the developments by the CALICE collaboration, e.g. [11, 12], and from the Phase 1 upgrade of the CMS HCAL [13]. The scintillator plus SiPM part is not further discussed anywhere in this thesis.

HGCAL's front-end electronics are particularly challenging [14]. The ultimate readout ASIC is called *HGCROC*. It will operate at a low noise level (<2000 e) and will cover a large dynamic range from 0.2 fC to 10 pC. A substantial novelty in calorimetry is that it is designed to also provide timing measurements with $\mathcal{O}(10\,\text{ps})$ resolution. In addition, HGCAL shall contribute to the L1 trigger of CMS through computing signal sums already in the front-end electronics with a maximum latency of 12.5 μs. These ambitious specifications must be realised within a compact ASIC because the available space for electronics and connectors in HGCAL is limited. Another limitation is defined by the cooling which caps the maximum power budget at 20 mW per readout channel.

Due to the anticipated degradation of the sensitive materials, HGCAL requires continuous energy calibration with minimum ionising particles throughout its lifetime, cf. Sect. 9.1. For this concept to function, the degradation of the sensors should however be kept to a reasonable minimum. In particular, the signal-to-noise ratio for minimum ionising particles should not go below one. Therefore, the full calorimeter system will be maintained at −35°C using an evaporative CO_2 cooling system [15]. With proceeding ageing of the silicon sensors, the bias voltage will steadily be increased up to 800 V towards the end of HGCAL's lifetime. Its final power consumption will amount to 125 kW per endcap.

When it is fully assembled, the upgraded calorimeter endcap will cover a region in pseudorapidity η from $1.5 < |\eta| < 3.0$. It will weigh 215 t per endcap and the acquisition of the required material will have costed more than 60 MCHF.

In summary, HGCAL is a necessary upgrade and in principle there are no major concerns regarding the particle detection technology. Yet, the project timeline for its realisation is relatively short and the engineering aspects are ambitious. In fact, the project was once described as "perhaps the most challenging engineering project ever undertaken in particle physics" [9].

4.3 Silicon as Active Material

Silicon is a semiconductor and is used as sensitive material in many applications in the detection of particles. Semiconductors alone are not suitable for the detection of ionising particles but the addition of dopants is necessary. With the doping, pn-junctions can be constructed to yield the equivalent to an ionisation chamber, a device that is sensitive to traversing charged particles. The passage of ionising particles through the silicon creates electron-hole pairs in its bulk that are subsequently collected by an electric field. The energy measurement is based on the collected charge carriers which cause a detectable signal at the electrodes. In general, the required energy of around 3.6 eV for creating the charge carriers in silicon is relatively low resulting in good energy resolutions. Figure 4.6a illustrates the main idea of using silicon for particle detection. In case of HGCAL, silicon is particularly preferred over other sensitive materials as it allows for a compact calorimeter design and provides fast signals. Furthermore, its robustness against radiation is acceptable.

The relevant concepts of using silicon as active material for particle detection are explained in this section, starting from the fundamental properties of semiconductors, followed by the working principle of the pn-junction diode and the energy measurement principle. At the end, the impact of radiation is discussed briefly. This discussion is kept to a minimum since radiation effects are not investigated in this thesis. Nevertheless, awareness of them is necessary for the understanding of some of HGCAL's design choices.

Broader introductions, than provided here, to semiconductors, their standalone working principles or their specific applications in the context of particle detection can be found in Refs. [16–18]. The impact of radiation on silicon is caused by many non-trivial effects and is a field of ongoing research. An overview of those is given in Ref. [19]. Other studies specific to the silicon sensors for HGCAL should also be mentioned: electrical properties of the HGCAL silicon sensors [20], radiation hardness of prototype sensors [21] or their intrinsic timing resolution close to 16 ps [22].

Fundamental Properties of Semiconductors

The atoms in a solid state material are typically arranged as a periodic lattice. By this means, the electron wavefunctions overlap in such a way that band structures in the allowed energy spectrum are created. Electrons bound to the atom are in the valence band. Free, movable electrons are in the conduction band. The gap in between both bands is called the *energy gap* or *forbidden band* and is depleted from any free charge carriers. While this energy gap is large for insulators and non-existent for conductors, it has an intermediate extent for semiconductors. In silicon, the energy gap amounts to approximately $E_g = 1.21$ eV. Thermal excitation of an electron in the valence band may transfer it into the conduction band rendering the semiconductor slightly

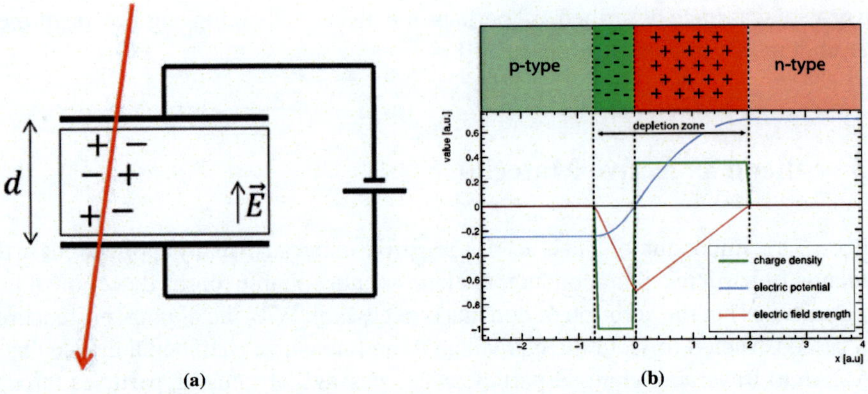

Fig. 4.6 **a** Working principle of silicon as an ionisation chamber for particle detection. **b** An instructive model of a pn-junction, as well as the electric potential and field strength therein

conductive. Not only electrons but also the absence of electrons, i.e. *holes*, in the lattice contribute to the semiconductor's conductivity. At thermal equilibrium, the latter generally depends on the temperature and on the so-called mobility $\mu_e(\mu_h)$ of the electrons (holes).

Pure semiconductors are called intrinsic. Only one out of 10^{12} atoms is typically ionised and the conductivity is minor. It can be enhanced when impurities are added into the material. Such semiconductors are then referred to as extrinsic or *doped*. The addition of phosphorus atoms (group 5 of the periodic system) into the silicon lattice (1 P:10^9 Si) creates additional levels in the energy gap close to the conduction band. An excess of freely moving electrons is the result and the doped silicon is characterised as being *n-type*. In this example, phosphorus is considered a *donor* material. Similarly, an excess of holes can be caused by adding levels in the energy gap close to the valence band which then attracts freely moving electrons. This can be achieved by introducing, for instance, a few boron atoms (group 3 of the periodic system) into the silicon lattice. In this case, boron serves as an electron *acceptor* and the doping is *p-type*.

The pn-Junction Diode

The equivalent to an ionisation chamber is created when n- and p-type doped silicon volumes are connected to form a pn-junction diode. Then, the freely moving electrons from the n-type region are attracted to the p-type region and vice versa for the holes. The region where this happens is called the *depletion zone* because it is depleted from freely moving charge carriers. Since the volume is initially neutral, this movement results in a distribution of charge densities in the depletion zone. The Poisson equation in Eq. 4.1 (and boundary conditions) may be applied to infer the corresponding electric potential $V(x)$ and with it the field strength in this zone. The difference in the electric potential on either side is called *contact potential* and amounts to $\mathcal{O}$ (1V) in typical applications with silicon.

$$\frac{d^2V(x)}{dx^2} = -\frac{\rho(x)}{\epsilon_0} \tag{4.1}$$

A simplified but instructive model assuming a rectangular distribution of charge carriers is shown in Fig. 4.6b. Using this model, the effective potential difference V_{eff} may be related to the capacitance of the junction C according to Eq. 4.2. The contact potential as well as an externally applied *bias voltage* contribute to V_{eff}. Besides the already introduced quantities, A is the junction's cross-section area and ρ is a junction-specific constant that depends on the concentration of the doping materials and the mobility of the electrons and holes.

$$C = A \cdot \left(\frac{\epsilon_0}{2 \cdot V_{\text{eff}} \cdot \rho}\right)^{0.5} \tag{4.2}$$

The application of a reverse bias voltage in the sense that positive polarity is applied to the n-type side and a negative one to the p-type increases the potential difference and enlarges the depletion zone. This enlarges the amount of generated charge in the silicon bulk and with it the diode's sensitivity to ionisation. In addition, it lowers the capacitance and with it the electronic noise in the readout. In practice, the maximum applied voltage is limited by the diode's resistivity that may cause sudden increases in the leakage currents (*breakdown*) at high enough voltages.

Large leakage currents I in the diode are generally undesired because they result in the reduction of the effective bias voltages and worsen the diode's noise behaviour. With increasing bias voltage and without breakdown occurring, the leakage currents increase until the depletion zone reaches its physical maximum. Moreover, the leakage current is highly dependent on the temperature, cf. Eq. 4.3, where k_b denotes the Boltzmann constant and T is the temperature in Kelvin.

$$I \propto T^2 \cdot \exp\left(-\frac{E_g}{2 \cdot k_b \cdot T}\right) \tag{4.3}$$

Charge Deposition and Energy Measurement

When plasma effects can be neglected that would otherwise affect the charge collection efficiency, the amount of created charge carriers in the depletion zone due to ionisation of an incident particle is proportional to its energy loss. The mean energy loss per thickness unit is described by the "Bethe equation", cf. Eq. 3.1. Yet, the distribution of deposited energy of ionising particles in $\mathcal{O}\,(100\,\mu\text{m})$ thick silicon is not Gaussian and the mean is driven by outliers. For the purpose of energy calibration of the diode, the most-probable energy loss (*MPV*) is often used instead. Landau [23] and Bichsel [24] have developed models for the MPV and have found that it reaches a constant plateau for increasing particle energies above $\beta \cdot \gamma > 100$ [25]. In this highly relativistic energy range, the *MPV* and with it the expected most-probable signal does not scale with the energy of the incident ionising particles. This will ultimately enable the energy calibration of the HGCAL diodes with non-showering, minimum ionising particles with rather arbitrary incidence energies.

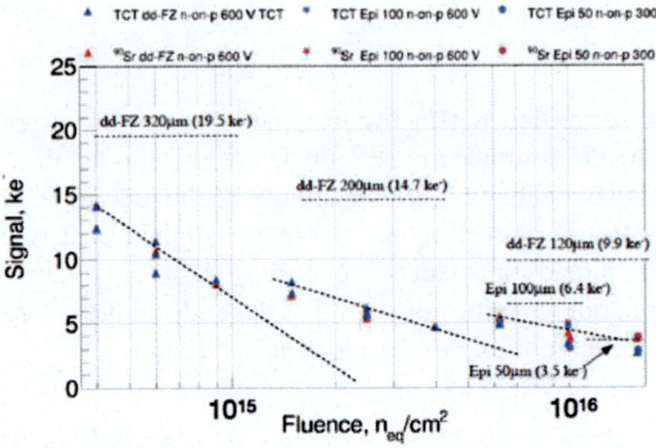

Fig. 4.7 Measured signal for different prototype silicon sensors after exposure to a certain neutron fluence. The range of fluences includes the maximum expected exposure after $3000\,\text{fb}^{-1}$ of integrated luminosity. Figure taken and modified from Ref. [21]

It should be remarked that the measured signal in the readout electronics is not produced by the original charge carriers but rather via induction. As consequence, signals may also be obtained from undepleted diodes. The signal rise time is at the order of a few nanoseconds, in principle allowing for timing inferences with resolutions at the order of $\mathcal{O}\,(10\,\text{ps})$.

Radiation Damage and Implications to HGCAL

Radiation may knock out atoms in the lattice and create point defects in the silicon. This leads to additional levels in the energy gap resulting in more trapping of free charge carriers and thus to a reduction of the efficiency in collecting the signal inducing charges. A common solution is the speed-up of the charge collection, hereby leaving less time for the trapping to occur. This can be achieved by increasing the electric field strength through the bias voltage. Potential changes in the depletion zone due to radiation may be limited by freezing out the underlying diffusion processes at cold temperatures. That explains why HGCAL is operated at −35°C. Despite these efforts, the effect of increasing leakage currents may not be prevented. In general, the corresponding increase ΔI per bulk volume V is proportional to the particle flux Φ_{eq} that the material was exposed to, cf. Eq. 4.4.

$$\Delta I = \alpha \cdot \Phi_{\text{eq}} \cdot V \tag{4.4}$$

Therefore, the silicon sensors placed in the regions of HGCAL with higher radiation levels are predominantly affected by increased leakage current and with it by the loss of energy resolution. The loss of signal was studied for HGCAL prototype sensors and one representative result is shown in Fig. 4.7. It demonstrates that sensors with smaller thicknesses and with it thinner depletion zones maintain a relatively higher signal after exposure to a certain level of radiation than thicker ones. For this reason, HGCAL will be equipped with sensors of varying active thicknesses of 120, 200 and $300\,\mu\text{m}$. By this means, the signal-to-noise ratio for minimum ionising particles will be kept above one throughout HGCAL's lifetime.

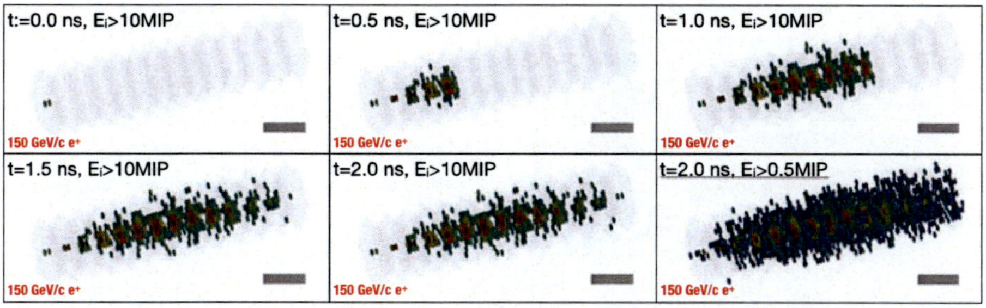

Fig. 4.8 Event "video" of a simulated 300 GeV/c positron-induced shower in a model of HGCAL's CE-E compartment. Timing measurements are assumed to be available for hits with energies above 10 MIP

4.4 5D Calorimetry

HGCAL can be considered a five-dimensional detector because it measures the energy density of particle showers therein both in space and in time. Hence, the derived calorimeter hits in HGCAL may ultimately be visualised as "videos" of the shower development. Figure 4.8 shows one example of such a video made from simulated calorimeter data with GEANT4.

The timing information provided by HGCAL will be essential in disentangling hits from the particle shower of interest and those from pile-up in the calorimeter, cf. the illustration in Fig. 4.9. The concept of using timing information for background rejection has already been proposed in the context of other design studies for future experiments, such as for the suppression of beam-induced backgrounds at CLIC [8, 26].

As it is shown in Fig. 4.10a, the expected Molière radius of electromagnetic showers in HGCAL is well below 3 cm. Given that the size of the readout cells in the CE-E is at least three times less than that, simultaneous and closely evolving electromagnetic showers can be resolved, cf. Fig. 4.10b. With the presence of pile-up, the association of calorimeter hits to a specific shower is realised by incorporation of all three spatial dimensions in an iterative clustering algorithm (*TICL*) inspired by Ref. [27]. The inclusion of the timing data into the clustering is currently being studied by the CMS collaboration. Due to the granularity and with a preliminary version of the clustering algorithm, a reasonable calorimetric performance can be sustained even in the presence of significant pile-up in the detector. For instance, Fig. 4.10c demonstrates the expected energy resolution for photon-induced showers in HGCAL for different pseudorapidities and pile-up scenarios. Implementation of online clustering algorithms are also foreseen both in the hardware- and in the high-level trigger at CMS. By this means, HGCAL will contribute to the event triggering of the CMS detector running at HL-LHC.

The spatial granularity will also be beneficial to fit HGCAL into the global event reconstruction using the Particle Flow paradigm with the CMS detector, cf.

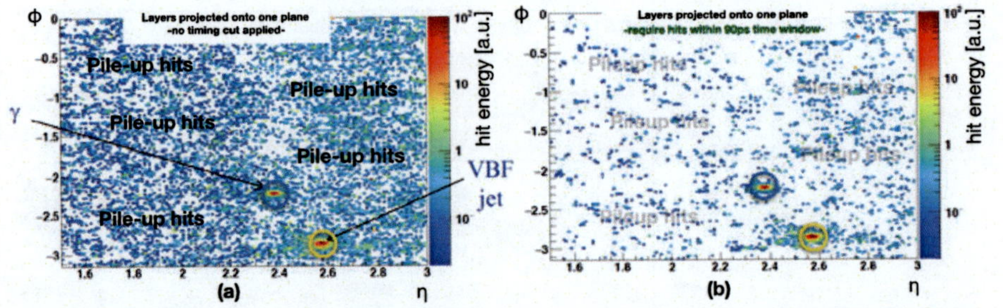

Fig. 4.9 Simulated signature of a Higgs boson from vector boson fusion (*VBF*) decaying into a pair of photons where one jet and one photon are measured within the same quadrant in HGCAL. All hits from all layers are projected into one plane. (a) All hits without application of any timing cut. (b) Hits within a 90 ps window around the time of the principal cluster are shown. Most of the undesired pile-up contributions are removed and the hits of interest are retained. Graphics taken and modified from Ref. [9]

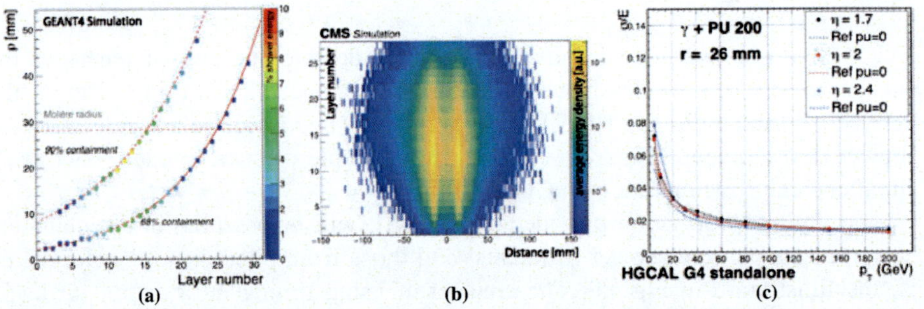

Fig. 4.10 (a) Shower radii (ρ) containing 68% and 90% of the energy deposited in an individual silicon layer by a photon shower as a function of the silicon layer. The colour-coded rectangles indicate the fraction of total energy deposited inside the 68% and 90% containment radii in each layer. (b) Energy deposited in HGCAL cells by pairs of unconverted photons for several thousand events. The colour code represents the energy density. (c) Fractional energy resolution (σ/E) as a function of E_T for unconverted photons using a region of radius 2.6 cm to sum the energy for different pseudorapidities and different pile-up levels. Figures and their captions are taken from Ref. [3]

Sect. 2.3.2. Furthermore, software compensation techniques analogous to the ones studied by the CALICE collaboration [28, 29] to enhance the energy resolution of hadronic showers may profit from the granularity. Finally, HGCAL will be able to complement the measurements from the muon chambers by tracking and tagging muons in the forward regions that are fully equipped with silicon.

Exploiting the wealth of information obtained with the upgraded endcap calorimeter is a major challenge for the reconstruction and analysis of its data [30]. It is anticipated that state-of-the-art image recognition techniques based on *deep learning* [31] will be helpful in this regard. Besides the ongoing evaluation of possible applications of such methods in terms of energy reconstruction or particle identification, the inclusion of the timing information in these tasks remains to be assessed in the future. In any case, a good understanding of HGCAL, in particular its calibration and the simulation of the showers therein, is necessary to trustworthily apply deep learning-based methods in the interpretation of its data, cf. Sects. 10.3.4 and 10.4.

4.5 Prototype Assembly in 2018

Most of the studies related to HGCAL in the past are based on simulation of its data. The simulation relies on assumptions on the electronic performance and on the validity of GEANT4 to properly model the energy depositions in the sensors. However, the former is not guaranteed since the foreseen electronics are custom-designed and the latter might not fully reflect reality. For instance, it is known that the simulation of energy depositions in thin silicon layers systematically deviates from measurements, cf. Fig. 13 in Ref. [32] and the clear statement[2] in Ref. [33]. Thus, an experimental proof of concept as well as first practical experiences in the operation of HGCAL under realistic conditions are essential to the successful progress of this upgrade project.

For this purpose, HGCAL prototype modules were assembled and installed in calorimeter-like configurations. Initial beam tests of such configurations took place in 2016 [34], there with a small number of 8–16 available prototype modules that were equipped with the SKIROC2 ASIC [35, 36]. This particular ASIC did not incorporate the relevant features of the ultimate ASIC (HGCROC). The small prototypes in 2016 were tested with minimum ionising particles and electromagnetic showers. After 2016, more prototype modules were assembled that were equipped with a new ASIC which featured prototype functionality of the final HGCROC design. Then in 2018, those modules were tested with particle beam of electrons and hadrons with momenta up to 300 GeV/c. The goal of the beam test campaigns was the validation of the prototype's functionality for calorimetric measurements. Especially, the encouraging results from 2016 should be confirmed with a fully equipped electromagnetic calorimeter prototype and the studies should be extended to hadronically induced showers using a prototype of the hadronic section.

In the following, the construction of the most recent HGCAL prototypes tested in 2018 is explained. More details can be found in a dedicated publication [37].

Prototype Silicon Sensors

Prototype silicon sensors were produced by Hamamatsu Photonics K.K. (*HPK*). Following the initial baseline, those sensors are 6″ in diameter and consist of 135 individual diodes (synonyms: pads, cells). Each prototype sensor houses five different geometries of the silicon diodes, cf. Fig. 4.11a. The area of full hexagonal cells amounts to approximately 1 cm^2. The smallest pad area is the so-called calibration pad. It has the lowest capacitance and with it a reduced noise level that should allow to detect clear signals from minimum ionising particles throughout HGCAL's lifetime.

[2] "The sensitive material of future advanced calorimeters is silicon, so accurate Geant4 predictions for energy deposition in silicon for all particle types and energies are required." [33].

Table 4.1 Properties of the 6″ prototype silicon sensors delivered by HPK for beam test module assembly in 2018. For a few modules, the depletion zone is only 200 μm thick

Property	Value
Physical thickness	320 μm
Depletion zone thickness	300 (200) μm
Primary doping	n-type
p-stop	none
Number of pads	135
Different pad geometries	5

The silicon sensor substrate is n-type with p-type implants on top. The depletion zone of the sensors is 300 μm thick at full depletion for 90 out of 94 HGCAL prototype modules that were available in October 2018. For the remaining four, the depletion zone is only 200 μm thick. Table 4.1 summarises the characteristics of the prototype sensors.

Prototype Readout Chip: SKIROC2-CMS

The SKIROC2-CMS ASIC [38] is the prototype readout chip for the HGCAL prototype modules. It houses 64 readout channels and features prototype functionality of the ultimate HGCROC. It covers a large dynamic range with dedicated readouts of the high- and the low gain as well as the time-over-threshold (*TOT*). The detected signals are shaped with a configurable time which was set to 40 ns during the beam tests. Prior to the reception of a trigger signal, waveforms are continuously sampled at 40 MHz with a 12-bit ADC into a 13-deep switched capacitor array (*SCA*). Furthermore, the SKIROC2-CMS is designed to measure the time-of-arrival (*TOA*) of the signals with a precision down to 50 ps. This is achieved through a discriminator and a TDC with 25 ps bins, and with a preceding fast-shaper processing the preamplified signals.

Construction of Prototype Modules

The HGCAL beam test prototype modules were made from stacked layers of several components bonded together with an epoxy glue. A baseplate, made from Cu, CuW or printed circuit board (*PCB*) material, provides mechanical support and thermal conductivity. It follows a layer of gold-plated Kapton for electric insulation and for routing the bias voltage to the sensor backplane. Glued on top of the sensor, the PCB is ∼1 mm thick and hosts four SKIROC2-CMS readout chips (*ROCs*). An onboard FPGA on the PCB serves for communication between the back-end data acquisition system and the ROCs. Only 32 out of the 64 available channels per ROC are connected to the silicon pads. The physical connections to the pads, the guard ring, and the Kapton foil are realised through multiple wire bonds per pad with stepped holes in the PCB.

Semi-Automated Assembly Process

The assembly of the prototype modules was performed by the University of California in Santa Barbara. The procedure was semi-automated whereby the placement

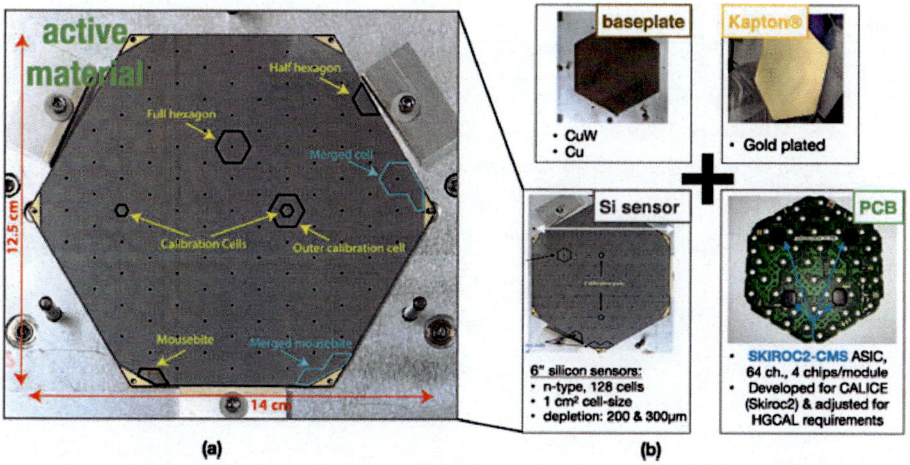

Fig. 4.11 **a** Prototype sensor design and different pad geometries. Merged cells were not incorporated into the 2018 modules that are subject in this thesis. **b** Overview on the beam test module assembly in 2018

and gluing of the different module layers was performed by a robotic gantry. The placement accuracy with this gantry was well below 100 μm. Six modules per day could be manufactured by this means. The followed procedure for the 6″ modules is analogous to the foreseen production of the ultimate 8″ HGCAL modules in the next years. After shipment to CERN, the modules were tested again in terms of their leakage currents and in terms of data integrity. The assembled modules were found to be properly functional prior to the beam test in October 2018.

HGCAL Prototype Calorimeters

Finally, 94 prototype modules were available and could be mounted on mechanical structures to construct HGCAL calorimeter prototypes. They consisted of two compartments, one mimicking the design of the CE-E and the second one representing the silicon-based part of the CE-H. Those prototypes are similar to the final HGCAL design but still showed some decisive differences. As example, both the longitudinal and transverse coverage of the beam test prototype was limited such that hadronic showers tended not to be fully contained in the prototypes. In order not to confuse the calorimeter performances derived from the beam tests with the ultimate HGCAL performance, those differences are indirectly emphasised by not referring to the prototypes' compartments as CE-E, nor as CE-H, respectively. Instead, they are called *EE* (= CE-E) and *FH* compartments (= CE-H) in this thesis, utilising the naming convention at the time of HGCAL's technical proposal [2]. Figure 4.12 depicts the prototype mechanics and structures of the sensitive layers made from the modules.

The EE compartment was made from up to 14 mini-cassettes hung into a hanging file structure. The cassettes were based on 6 mm thick copper cooling plates on which two modules were screwed back-to-back. Grooves inside the plates allowed for active maintenance of a constant temperature of around 28°C. 4 m thick lead

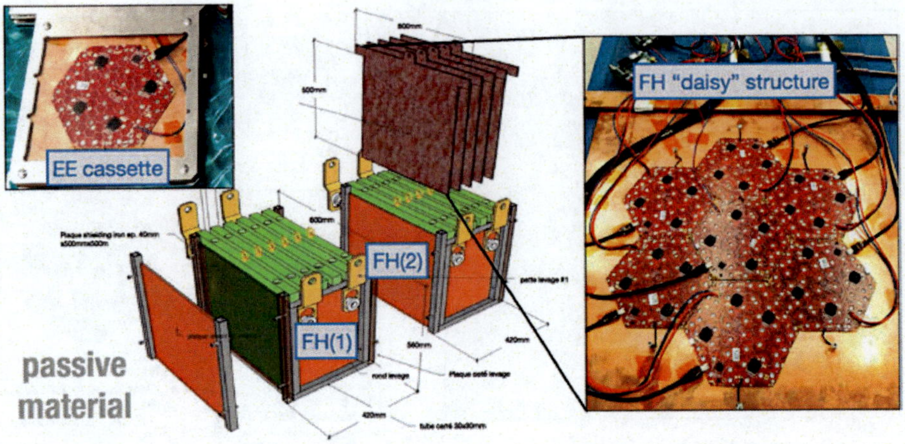

Fig. 4.12 Mechanics and structure of the sensitive layers made from the silicon modules forming the HGCAL calorimeter prototypes for the beam tests in 2018. FH(1) and FH(2) denote the mechanically separated hanging file structures that make the FH compartment

absorbers served as main passive material in the cassettes generating the particle showers. They were closed by thin aluminum plates and a polyester film. Interposer boards were installed on top for the communication to the back-end data acquisition system and for the supply of the bias voltage. Noise due to improper grounding was initially an issue during the assembly. It was found that the the best solution was to connect the copper cooling plates to the ground of the PCBs.

Again, 6 m thick but larger copper plates were the basis for the construction of the FH. Up to seven modules were installed in a daisy-like structure on each of the plates, cf. Fig. 4.12, making one sensitive layer in the FH compartment. The special solution for the grounding of the central module in the daisies is documented in Ref. [37]. Cooling pipes were installed on the respective other side of the copper plates. Subsequently, the layers were inserted into one of the two hanging file structures for the FH compartment. 4 m thick stainless steel plates were used as main passive material. In total, the assembled FH compartment weighed more than one metric ton.

References

1. Barney D, Sakuma T (2017) Sketchup images highlighting the sub-detectors. http://cds.cern.ch/record/2628519
2. Contardo D, et al. (2015) Technical proposal for the phase-II upgrade of the compact muon solenoid. CERN-LHCC-2015-10 (CMS-TDR-15-02). https://cds.cern.ch/record/2020886
3. CMS Collaboration (2017) The phase-2 upgrade of the CMS endcap calorimeter. CERN-LHCC-2017-023 (CMS-TDR-019). https://cds.cern.ch/record/2293646
4. CMS Collaboration (2019) How CMS weeds out particles that pile up. https://cms.cern/news/how-cms-weeds-out-particles-pile. Accessed 24 Sep 2019

5. Holzner A (2012) 78 reconstructed vertices in event from high-pileup run 198609. http://cds.cern.ch/record/1479324
6. Zhang L, et al. (2005) Performance of the monitoring light source for the CMS lead tungstate crystal calorimeter. IEEE Trans Nucl Sci 52:1123–1130. https://doi.org/10.1109/TNS.2005.852661
7. Anfreville M, et al. (2008) Laser monitoring system for the CMS lead tungstate crystal calorimeter. Nucl Instrum Meth A 594:292–320. https://doi.org/10.1016/j.nima.2008.01.104
8. Marshall JS, Münnich A, Thomson MA (2013) Performance of particle flow calorimetry at CLIC. Nucl Instrum Meth A 700:153–162. https://doi.org/10.1016/j.nima.2012.10.038
9. Barney D (2018) The high granularity calorimeter upgrade project for CMS." https://indico.cern.ch/event/718124/. CERN Detector Seminar, April 2018
10. Barney D (2019) Overview slide of CE with main parameters. https://cms-docdb.cern.ch/cgi-bin/PublicDocDB/ShowDocument?docid=13251. Accessed 24 Sep 2019
11. Simon F, Soldner C (2010) Uniformity studies of scintillator tiles directly coupled to SiPMs for imaging calorimetry. Nucl Instrum Meth A 620:196–201. https://doi.org/10.1016/j.nima.2010.03.142
12. Blazey G, et al. (2009) Directly coupled tiles as elements of a scintillator calorimeter with MPPC readout. Nucl Instrum Meth A 605:277–281. https://doi.org/10.1016/j.nima.2009.03.253
13. CMS Collaboration (2012) CMS technical design report for the phase 1 upgrade of the hadron calorimeter. CERN-LHCC-2012-015. https://cds.cern.ch/record/1481837
14. Lobanov A (2019) Electronics and triggering challenges for the CMS high granularity calorimeter. JINST 13. https://doi.org/10.1088/1748-0221/13/02/c02056
15. Verlaat B, Van Beuzekom M, Van Lysebetten A (2008) CO_2 cooling for HEP experiments. https://doi.org/10.5170/CERN-2008-008.328
16. Simon SH (2013) The Oxford solid state basics. Oxford University Press, Oxford
17. Leo WR (1993) Techniques for nuclear and particle physics experiments, 2nd edn. Springer, Berlin
18. Hartmann F (2008) Evolution of silicon sensor technology in particle physics. Springer, Berlin
19. Moll M (1990) Radiation damage in silicon particle detectors. PhD thesis, University of Hamburg, Germany. http://www-library.desy.de/cgi-bin/showprep.pl?desy-thesis99-040
20. Pree E (2018) Development of large area silicon sensors for the high granularity calorimeter at CMS. PhD thesis, Technische Universität Wien Fakultät für Physik, Austria. http://katalog.ub.tuwien.ac.at/AC15192834
21. Rivera EC (2017) Advanced silicon sensors for future collider experiments. PhD thesis, IFCA (CSIC-UC), Spain. https://cds.cern.ch/record/2291517
22. Apresyan A (2016) Investigation of fast timing capabilities of silicon sensors for the CMS high granularity calorimeter at HL-LHC. https://doi.org/10.1109/NSSMIC.2016.8069873
23. Landau L (1944) On the energy loss of fast particles by ionization. J Phys (USSR) 8:201–205. https://cds.cern.ch/record/216256
24. Bichsel H (1988) Straggling in thin silicon detectors. Rev Mod Phys 60:663–699. https://doi.org/10.1103/RevModPhys.60.663
25. Patrignani C, et al. (Particle Data Group) The review of particle physics. Chin Phys C 40:100001. https://doi.org/10.1088/1674-1137/40/10/100001
26. Burrows PN, et al. (2018) The compact linear collider (CLIC) - 2018 summary report. https://doi.org/10.23731/CYRM-2018-002
27. Rodriguez A, Laio A (2014) Clustering by fast search and find of density peaks. Science 344(6191):1492–1496. https://doi.org/10.1126/science.1242072
28. Adloff C, et al. (2012) Hadronic energy resolution of a highly granular scintillator-steel hadron calorimeter using software compensation techniques. JINST 7:P09017. https://doi.org/10.1088/1748-0221/7/09/p09017
29. Repond J, et al. (2018) Hadronic energy resolution of a combined high granularity scintillator calorimeter system. JINST 13:P12022. https://doi.org/10.1088/1748-0221/13/12/p12022

30. Diamante L (2019) What's in store for the CMS detector over the next two years? https://home.cern/news/news/experiments/whats-store-cms-detector-over-next-two-years. Accessed 18 Nov 2019
31. LeCun Y, Bengio Y, Hinton G (2015) Deep learning. Nature 521:436–444. https://doi.org/10.1038/nature14539
32. Allison J, et al. (2016) Recent developments in Geant4. Nucl Instrum Meth A 835:186. https://doi.org/10.1016/j.nima.2016.06.125
33. Incerti S, et al. (2017) Recent progress of Geant4 electromagnetic physics for calorimeter simulation. JINST 13:C02054. https://doi.org/10.1088/1748-0221/13/02/c02054
34. Akchurin N, et al. (2018) First beam tests of prototype silicon modules for the CMS high granularity endcap calorimeter. JINST 13:P10023. https://doi.org/10.1088/1748-0221/13/10/p10023
35. Bouchel M, et al. (2007) SKIROC: a front-end chip to read out the imaging silicon-tungsten calorimeter for ILC. In: IEEE nuclear science symposium conference record. https://doi.org/10.1109/NSSMIC.2007.4436517
36. Suehara T, et al. (2018) Performance study of SKIROC2/A ASIC for ILD Si-W ECAL. JINST 13:C03015. https://doi.org/10.1088/1748-0221/13/03/c03015
37. Acar B, et al. (2021) Construction and commissioning of CMS CE prototype silicon modules. JINST 16:T04002. https://doi.org/10.1088/1748-0221/16/04/T04002
38. Borg J, et al. (2019) SKIROC2_CMS an ASIC for testing CMS HGCAL. JINST 12:C02019. https://doi.org/10.1088/1748-0221/12/02/c02019

Chapter 5
Strategy

The main contribution of this thesis to the upgrade of the CMS endcap calorimeter is the experimental verification of the silicon-based HGCAL design using the prototypes described in Sect. 4.5. For this purpose, the assembled prototypes were tested with particle beams at the DESY (Hamburg) and CERN SPS beam test facilities.

A significant part of the work for this thesis was done for the development and preparation of the experimental infrastructure required for the experiments with the HGCAL prototypes at the two aforementioned beam test facilities. Furthermore, the reconstruction of the HGCAL prototype data and of the involved beam characterising detectors had to be designed and implemented. The corresponding developments and procedures are documented in Chaps. 6 and 7.

Given this infrastructure and the reconstruction software, numerous beam tests and other characterisation measurements could be carried out successfully in 2017–2018. The analysis of the data and the main results are presented in Chaps. 8–10. The strategy hereby can be categorised twofold:

1. With the data from sensor characterisations and beam tests, the qualification of the functionality of the silicon-based HGCAL prototype modules is investigated.
2. After the modules' proper functionality is established, the calorimetric performance of the assembled HGCAL prototype calorimeters is validated with particle beams. In this context, the agreement of the obtained performance indicators to GEANT4 simulation needs to be confirmed because the ultimate HGCAL design is based on simulation with GEANT4.

Also during its operation at HL-LHC, HGCAL data will have to be simulated for the production of Monte Carlo samples for physics analyses. However, sequential simulation of particles interacting with material, such as with GEANT4, is time consuming. It demands a significant amount of computing resources which would likely exceed the foreseen computing budget at HL-LHC. Ongoing R&D is investigating the possibility to use state-of-the-art machine learning techniques for the fast simulation of calorimeter showers that has the potential to reduce the computing needs in

T. Quast, *Beam Test Calorimeter Prototypes for the CMS Calorimeter Endcap Upgrade*, Springer Theses, https://doi.org/10.1007/978-3-030-90202-5_5

this regard. A corresponding application for the generation of showers in the HGCAL topology defines the third strategy item which is studied in this thesis:

3. State-of-the-art machine learning methods are used for the fast simulation of electromagnetic showers in HGCAL prototypes. The method is studied in terms of its accuracy in the generation of calorimeter data, its speed and applicability to realistic calorimeter data.

In the subsequent sections, sets of criteria are defined for each of the three strategy items. Those criteria will be addressed in the course of this thesis.

5.1 Qualification of Prototype Silicon Modules (MQ)

Module Components and Functionality (MQ-FUNC)

The first set of requirements related to the silicon-based module qualification addresses the characteristics of the module components and verifies the basic functionality of the fully assembled modules:

- ☐ **MQ-FUNC.1** At the typical module bias voltage of 200–800 V, the per-cell leakage current of the bare prototype silicon sensors should be at the order of a few nanoamperes.
- ☐ **MQ-FUNC.2** The silicon sensors should be fully depleted around the expected depletion voltage, which is around ∼200 V for 300 μm sensors. Especially, the per-cell capacitance should decrease with increasing bias voltage until full depletion is reached.
- ☐ **MQ-FUNC.3** Operated on the fully assembled modules, the SKIROC2-CMS ASICs should give consistent data that corresponds to the intended data format. The pedestal values should be reasonable.
- ☐ **MQ-FUNC.4** The noise level of the modules should be manageable and can, to some extent, be corrected for. After such corrections, the intrinsic noise level should not exceed a fraction of the signal induced by minimum ionising particles (*MIPs*) in both the high and the low gain.
- ☐ **MQ-FUNC.5** The lowest relevant energy depositions due to MIPs should be detectable with efficiencies close to 100%.
- ☐ **MQ-FUNC.6** Only a few isolated cells should exhibit low MIP efficiency.
- ☐ **MQ-FUNC.7** In order to use the HGCAL prototype to track MIPs, the calorimeter hit fake rate should be below 1%. This corresponds to approximately one fake calorimeter hit per module and per readout which is considered to be treatable in any MIP-tracking algorithm.
- ☐ **MQ-FUNC.8** Sharing of charge between neighbouring cells is not expected because the extent of the charge diffusion cloud in the silicon is much smaller than the pad dimension. Hence, MIP signals should be confined to single cells.

If all the requirements above are fulfilled, the prototype modules are considered operational for subsequent tests with showering particles.

Energy Calibration of the Modules (MQ-CALIB)
In that case, the assessment proceeds with the investigation if the energy scale can be calibrated using MIPs and if the gains can be intercalibrated with particle showers. Moreover, it needs to be verified that the modules provide a trustworthy inference of shower energy densities. For this, the following criteria are studied:

- ☐ **MQ-CALIB.1** The response to MIPs, evaluated for as many cells as possible, should not vary more than 5% for all cells read out by the same readout chip. This would justify the usage of chip-averaged MIP calibration constants in the final HGCAL design during its operation.
- ☐ **MQ-CALIB.2** The signal-to-noise ratio (*SN*) should scale according to the area of the silicon pads. In particular, the calibration pads should have significantly higher SN than full hexagonal pads. By this means, they can be used for energy calibration even towards the end of HGCAL's lifetime when the signal will have decreased and the noise increased due to increasing radiation damages.
- ☐ **MQ-CALIB.3** The high and low gains in the SKIROC2-CMS should exhibit the designed linearity in their signal amplitudes prior to the saturation of the high gain. This requirement is specific to the tests with the HGCAL prototype and has no implications for the ultimate HGCAL design.
- ☐ **MQ-CALIB.4** At high energy densities, the time-over-threshold should be usable to reconstruct the deposited energy and to effectively cover the full range of expected shower energy densities.

TOA Performance (MQ-TOA)
So far, the outlined qualification strategy has been targeted on the measurement of energy depositions. Yet, the HGCAL prototype also comprises a major novelty in calorimetry, namely timing capabilities with $\mathcal{O}$(50ps) resolution in form of the time-of-arrival measurement (*TOA*). Investigations of the SKIROC2-CMS's TOA performance in beam tests have not yet been published. A preliminary assessment is presented in this thesis:

- ☐ **MQ-TOA.1** It should be possible to calibrate the TOA using an external time reference like it could be done during HGCAL's operation in the Phase 2 CMS detector.
- ☐ **MQ-TOA.2** The timing resolution provided by the prototype SKIROC2-CMS ASIC should be close to its design value of 50ps for the highest signal sizes. This would validate the analogous TOA measurement principle in the ultimate HGCROC design.

PCB Tomography (MQ-PCB)
As preparation for the calorimeter performance studies in this thesis, a basic understanding of the prototype's passive material as input to the simulation model is essential. The thicknesses, compositions and densities of the large absorbers are provided by the manufactures and thus they are rather well known. By comparison,

the prototype modules' PCB is a heterogeneous element with many electronic elements of unknown material budget. Therefore, a tomography of the prototype PCB is performed in addition to the main module qualification strategy:

- ☐ **MQ-PCB** It should be validated that the PCB may be assumed as a homogeneous element in the simulation model.

5.2 Performance Validation of Calorimeter Prototypes (PV)

After integration of the qualified modules into HGCAL prototype calorimeters, their performance is validated in test beams of muons as MIPs, electromagnetically showering particles and hadronically showering particles. While the derived performance indicators in the beam tests are not fully applicable to the final HGCAL design, e.g. due to the limited size of the HGCAL prototypes or undefined experimental conditions, the prototypes should perform well enough to reconstruct typical characteristics of the showering particles. When possible, the obtained results are compared to detailed GEANT4 simulations of the prototypes and a basic agreement between them is to be expected.

Performance Validation with MIPs (PV-MIP)

- ☐ **PV-MIP.1** Exploiting HGCAL's granularity and the low enough noise level, the prototype calorimeter should be usable for the tracking of MIPs therein.
- ☐ **PV-MIP.2** MIP-like signatures should be identifiable in rather busy, i.e. pile-up-like, environments.

Performance Validation with Positrons (PV-POS)

- ☐ **PV-POS.1** The longitudinal development of electromagnetic showers should be in agreement between the simulation and the beam test measurements. In particular, the longitudinal shower depth should scale logarithmically with the incident positron energy. The behaviour may be exploited to disentangle pile-up contributions from actual electromagnetically showering particles in the Phase 2 CMS detector.
- ☐ **PV-POS.2** The mean visible energy measured by the prototype calorimeter should be proportional to the incident positron energy within the energy range accessible in the beam tests.
- ☐ **PV-POS.3** The energy resolution as a function of the positron energy is a key property of the calorimeter design. It should be in good agreement with the simulation, especially at high energies. The constant term for a CE-E prototype with 28 sensitive layers should be below 1 % consistent with the ultimate HGCAL design.
- ☐ **PV-POS.4** Using the transverse granularity, the pointing resolution to reconstruct the shower axis should amount to a fraction of a millimetre. By this means, the

assignment of energy clusters in HGCAL to reconstructed particle trajectories in the CMS tracker using Particle Flow would be supported.

- ☐ **PV-POS.5** Also beneficial to Particle Flow-based event reconstruction in the CMS detector, the resolution to reconstruct the shower axis angle should be below the level of 10 mrad at high energies.

Performance Validation with Charged Pions (PV-PION)

- ☐ **PV-PION.1** The primary interaction of incident hadrons should be reconstructed reliably exploiting the longitudinal segmentation of the HGCAL prototype.
- ☐ **PV-PION.2** The energy resolution for hadron-induced showers should be improved when compensation of the different energy responses to electromagnetic and hadronic shower constituents is applied. The compensation procedure relies on the granularity in the measurement of the particle showers.
- ☐ **PV-PION.3** The energy resolution and linearity of hadron-induced showers should exhibit basic agreement to the simulation. In this connection, the level of agreement is not as stringent as for the electromagnetic showers due to experimental deficits.

Hit Energy Reconstruction Similar to the Final HGCAL Design (PV-REC)
For the assessments with the showering particles, the resulting calorimetric performances of the HGCAL prototype should not be impaired significantly when alternative energy reconstruction methods are used that are more similar to the final HGCAL design:

- ☐ **PV-REC.1** The calorimetric performances should be unaffected when chip-averaged MIP calibration constants are used in the inference of hit energies.
- ☐ **PV-REC.2** The calorimetric performances should be unaffected when only the low gain and the TOT information are used in the inference of hit energies.

Electron-Pion Classification with Machine Learning (PV-EPI)
Finally, HGCAL is advertised as a detector that is predestined for the application of sophisticated, machine learning-based methods on its data. This hypothesis is investigated by deploying a machine learned classifier to separate electron- from pion-induced cascades in the prototype's EE compartment and applying it on beam test data. A successful, state-of-the-art classifier should have the following attributes:

- ☐ **PV-EPI.1** The classifier should be able to separate electromagnetic and charged hadron-induced signatures also when it is blinded to the measurements in the hadronic calorimeter compartment.
- ☐ **PV-EPI.2** The classifier should be applicable reliably to realistic calorimeter data.

5.3 Fast Calorimeter Simulation (FSIM)

State-of-the-art machine learning methods are used for the fast simulation of calorimeter showers in the HGCAL topology. The study in this thesis is intended as a proof of principle of the applied methodology. It focusses on electromagnetic showers that impinge on prototypes of the CE-E compartment.

The developed method for fast simulation of HGCAL's granular calorimeter data with electromagnetic showers is considered appropriate if the following criteria are met:

- ☐ **FSIM.1** The fast simulated showers should obey the imposed characteristics that are injected into the generation process, namely their energies and impact locations.
- ☐ **FSIM.2** Reduced to typical calorimeter observables, the fast simulated shower data should be in good agreement to the reference dataset, e.g. from detailed GEANT4 simulation, both in their distributions and also in their pairwise correlations.
- ☐ **FSIM.3** The fast simulation approach should allow for interpolation to phase spaces for which it has not been set up explicitly.
- ☐ **FSIM.4** Compared to sequential shower simulation with GEANT4, the fast simulation approach should be significantly faster in the generation of calorimeter data.
- ☐ **FSIM.5** The fast simulation approach should not only be applicable to one particular calorimeter configuration or energy range but should also be applicable to others.
- ☐ **FSIM.6** Ideally, the fast simulation approach should be applicable to realistic beam test data. This may demonstrate that the ultimately achieved correspondence of the generated calorimeter data to realistic calorimeter data is better than with the default GEANT4 simulation.

Chapter 6
Experimental Infrastructure

Experimental data is the basis for the qualification and performance validation of the HGCAL prototypes presented in this thesis. This data was recorded with single prototype modules and with various prototype calorimeter configurations being exposed to particle beams. Besides that, measurements on the electrical properties of prototype silicon sensors prior to their assembly to modules were made.

This chapter deals with the infrastructure necessary for these experiments. Special emphasis is put on the contributions to the infrastructure made as part of this work. This includes the development of the data acquisition system used in the beam tests, the incorporation of several beam characterising detectors into the setups that support the subsequent analyses of the calorimetric performance of the HGCAL prototypes and the exploration of the DESY (Hamburg) beam test facility as an alternative experimental site during shutdowns of the LHC and its injectors.

Section 6.1 describes the system used for the electrical sensor characterisation assembled and commissioned at CERN. Both the novel switch- and probe-card system and the measurement procedure are highlighted. A dedicated data acquisition system (*DAQ*) was required for the readout of the prototype calorimeter's data (*events*). The DAQ hard- and software of the beam test prototypes and the strategy for offline event synchronisation between all involved detectors are addressed in Sect. 6.2. Detailed information on the DESY beam test facility and the corresponding HGCAL prototype test setup can be found in Sect. 6.3. The majority of the beam tests have taken place in area PPE172 at the SPS North Area Beam Test Facility at CERN. Principles of the beam generation, an outline of this experimental area, the readout of beam characterising detectors and a brief summary of the tested HGCAL prototype configurations are explained in Sect. 6.4. This chapter concludes with Sect. 6.5 giving an overview on the conducted measurement programme by listing the accumulated datasets that are used in this thesis.

T. Quast, *Beam Test Calorimeter Prototypes for the CMS Calorimeter Endcap Upgrade*, Springer Theses, https://doi.org/10.1007/978-3-030-90202-5_6

6.1 Electrical Sensor Characterisation

Before their assembly to beam test modules, the 6″ HGCAL prototype silicon sensors underwent electrical qualification in terms of their per-diode dark current up to 1000 V and of their per-diode capacitance to voltages above full depletion. For this purpose, all diodes on a sensor need to be tested individually as it is foreseen in the final HGCAL design. In particular, the bias voltage must be applied simultaneously to at least all diodes surrounding the central volume under test in order to circumvent the extension of electric field lines into neighbouring diodes. Measurements with seven contacting needles would fulfill this requirement but are considered impractical, namely too time consuming, for sensor quality control during future mass production.

Instead of sequentially stepping such a set of seven needles across the whole sensor, the EP-LCD group at CERN has developed a switch- and probe-card-based system for silicon sensor characterisation. Spring-loaded pogo pins mounted on the probe-card connect all diodes, routed through the switch card, to ground at once. The latter consists of a 512-channel multiplexer that provides automated switching, connecting one pogo pin at a time to an ammeter for leakage current measurements or to an LCR meter for capacitance measurements, respectively. The cards are mechanically attached to the probe station. The sensor is held by a vacuum provided by a chuck that also provides the bias voltage to the sensor's backplane. Figure 6.1 shows the setup's principal components. As it is typical for such a system, they are located inside a dark box minimising light-induced dark currents in the silicon. Two USB microscopes aid visually with the sensor-to-probe-card alignment and with the manual needle-to-guard ring contact procedure. In the following, the main characteristics of the developed probe and switch card are summarised whereby details are mainly quoted from Ref. [1]. Afterwards, practical aspects of the sensor measurement procedure are described.

6.1.1 Switch- and Probe-Card System

Figure 6.2a shows a computer-aided design drawing of the two developed core components: the probe card together with the switch card plugged in on top. The former is the passive device in the measurement whereas the latter can be understood as the active element.

Although the probe card's exact design depends on the sensor geometry, cf. Fig. 4.11, its working principle remains unchanged: Spring-loaded pogo pins with a travel range of 1.4 mm are mounted to the probe card and serve as electrical bridges between the sensor's contact points and the switch card. Their tip radius amounts to 250 μm and in combination with the soldering as well as the card-to-sensor alignment accuracy results in a positioning precision that exceeds the width of the sensor guard ring contacts that amounts to a few hundred microns. Hence, an additional contact needle is used to contact the guard ring in this setup.

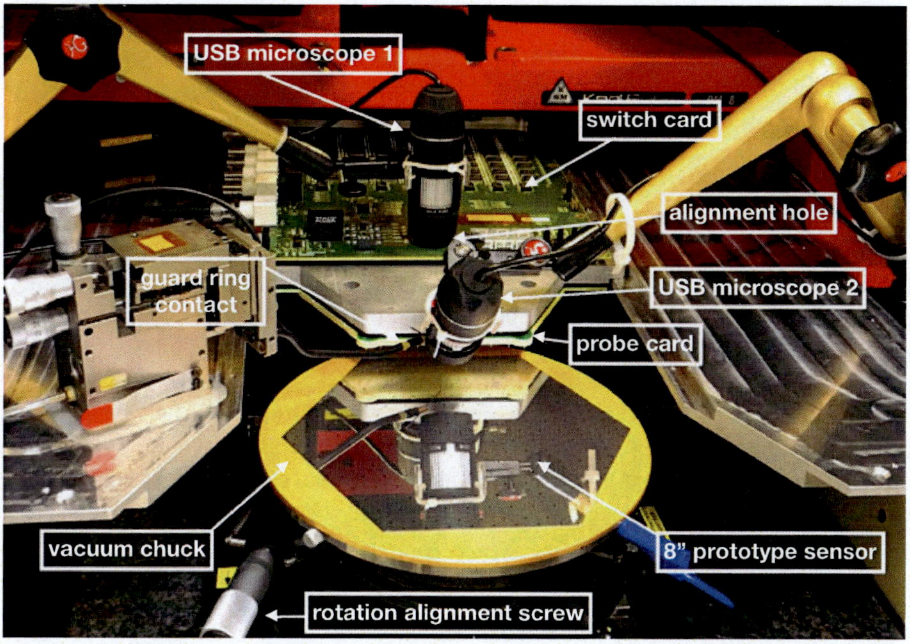

Fig. 6.1 Photograph of the system for electrical sensor characterisation of HGCAL (prototype) sensors at CERN taken in 2018. The setup is analogous for the tests with the 6″ prototype sensors

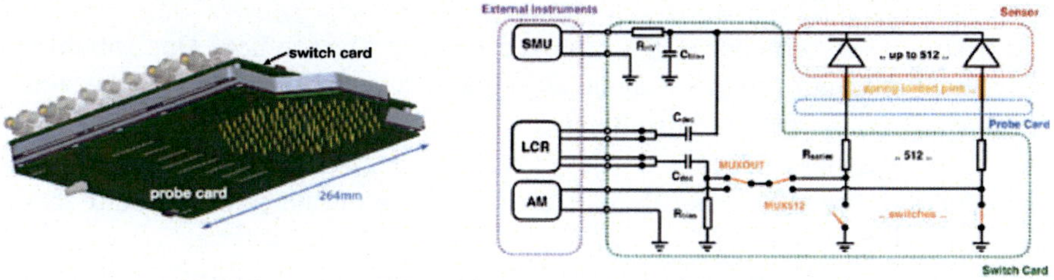

(a) (b)

Fig. 6.2 **a** CAD drawing of the probe card for the 6″ prototype sensors with the switch card plugged in on top, taken from Ref. [2]. **b** Schematic circuit diagram of the probe card switching matrix system, taken from Ref. [1]. The SMU symbolises the power supply (Keithley 2410), LCR the LCR meter (Keysight E4980A) and AM the ammeter (Keithley 6487)

10 kΩ resistors are installed in series to the pogo pins inside the switch card for protection against high electrical currents that may occur in the event of breakthrough diodes. The card's core component consists of a hierarchical, three-level multiplexer with 512 channels ("MUX512") for software-controlled switching between the diodes under test. Deploying this multiplexing scheme, the switching time from one diode to another is approximately 100ms. A separate multiplexer ("MUXOUT") controls the connection to the leakage current and capacitance measurement circuits without the necessity of manual intervention. Figure 6.2b shows the schematic of this

switching matrix system. In addition, the switch- and probe-card system allows for environmental monitoring via integrated temperature and humidity sensors. Communication to the components is possible through an on-board microcontroller and an USB interface. The remaining elements shown in the schematic are auxiliary components especially relevant to the capacitance measurement. For instance, a set of 1 μF capacitors (C_{dec} in Fig. 6.2b) is used to decouple the LCR meter from the DC voltage needed to bias the sensor. For more details, see Ref. [1].

In summary, the implementation of this switch- and probe-card system allows for automated, fast electrical qualification of HGCAL silicon sensors during their prospective mass production. Contributions to the leakage current by the system are at the order of 10 pA and can be neglected in the context of the targeted sensor quality assurance. On the other hand, substantial parasitic capacitances at the order of 100 pF at 10 kHz are introduced due to the complexity of the circuitry. When the parasitic capacitances are known, they can be subtracted from the measurement by applying an *open correction* [3]. After open correction of the data, a precision of 0.2 pF in the capacitance measurement can be expected [1].

6.1.2 Measurement Procedure

The procedure for electrical characterisation of HGCAL prototype sensors consists of well-defined steps.

- At first, the sensor under test is removed from its storage box. One should take particular precaution at this to avoid any contact with another sensor in the storage box separated by foils because it may lead to undesired, mechanical damages.
- It follows a visual inspection of the sensor's guard ring and the search for visible defects on its surface. Any obvious dust particles on the sensor should also be removed at this point.
- Using a vacuum pick-up tweezer, the sensor is placed onto the vacuum chuck inside the probe station. This step requires good synchronisation in turning on the chuck's vacuum with the release of the sensor by the tweezer. Sliding of the sensor on the chuck and simultaneous strain from both vacuums should be averted.
- Next, the probe card and the sensor wafer are aligned. This is done with the aid of the alignment hole and USB microscope 1, cf. Fig. 6.1. Figure 6.3 illustrates the situation before and after a successful alignment. Ideally, the star-like structure is situated centrally with respect to the alignment hole. Moreover, the vertical edge should be parallel to the picture frame which can be assessed by checking if the other cells are equally focussed when moving the chuck out and back in.
- The probe card is lowered to attain contact of the pogo pins to the sensor. A low forward bias voltage, typically of 1 V, can be applied to the sensor to assess the connectivity of all cells to the system. Currents through floating pins are zero, whereas they amount to $\mathcal{O}$ (100 nA) at 1 V if a pin is correctly connected to a diode.

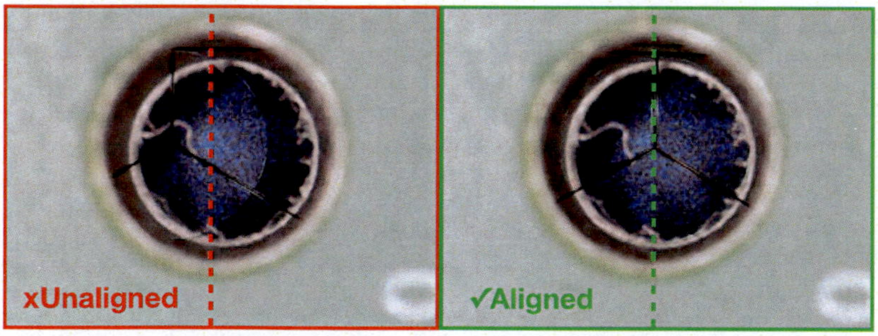

Fig. 6.3 Sensor-to-probe-card alignment using the alignment hole and the USB microscope

The chuck is lowered again and the previous step is iterated when a substantial number of pins is found to be disconnected.

- Prior to the closure of the dark box, the contact needle to the guard ring is positioned with the help of USB microscope 2.
- Measurements of the leakage currents (*IV*) and of the capacitances (*CV*) for different bias voltages are run automatically by a LabVIEW-based data acquisition system [4]. A typical IV + CV scan for 150 cells with 15 voltage steps takes less than 2 h and does not necessitate any intervention by the user. Voltage scans are stopped automatically as soon as a configurable compliance level in the total leakage current, usually around 1 mA, is reached.
- After the measurement, the sensor is extracted from the probe station. First, the needle contacting the guard ring is detached followed by the raise of the probe card. A visual inspection of the guard ring and of cells that have shown indications of breakthrough concludes the characterisation procedure.

Altogether, the typical duration for such an electrical characterisation took less than 3 h per tested sensor in spring 2017. This time was dominated by the length of the automated IV and CV measurement. Of course, the latter depends on the level of detail in the voltage scan and on the number of diodes under test on the sensor. In general, users are advised to perform the described preparatory and revising measures as well as the measurement itself in twos because of the delicacy in the mechanical interactions with the sensor and due to the utilisation of high voltages up to 1000 V.

6.2 Data Acquisition System for Beam Tests

In beam tests, data from the HGCAL prototype calorimeters and other involved detectors are recorded, monitored and stored to disk. This task is performed by the data acquisition system (*DAQ*). Since the arrival of HGCAL prototype modules equipped with the SKIROC2-CMS ASIC [5], the initial system for the beam tests in 2016 [6, 7] could not be used anymore. Instead, a new DAQ had been developed,

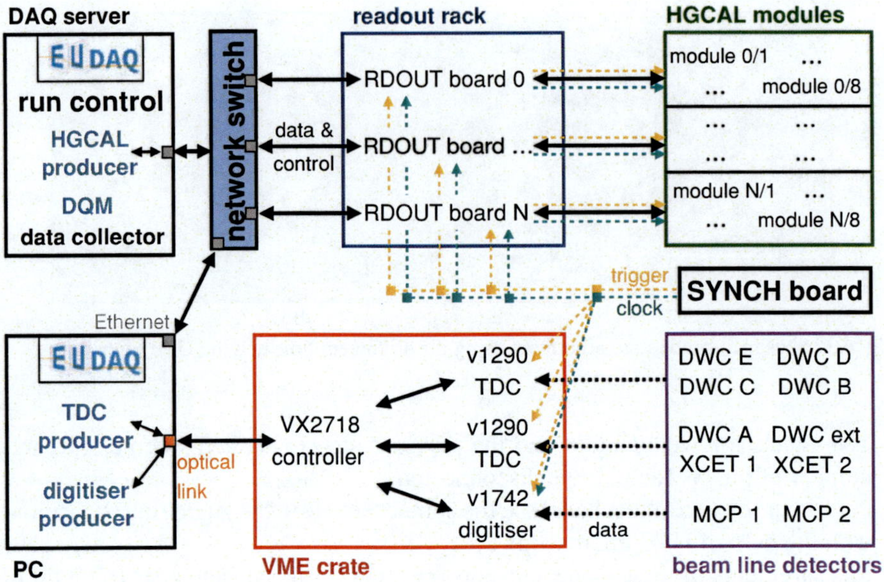

Fig. 6.4 Schematic of the data acquisition system as it was operated during the October 2018 beam test at the H2 beam line at CERN. For better visualisation, not all connections and communication paths are indicated. The cyan-coloured instances on the left-hand side symbolise EUDAQ-based applications implemented specifically for the HGCAL prototype beam tests

tested and qualified in 2017 and was successfully operated throughout the HGCAL beam tests in 2017 and 2018. Its core components corresponding to the, so far largest, beam test in October 2018 are schematically presented in Fig. 6.4. The depicted elements are described in the following.

Section 6.2.1 deals with the scalable HGCAL DAQ hardware that has been used for the readout of more than 90 prototype modules. The hardware incorporated for the readout of auxiliary beam characterising detectors is described in Sects. 6.3.2 and 6.4.3. The DAQ program is written within the EUDAQ framework [8, 9]. Its applications are executed on separate computing instances which communicate through a network. Section 6.2.2 summarises the EUDAQ framework and highlights relevant developments in the context of beam tests with the HGCAL prototype. In particular, the extension of the integrated online data quality monitoring (*DQM*) is discussed further in Sect. 6.2.3. Finally, the strategy for event synchronisation as well as its validation among all involved detectors is explained in Sect. 6.2.4.

More details are given in a publication dedicated to this DAQ [10].

6.2.1 HGCAL DAQ Hardware

Custom hardware has been designed and produced for the configuration and readout of the SKIROC2-CMS ASICs. The local, on-board communication to them is governed by an FPGA on the prototype module PCBs. Via an interposer board (not shown in Fig. 6.4) and HDMI sockets, the connection to the back-end DAQ is achieved.

The back-end consists of multiple customised PCBs: one synchronisation (*SYNCH*) board and multiple readout (*RDOUT*) boards. Both board types make use of existing components such as the optical receiver modules (*ORM*) which has already found application in the CMS triggering system [11].

A maximum of eight HGCAL modules can be connected to one RDOUT board. Its functions comprise control of the ASICs and acquisition of their data, clock- and trigger-signal forwarding as well as high and low voltage distribution to the modules. At the beginning of each data-taking unit (*run*), an incorporated Raspberry Pi computer (*RPI*) performs the ASICs' configuration. After reception of a trigger signal, data is collected by four data ORMs and merged into one control ORM. Inside the latter ORM's RAM, a boolean flag is set to signal that the accumulated data is ready to be polled by the DAQ server. Afterwards, the RPI resets the ASICs and reinitialises their data acquisition while the server reads the accumulated data. Once the reading is finished, the RPI emits a "Readout Done" signal to the SYNCH board to indicate the end of the readout cycle.

The SYNCH board is responsible for synchronous data taking. It receives physical trigger signals, e.g. from scintillator signal coincidences in the beam line, and forwards it to the RDOUT boards only when all connected boards are ready. Trigger signals are vetoed if at least one of the connected RDOUT board remains busy, meaning it has not sent its "Readout Done" signal. A global busy signal and copies of the trigger are available through LEMO connectors on the SYNCH board which are essential for interfacing to external DAQ hardware. Furthermore, clock-signal generation and provision to all connected RDOUT boards is performed by the SYNCH board.

In practice, this HGCAL prototype DAQ system has reached acquisition rates close to 50 Hz. Given the data size of $\sim$3.8 kB per ASIC and trigger, the system eventually accumulated approximately 70 MB s^{-1} of data during its deployment in the latest beam test in October 2018.

6.2.2 EUDAQ Framework

EUDAQ is a software framework written in C++ that provides essential ingredients to a typical DAQ system. Its concept allows for high flexibility making it particularly suitable for beam test applications where devices under test, and with it the corresponding DAQs, are usually exchanged frequently [12]. Moreover, EUDAQ is a top-level framework in which the user mainly interacts with the run control that steers all other components. Its control flow follows the principle of a finite state machine rendering its utilisation intuitive.

Apart from the aforementioned run control, its core parts are log- and data collectors as well as functionality for data quality monitoring. By design, all components are kept independent from the user-specific applications. The latter are called producers and serve as the interface between the user's hardware and EUDAQ's core. IP/TCP sockets are employed for communication between all DAQ components. Consequently, core functionality and the producers do not need to run on the same physical computing node.

By 2018, EUDAQ has found broad application among many R&D groups [13]. Since 2017, it has also been established successfully in the beam tests of HGCAL prototype modules with the SKIROC2-CMS ASIC. For its deployment, a principal task was the establishment of a producer that communicates to the RDOUT boards and polls its data. As part of this work and independent from the HGCAL prototype's readout, producers for interfacing CAEN v1290 TDCs [14] and a CAEN v1742 digitiser [15] have been developed.

The underlying EUDAQ version has been updated regularly as soon as new versions were publicly available. While v1.6 and v1.7 [8] were the basis for most of the earlier beam test campaigns, v2 [9] was used during the most recent beam test in October 2018. Distinct upgrades available in the v2 release such as the possibility of defining sophisticated online event synchronisation schemes based on trigger timestamps within a custom data collector were not used but are potentially useful for future tests.

6.2.3 *Online Data Quality Monitoring*

Especially during beam tests of novel detector prototypes, data integrity cannot be assumed. In order to exploit the precious experimentation time with the particle beam effectively, any symptoms of problems with the accumulated dataset for future offline analysis should be identified as soon as possible. With this in mind, special emphasis has been put into the creation of a fast and online data quality monitoring (*DQM*) for the HGCAL prototype tests through extension of the EUDAQ's default DQM.

Monitored quantities computed from the HGCAL prototype calorimeter's data can be categorised twofold: First, a graphical collection related to the integrity of the data from each module is constructed. Second, global calorimeter shower characteristics are estimated giving feedback on the overall detector performance and on the quality of the particle beam.

In both cases, the procedure starts with raw data written to disk by the native EUDAQ data collector. An event in the recorded file is read, is converted and its relevant content is wrapped into predefined EUDAQ classes at hand. Since these EUDAQ classes have originally been intended for application to data structures from tracking detectors, rather abstract conventions have been decided on for this conversion. As example, data collected from a readout channel is wrapped into an

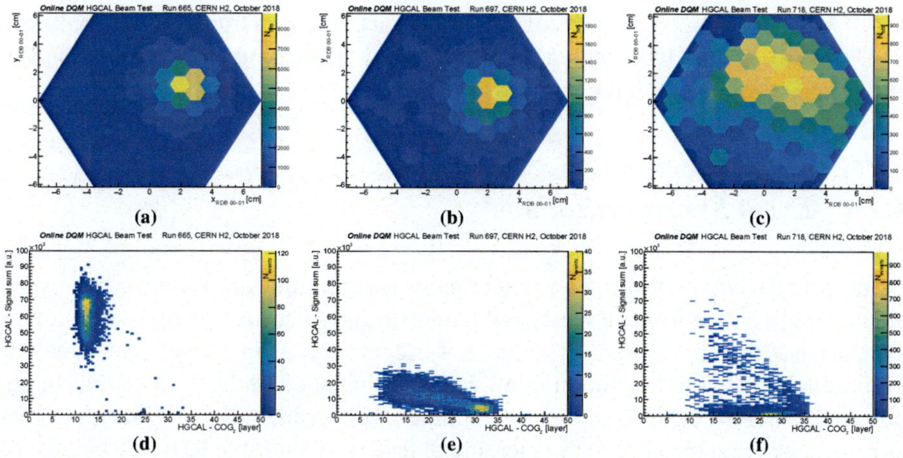

Fig. 6.5 Online DQM: (Top) Number of hits of channels with estimated signals above a configurable threshold. (Bottom) Two-dimensional distribution of the summed calorimeter signal and the estimated longitudinal centre of gravity using low gain only. Examples taken from runs with **a**, **d** 200 GeV/c positrons, **b**, **e** 50 GeV/c pions and **c**, **f** 200 GeV/c muons are shown

EUDAQ pixel object with its column and row indexes set as the associated ASIC and channel indexes. ADCs for different time samples and gains enter as so-called EUDAQ frames [8].

After this metamorphosis, data is forwarded to and is visualised by the DQM. Figure 6.5 gives some examples of monitored quantities during data taking. Signal occupancy visualisations like the ones shown for Fig. 6.5a–c are constructed for every module under test and for various signal definitions. This way, not only data integrity problems under different aspects can be identified easily but also the beam impact onto the calorimeter prototype can be assessed. Graphics like Fig. 6.5d–f are obtained after a simplified signal reconstruction of the calorimeter data, cf. Sect. 7.2. In this context, only the low gain information is used, the signal amplitude is taken from a configurable time sample and the pedestals are estimated event-by-event with the signal-free (by design) time sample zero. Estimates of calorimetric observables whose distributions and correlations are typically different for electromagnetic and hadronic showers as well as for minimum ionising particles are computed subsequently. Under the assumption of a reasonable prototype performance, the hereby monitored distributions yield an immediate impression of the beam composition during data taking.

Similar to the outlined procedure for the HGCAL prototype data, the DQM was extended further to also monitor data from beam characterising detectors. Details are given in Sects. 6.3.2 and 6.4.3.

In summary, the online monitoring constitutes a full, simplified analysis of the accumulated prototype calorimeter data enabling immediate conclusions on the prototype's performance and on the quality of the beam. A possible improvement for the future is the DQM rate which was empirically limited to $\mathcal{O}(1\,\text{Hz})$ for a data volume of 1.4 MB per event. One likely cause is the fact that EUDAQ foresees only one CPU thread for the DQM including all necessary unpacking, conversion, analysis and visualisation of this data.

6.2.4 Event Synchronisation

By default, EUDAQ constitutes a synchronous DAQ framework meaning that events are only built if data from all registered producers have been received. Event number synchronisation refers to the assignment of trigger indexes in a given data stream to event numbers that are common to all detectors included in the data taking. In this context, data from HGCAL prototype modules are considered to originate from one stream, neglecting the underlying merging of data from multiple RDOUT boards. As it is stated later in this chapter, other data streams from external hardware have been formed by TDCs, digitisers, MIMOSA26 chips or even other calorimeter prototypes.

Synchronisation was mainly realised through the physical copy of the trigger signal by the SYNCH board that was fed into all other trigger-operating hardware in the setup. However, for some runs, the number of recorded triggers among the distinct data streams differs, and subsequent offline corrections are mandatory. The occurring symptomatic can be interpreted as follows:

1. At the beginning of a data taking run a trigger copy was issued to the external hardware despite the HGCAL readout not being fully ready. In consequence, the external hardware registered one trigger too much at the beginning of a run.
2. Trigger copies by the synch board were either not made or were not registered by the external detectors, cf. *event 3* in Fig. 6.6a. Randomly, the readout of external hardware lacked a few triggers.

The most stringent treatment of such desynchronisations would be the rejection of all affected runs for which the number of recorded triggers disagree among the participating detectors. Following this principle, only a small fraction of the recorded data could be analysed in connection with the study on position resolution with the 2016 HGCAL prototype calorimeter [7]. In 2016, no additional information was available that could have been useful for event synchronisation after the beam test.

Given this experience and in order to circumvent drastic measures like the one described above, the timestamp of each trigger registered by the respective electronics was also recorded from 2017 onward. These timestamps are now used offline to compute time intervals Δt_X between a given trigger i with respect to the previous one, $i-1$, for each detector X.

$$\Delta t_X := t_{X,i} - t_{X,i-1} \tag{6.1}$$

Empirically, Δt_X is found to be continuously distributed in the range from 15ms to 30ms during the HGCAL beam tests, consistent with the overall DAQ rate of 40 Hz to 50 Hz. For longer interruptions of the trigger signal, e.g. in between spills,

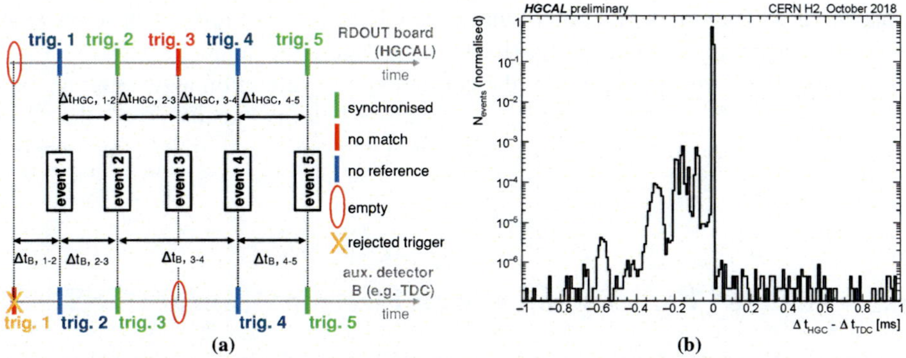

Fig. 6.6 **a** Illustration of the offline event synchronisation validation and correction algorithm. It relies on the presence of trigger timestamps from each electronics involved in the data taking. Only synchronised triggers (green) are kept for offline analysis. By convention, the trigger number obtained from the HGCAL prototype calorimeter data stream defines the matched event index. Possible additional triggers at the beginning of a run in the auxiliary detectors or missed triggers during the run are corrected for. **b** Exemplary distribution of $\Delta t_{\text{HGC}} - \Delta t_{\text{TDC}}$ between the RDOUT boards and the TDCs during the October 2018 beam test

trigger time intervals may even reach a couple of 10 s. After computation of all Δt_X, triggers in different data streams are aligned as illustrated in Fig. 6.6a. The applied algorithm validates the event synchronisation and corrects for the deficits causing event desynchronisation as outlined above when necessary. By convention, the trigger time interval provided by the RDOUT boards (Δt_{HGC}) serves as the reference and thus their associated trigger number defines the matched, common event index.

All parameter values occurring in this procedure are optimised for the time structure of test beams delivered by the SPS at CERN. A given event is considered synchronised if the absolute trigger time interval difference between HGCAL, i.e. the RDOUT boards, and a detector X does not exceed 1 ms, cf. Eq. 6.2.

$$|\Delta t| = |\Delta t_{\text{HGC}} - \Delta t_X| \overset{!}{<} 1\,\text{ms} \tag{6.2}$$

Given the average time interval between to events amounted to 20–25 ms, intermediate values of $-100\,\text{ms} < \Delta t < -15\,\text{ms}$ can be related to cause 2 and lead to a skipping of HGCAL triggers in the synchronisation. On a larger scale, a full run is considered synchronised if events at the edge of a longer interruption in the triggering are aligned, i.e. by definition, if Δt does not exceed 200ms. The latter chosen value corresponds to the eight- to tenfold of the observed DAQ rate with the HGCAL prototypes and is much less than the spill period at the SPS of $\mathcal{O}$ (10s).

Figure 6.6b shows a representative distribution of trigger time interval differences recorded between one of the involved TDCs and the HGCAL prototype data stream during the October 2018 beam test at CERN. Although the TDCs and HGCAL operate with different clocks in this example, $|\Delta t|$ is less than a fraction of a millisecond for the majority of readouts. This fact is consistent with synchronicity between both data streams. The minor fraction of entries smaller than zero hints at differences in the timing accuracies between the RDOUT boards and the deployed TDC.

Notably, after the application of the correction and validation algorithm, no sign of unsynchronised event numbers has been detected during the analysis of the beam test data. In conclusion, the presented event synchronisation strategy has worked well in practice.

6.3 Module Characterisation with Electron Beam at DESY

During the scheduled long shutdowns of the LHC and its injectors, test beam experiments are not possible at CERN. On that account, the feasibility of performing HGCAL prototype testing with particle beams at other facilities during such periods needs to be assessed. Fermilab has already been proven to be a suitable, alternative site in this regard during the initial HGCAL beam tests in 2016 [7].

The potential of HGCAL beam tests at DESY (Hamburg) [16, 17] has been explored for the first time in March 2018 at the DESY II beam test facility. The DESY II beam line provides electrons and positrons up to a maximum momentum of 6 GeV/c rendering the expected energy density inside an initiated electromagnetic shower too low for characterisation of the SKIROC2-CMS's entire dynamic range. Therefore, detailed measurements with a fully equipped HGCAL calorimeter prototype were not foreseen at this time. Instead, one of the test's main purposes was defined to be the qualification of a small sample of prototype modules in terms of their noise behaviour and of their response to minimum ionising particles (*MIPs*). In addition, the effective inclusion of the DATURA beam telescope [18] into the data taking was targeted to enable detailed studies of potential sensor interpad effects in the response to MIPs.

In Sect. 6.3.1 the beam generation principles at this facility and its central features are highlighted. Afterwards, the first HGCAL prototype test setup in its T21 area is described in Sect. 6.3.2.

6.3.1 DESY II Beam Test Facility

Figure 6.7 depicts the generation principle of test beams at the DESY II beam test facility. It starts with the circulation of electrons and positrons inside the 292.8 m long DESY II synchrotron. Particles coming from LINAC II and PIA [19] are injected at 450 MeV/c and are accelerated to 6.3 GeV/c for subsequent transfer to PETRA III [20]. The sinusoidal magnet cycle lasts 80 ms corresponding to an oscillation of the circulating particles' momentum at a frequency of 12.5 Hz.

Generation of test beam at DESY II relies on the presence of two targets. The first target is made of 6 μm thick carbon fibres positioned inside the beam. Bremsstrahlung photons are created and impinge on a second target, typically made of copper or

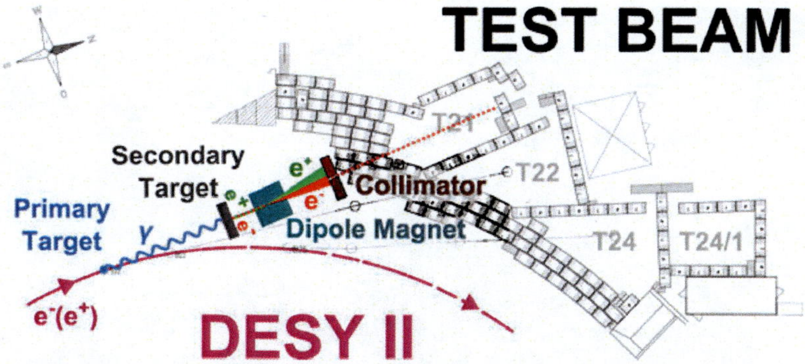

Fig. 6.7 Illustration of the test beam generation and the facilities at the DESY II ring accelerator. Figure taken from Ref. [17]

aluminum plates. There, photon conversion into electron-positron pairs occurs. It follows a sequence of vacuum pipes guiding the particles through beam optics for selection of polarity and momentum. At the end of this chain, electrons and positrons with momenta ranging from 1 GeV/c to 6 GeV/c are available. Their rate decreases with higher momentum due to the qualitative ${E_\gamma}^{-1}$-dependence in the spectrum of emitted Bremsstrahlung photons [21] at the primary target. From detailed simulation [22], the beam's relative momentum spread is estimated to be 13% at 1 GeV/c going down to 1.5% at 6 GeV/c.

Particle beams are generated practically continuously and are delivered to three independent areas: T21, T22 and T24. Each one is surrounded by thick concrete walls guaranteeing radiation levels below 1 mSv/a in the control rooms outside of these areas despite the presence of beam. The latter can be turned off for each area independently through shutters made of a 40 cm thick lead blocks. The tests of HGCAL prototype modules at DESY in March 2018 took place inside the T21 area.

6.3.2 Experimental Setup in the T21 Area

Like the other test beam areas at DESY, the T21 area provides necessary infrastructure for prototype tests with particle beam. 230 and 400 V power fuses, 1 Gbits^{-1} ethernet connections for remote operation and communication to the local DAQ inside the area are available.

Most prominently, the EUDET-based DATURA beam telescope [18] is installed in the T21 area allowing for tracking of beam particles with micron precision. As it is depicted in the photograph of the setup in Fig. 6.8, the telescope consists of two arms each comprising three tracking planes equipped with MIMOSA26 chips. The MIMOSA26 is a pixelated, monolithic silicon detector ASIC with 18.4 μm square pitch. 1152 columns and 576 rows span a sensitive area of 21.2 mm $\times$ 10.6 mm.

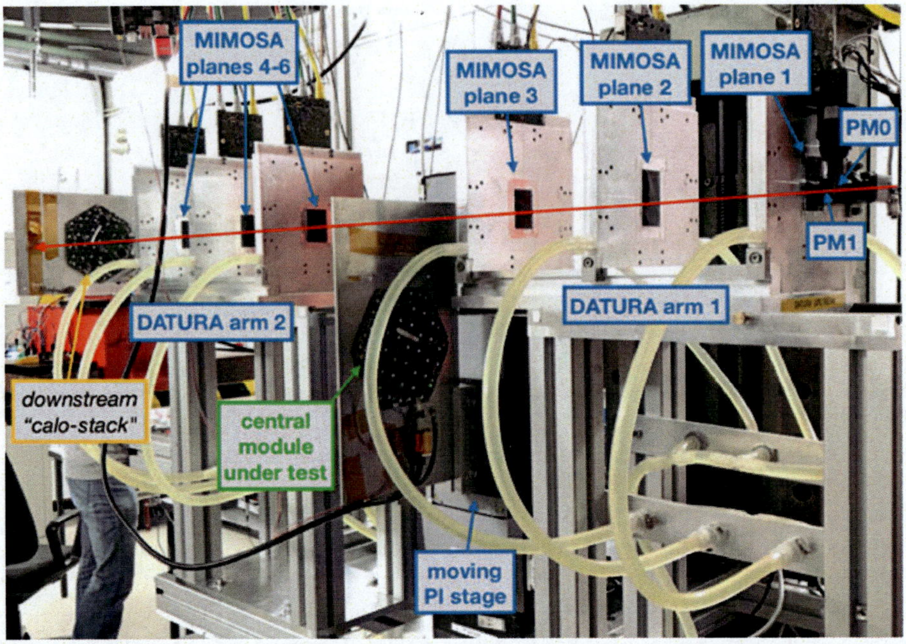

Fig. 6.8 Photograph of the HGCAL prototype module test setup at the DESY T21 area in March 2018. The main module under test was placed in between the DATURA telescope arms for optimal tracking resolution at its location. Additional modules together with passive absorber material are placed further downstream for calorimetric measurements which are not discussed in this thesis

The ASIC is read out with a rolling shutter, whereby the integration time amounts to 115.2 μs [18]. Its particle detection efficiency is reported to be 99.5% while a low fake rate of 10^{-4} per pixel [23] is maintained. A dedicated EUDAQ producer runs on a separate computer inside the area, performs the configuration and the readout of the MIMOSA26 chips. Moreover, the default EUDAQ online monitor already implements an online visualisation of the corresponding data [12]. The MIMOSA26's overall material budget is kept to a minimum: Their sensitive thickness amounts to only 50 μm complemented by 2x25 μm thick, light-shielding and insulating Kapton layers. Telescope operation was conducted at a discriminative signal-to-noise threshold of $\epsilon_n = 6$ [24] which results in a pointing resolution of 3.24 μm at each plane [18]. With such a performance, tracking precisions down to $\sim$2 μm have been reported [18]. It must be emphasised that this number, i.e. the tracking precision at the device under test (DUT), highly depends on the exact arrangement of the planes, on the beam momentum and on the DUT's material budget.

Figure 6.9a, b depict the two configurations relevant in this work. In configuration 1, a bare prototype PCB was placed in the middle between the telescope arms for detailed assessment of its material budget. In configuration 2, a prototype HGCAL module was operated in between the arms for the study of efficiencies in the detection of electrons as MIPs as a function of their impact position. In this configuration, the

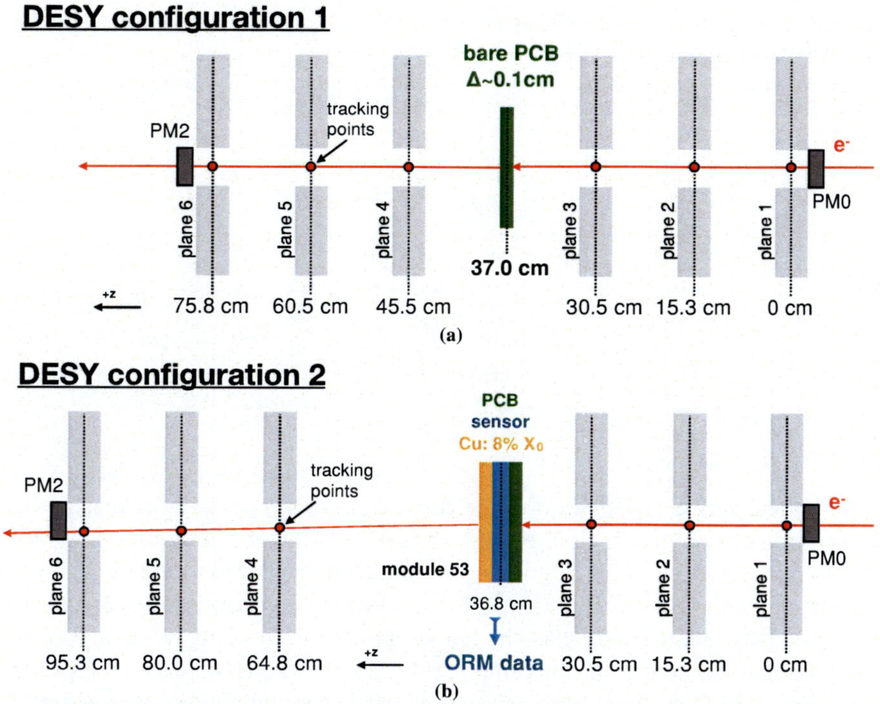

Fig. 6.9 Sketch of the two experimental configurations during the HGCAL beam test at DESY in March 2018 that are investigated in this work. PM0 and PM2 represent scintillators read out by photomultipliers that were used for triggering. The other two scintillators (PM1 and PM3) were placed in the beam but are not part in the trigger. The "calo-stack" from Fig. 6.8 is not shown

downstream arm had been moved further downstream to minimise its distance to the modules forming the "calo-stack" (Fig. 6.8) and thus to optimise the track resolution at its location. Note that calorimetric measurements from the "calo-stack" are not discussed in this thesis and these modules are not shown in Fig. 6.9b.

For both configurations, the DUTs in between the telescope arms were mounted on a moving stage which enables automated movement in transverse direction with respect to the beam. In principle, it allows for characterisation of the entire module area ($\sim$15 cm diameter) because otherwise the experimental acceptance would be limited to the trigger acceptance of roughly 2 cm $\times$ 1 cm defined by the photomultiplier scintillators (PM) up- and downstream the telescope.

For initiating the readout, the coincidence of signals from PM0 (upstream) and PM2 (downstream) was formed inside a Trigger Logic Unit (TLU) [25] which in turn generated the main trigger. Then, that signal was forwarded to the MIMOSA26 DAQ and the SYNCH board. Event synchronisation is achieved mainly at hardware level inside the TLU, namely through the provision with their respective busy signals. During readout of the data from the MIMOSA26 and/or SKIROC2-CMS, any generated trigger is vetoed by the TLU.

Table 6.1 DAQ rates during the HGCAL beam test in the T21 area at DESY in March 2018. Spills are provided continuously at the DESY II beam line

Beam momentum (GeV/c)	HGCAL	MIMOSA26	DAQ rate (Hz)
1.0	✓	✓	50
1.4	✓	✓	45
3.0	✓	✓	37
3.6	✓	✓	30
4.4	✓	✓	25
5.4	✓	✓	20
6.0	✓	✓	12.5
3.0	–	✓	1300

As anticipated, a modulation in the data acquisition rate with increasing beam momentum was obtained, cf. Table 6.1. In fact, the HGCAL prototype DAQ is too slow to record test particles originating from the same phase in the DESY II accelerator's magnet cycle, cf. Fig. 3 in Ref. [17], leading to quantisation effects in the observed DAQ rate. In particular, the rate dropped from initially 50 Hz at low beam momentum, equivalent to the maximum DAQ rate with the HGCAL prototype, down to 12.5 Hz at high test particle momenta, equal to the momentum modulation frequency of particles in DESY II. The DAQ rate levelled around 1.3 kHz when the HGCAL prototype's DAQ was not included in the data taking.

6.4 Calorimeter Prototype Tests with Beam at CERN

Proton-proton collisions at the LHC have been recorded at centre-of-mass energies up to $\sqrt{s} = 13$ TeV with the CMS detector until the second long shutdown at the end of 2018. Typical analyses of this data focus on reconstructed physics objects with high transverse momentum (p_T), usually at the order of a couple of 10 GeV/c. For instance, as integral part in the observation of top quark associated Higgs boson production [26], electrons[1] with $p_T > 15$ GeV/c are considered exclusively in a specific analysis where the Higgs boson decays into a pair of bottom quarks ($H \rightarrow b\bar{b}$) together with at least one of the top quarks decaying leptonically [27]. Given the increase in pile-up and the advance to $\sqrt{s} = 14$ TeV at HL-LHC, selection thresholds like such are not expected to be lowered in the future.

At a pseudorapidity of $|\eta| = 2$, $p_T = 15$ GeV/c is equivalent to 56 GeV/c in absolute momentum. Tests with electrons at that energy realm are only contingent at CERN's North Area where fixed target experiments and prototype tests are supplied

[1] Both electrons with negative and with positive polarity, i.e. positrons, are implied. Both types are referred to as "electrons" for simplicity in this section.

with primary proton beam up to 450 GeV/c delivered by the Super Proton Synchrotron (*SPS*) [28]. Like it had been for other CMS-related beam tests in the past, e.g. of the currently installed CMS HCAL [29], area PPE172 at the H2 beam line [30] was the main site for the HGCAL prototype tests until now. The so far largest test of an HGCAL prototype calorimeter equipped with more than 90 prototype modules was conducted there in October 2018.

Section 6.4.1 briefly describes the non-trivial generation of electron and hadron beam and its subsequent transport to this test area. Detailed studies on possible beam deteriorations and their impact on the calorimeter prototype qualification results are beyond the scope of this work. The following Sect. 6.4.2 deals with the October 2018 experimental setup in the PPE172 area. As part of this work, the incorporation of the available beam characterising detectors into the HGCAL prototype tests have been realised. Key concepts and developments in this regard are presented in Sect. 6.4.3. Finally, Sect. 6.4.4 focusses on the HGCAL prototype calorimeter configurations from which the results in Chaps. 9 and 10 are largely obtained.

6.4.1 SPS North Area Beam Test Facility

Generation of test beams at the North Area starts with protons from the Super Proton Synchrotron (*SPS*). The SPS is a 7 km long circular accelerator whose purpose is the acceleration of protons (and heavy ions) from the Proton Synchrotron, cf. Fig. 2.4. Subsequently, the protons are either injected into the LHC or are delivered to the experiments located at the SPS. In most of its operation time, the SPS beam is ejected towards the North Area targets *T2*, *T4* and *T6*. Depending on the SPS cycle, spills occur roughly once or twice per minute, they last approximately 5 s and contain $\mathcal{O}(10^{13})$ protons each. On average, about 3000 spills of protons are provided by the SPS per day.

At the three targets, a broad spectrum of secondary particles is created when they are exposed to the proton beam. The beam lines work as magnetic spectrometers that provide user-specific particle beams in terms of the particle type, the polarity and the momentum with a 2% uncertainty. The beam lines H2 and H4 are served by one common target (T2), which introduces correlation in the availability of beam momenta and particle types between them. For its high melting point and for its similarity in its nuclear interaction- (λ_n) and radiation length (X_0) per thickness unit, the target is made of beryllium. The latter attribute guarantees that the production of electrons and hadrons due to incident protons occur at comparable rates at the target. Qualitatively, production of hadrons is based on nuclear interactions of the incident proton with beryllium nuclei. It is a one-step process. By contrast, electrons are primarily produced by electron-positron pair production of photons from a previous π^0-decay. In other words, electron production consists of multiple steps. Consequently, the production rates for both hadrons and electrons generally scale differently with the beryllium's thickness. 500 mm has been used for the HGCAL

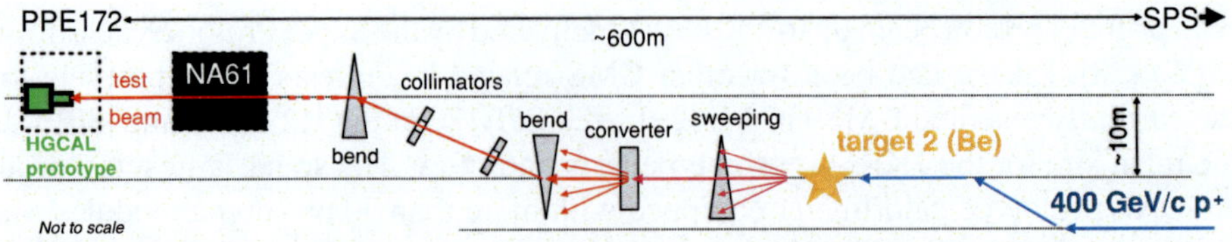

Fig. 6.10 Sketch of the H2 beam line operated with electron beam. The length of the beam line from the SPS to the test area (PPE172) amounts to approximately 600 m. It comprises various targets, collimators, di- and quadrupole magnets whose operating parameters can be modified depending on the user-specific requirements to the beam. The NA61 experiment [34] is located in front of PPE172 and adds some amount of passive material to the beam line

prototype tests in October 2018 yielding a good compromise in the abundance of hadrons and electrons in the secondary beam.

The exact secondary beam's particle composition with incident 400 and 450 GeV/c protons on beryllium has been investigated in numerous works, e.g. in [31–33]. At negative beam polarity and especially at high momentum, it is found that the relative fraction of anti-protons (p^-) and K^- is suppressed. In turn, it is expected that π^- are predominant in hadron beam with negative polarity. As an example, the fraction of anti-protons to negative pions was determined in Ref. [31] to be $p^-/\pi^- < 10^{-3}$ at 300 GeV. Hadron beams with negative polarity are thus named "pion beams" for simplicity throughout this thesis.

Further purification of hadron beams can be achieved through insertion of additional converter material with high ratios in X_0/λ_n, such as lead, into the beam line. Then, electrons in the beam lose significant amount of momentum and subsequently are not transported further through the beam line magnets, whereas hadrons traverse this material with only little interaction.

A possible implementation for electron beam purification is sketched in Fig. 6.10. Behind the primary target, a sweeping magnet distorts all charged, secondary particles from the nominal path allowing only neutral particles, e.g. neutral pions, to pass. Afterwards, a converter, similar to the one for hadron purification in a hadron beam, causes the neutral particle's decay cascade to electron-positron pairs.

Selection of the momentum is done in the vertical plane via sets of dipole magnets and collimators. Due to the bending in the magnetic field, the energy loss from synchrotron radiation (P_S) is not negligible for high-momentum electrons while it is practically irrelevant for hadrons, cf. the scaling law given in Eq. 6.3 for particles with momentum p, mass m travelling inside a magnetic field with perpendicular field strength B.

$$P_S \propto \frac{p^2 B^2}{m^4} \tag{6.3}$$

Synchrotron losses have two effects: First, electron beams can be purified further through tuning of the magnet currents by the expected electron energy losses. In this case, hadrons, unaffected by this energy loss, would fall out the momentum acceptance for transportation along the beam line. Also, the configured electron momentum

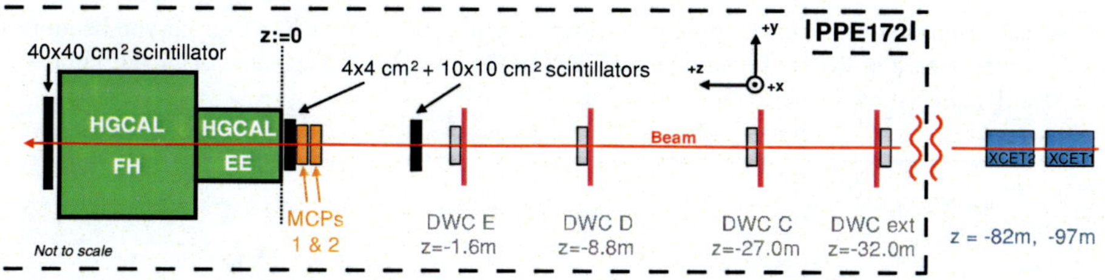

Fig. 6.11 Sketch of the experimental setup inside the PPE172 area at the H2 beam line during the HGCAL prototype test in October 2018

does not necessarily correspond to the real momentum of incident electrons in the test area. This is particularly critical in the context of energy linearity characterisation with the calorimeter prototype in Sect. 10.2.2.

For specification of the beam particles' momenta and their type, most of the magnet and collimator settings along the ~600 m long beam line can be controlled manually by the user. In addition, it enables tuning of the transverse beam size and, to some extent, it allows for control of the particle rates. Ultimately, the beam is designed to have a relative momentum acceptance of less than 2% [30] when arriving at the experimental site. However, the NA61 experiment [34] is located in front of the PPE172 area which added a few percent of a radiation length of passive material in the beam line that could not be removed during the HGCAL beam tests. Preliminary simulation studies on the beam transport in the H2 line indicate a systematic shift at the order of 0.5% on the average beam momentum when accounting for the relevant parts of NA61.

6.4.2 Experimental Setup in the PPE172 Area

Like at the DESY beam test facility, necessary infrastructure for prototype testing, such as connections to fast ethernet, electrical power, and supply of nitrogen gas, is provided. Unlike to the facility at DESY however, the entire North Area hall is a radioactive controlled area and the usage of a personal dosimeter is mandatory. Figure 6.10 illustrates the experimental setup inside the PPE172 area like it was during the most recent HGCAL prototype test in October 2018. The beam enters the area and immediately meets the first delay wire chamber (*DWC*) which was located approximately 32 m upstream of the HGCAL prototype. Vacuum pipes were installed covering most of the particles' flight distance to the calorimeter prototype in order to minimise the influence by air on the beam quality.

As explained in Sect. 4.5, the HGCAL prototype calorimeter consists of two compartments, referred to as EE, representing the electromagnetic, and referred to as FH, mimicking the silicon-based hadronic parts of the final calorimeter design. A com-

Table 6.2 Scintillator coincidence logic deployed for the principal HCGAL prototype beam test in the PPE172 area at the H2 beam line at CERN in October 2018. The $4\times4\,cm^2$ scintillator was eventually replaced by a smaller one of $2\times2\,cm^2$

Beam type	$10\times10\,cm^2$	$4\times4\,cm^2$	$40\times40\,cm^2$
Positrons	YES	YES	VETO
Charged pions	YES	YES	–
Muons	YES	–	YES

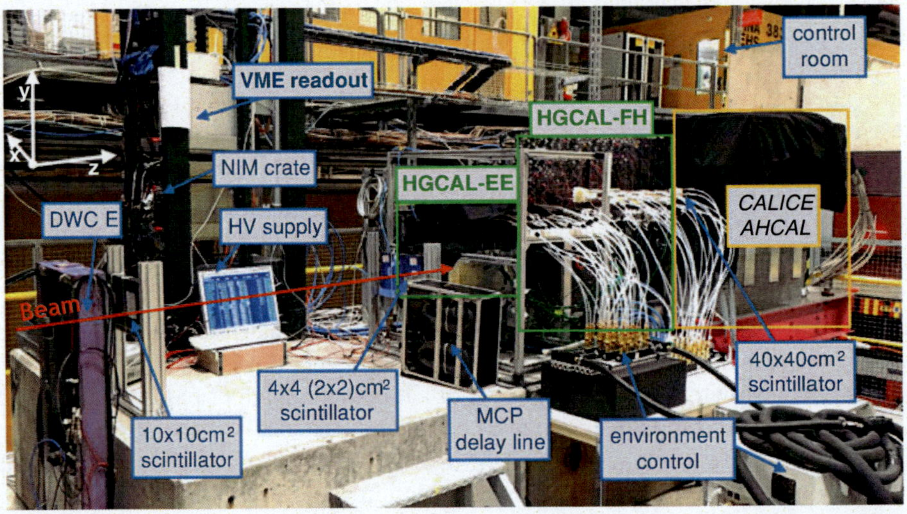

Fig. 6.12 Photograph of the HGCAL calorimeter prototype test setup in the PPE172 area at the H2 beam line at CERN in October 2018. The calorimeter is made of two compartments labelled EE and FH referring to the electromagnetic and silicon-based hadronic parts of the final design. Data taken with the CALICE AHCAL are not analysed in this thesis

mon right-handed coordinate system is defined such that the z-axis coincides with the beam direction and the y-axis points away from the hall floor.

At each spill, thousands of particles reach the test area. Readout of the detectors was triggered by coincidence of diverse scintillator signals. The scintillators were placed at various locations, had distinct sizes and defined the beam acceptance used for the tests. The coincidence logic was chosen according to the desired particle content in the beam, cf. Table 6.2. As example, given the eventual depth of the HGCAL prototype calorimeter of more than 50 X_0, hadron contamination in the recorded showers from electron beams could be reduced by demanding the coincidence of the two scintillators upstream the EE while the scintillator behind the FH functioned as a veto.

Figure 6.12 is a photograph of the entire setup illustrating also the area around the prototype calorimeter. In contrast to previous HGCAL prototype beam tests in that area, the setup itself was not installed on a movable table but remained stationary on a concrete platform. The main reason for that was the synchronous data taking together with the CALICE AHCAL engineering prototype [35] which rendered the combined setup too large to fit on any available table. Being a proxy for the scintillator-based part of HGCAL, CALICE AHCAL's data have been taken and indeed have been synchronised successfully following the strategy described in Sect. 6.2.4. Nevertheless, owing to the different methodology in their reconstruction and calibration, analysis of AHCAL's data is beyond the scope of this thesis.

The high voltage supply allowed for remote monitoring of applied bias voltages and leakage currents summed over several HGCAL prototype modules connected to the same supply channel in groups of up to 8 modules per channel. The environmental control provided insight to relative air humidity and temperature levels at various locations inside the EE and FH compartments. Synchronised analysis of these quantities with the HGCAL prototypes was not done which leaves room for further improvement in future tests. In consequence, detailed analysis of environmental data is not part of this thesis either. The VME crate shown in Fig. 6.12 is essential to the readout of the beam characterising detectors that are discussed in the next section.

6.4.3 *Operation and Readout of Beam Characterising Detectors*

For the beam tests with the HGCAL prototype calorimeter at CERN, the setups have partially been complemented by the readout of the following beam characterising detectors listed below:

- Up to four delay wire chambers for particle tracking,
- two microchannel plate plus PMT detectors for precise timing measurements,
- two threshold Cherenkov detectors intended for hadron identification.

Data collection from these detectors is performed by dedicated VME modules that communicate via a VME bus and an optical link to a computer running distinct EUDAQ producers, cf. Figs. 6.4 and 6.13a. This section describes these auxiliary detectors' working principles and provides information that might be helpful for re-integration at future HGCAL beam tests.

Delay Wire Chamber (DWC)

In first approximation, a DWC's working principle is identical to the traditional multiwire proportional chambers proposed by Charpak in 1968 [36]. The delay wire chambers in the H2 beam line, one example is depicted in Fig. 6.13b, consist of a chamber filled with a 50:50 mixture of argon and carbon-dioxide gas. 20 μm thin cathode wires are installed inside that chamber and are operated at a high voltage of usually around 2800 V [37]. When an ionising particle enters the gas volume, it generates a charge cloud which is accelerated towards the cathode wires. Precise inference of the incident particle's impact position is feasible through measurement

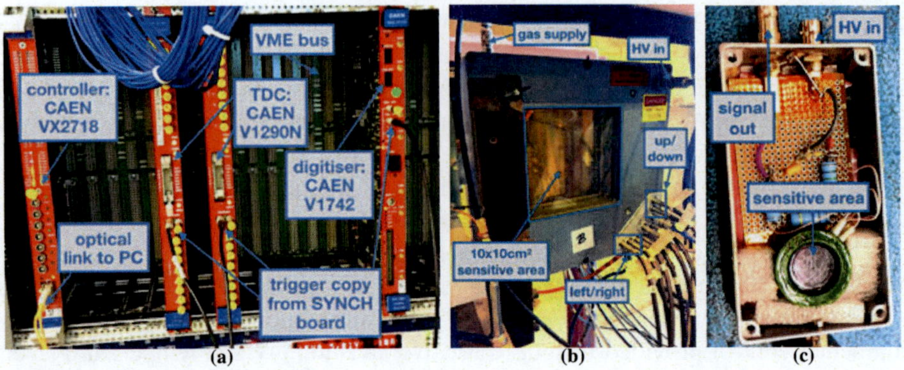

Fig. 6.13 **a** VME modules used for the readout of the beam characterising detectors. The CAEN VX2718 controller is the bridge between the modules and the DAQ computer. **b** Representative photograph of (the unused) delay wire chamber B in the PPE172 area. **c** Photograph of one of the two MCPs. During operation, its case is closed to minimise light-induced signal distortions. All photos have been taken during the preparation for the HGCAL prototype test in October 2018

of the collected charge at each wire and subsequent computation of a charge-weighted centre of gravity.

The major difference in the DWC's functionality to this original concept is the fact that DWCs make use of a 600 ns delay line connected to all wires for the purpose of impact position interpolation between the strands. This concept was first proposed in 1985 at CERN and has facilitated the readout of these devices ever since [38]. Instead of reading out the signal of each cathode individually, one has to record the relative signal delay for the accumulated charge propagating along the delay line to reach either of its ends (left/right, down/up), see Fig. 3 in Ref. [37]. Given the measured time of arrivals T_i, the computation of Cartesian impact points makes use of the relationships in Eq. 6.4.

$$\begin{aligned} x_{\text{DWC}} &= \alpha_x \cdot \left(T_{\text{left}} - T_{\text{right}}\right) + \beta_x \\ y_{\text{DWC}} &= \alpha_y \cdot \left(T_{\text{down}} - T_{\text{up}}\right) + \beta_y \end{aligned} \tag{6.4}$$

In these equations, the parameters $\beta_{x(y)}$ can be set to zero. Their exact values are subject to subsequent alignment of the DWCs, cf. Fig. 7.12a. In principle, $\alpha_{x(y)}$ are to be derived from calibration using a portable test generator device [37]. Unfortunately, this device is not available anymore. Instead, the recommended values of $\alpha_x = \alpha_y = 0.2\,\text{mm ns}^{-1}$ are taken.

Although the active region of the DWCs spans an area of 10 cm × 10 cm, the DWC operation manual [37] states that the linear region, i.e. the region in which Eq. 6.4 is valid, is confined to rather 8cm × 8cm. Moreover, the manual claims the following expected performance: Particle rates of more than 200 kHz can be sustained, the pointing resolution is around 0.2 mm and the particle detection efficiency is higher than 99%. The latter two statements are revisited and verified experimentally in Sect. 7.4.

Six DWCs are installed in the PPE172 area. Their positioning is influenced by mechanical and logistical constraints. All of them had been qualified with oscilloscope measurements and parasitic muons before the main HGCAL beam test. Hereby, two of them were identified to give no signals. Thus, the two malfunctioning DWCs were moved to positions A and B, which are located in between DWC ext and DWC C, cf. Fig. 6.11, and were not read out during the test.

The required connector for low voltage supply, cf. Fig. 5 in Ref. [37], to DWC E was not available and a custom made connector had to be assembled. Other than that slight complication, the DWCs' connection to the DAQ was straightforward: Up to 35 m long coaxial cables were used to connect the DWCs to the NIM crate. Neither signal loss nor delays were an issue.

Arriving at the NIM crate, signals from either end of the DWCs' delay lines were discriminated at −30 mV and then fed into a TDC for timing measurement. A CAEN v1290N module [14] was used because of its timing resolution well below 1 ns, i.e. the required precision to achieve 0.2 mm pointing resolution with the DWCs. Its 16 input channels in total are sufficient for collecting data from four DWCs. The readout was initiated by the trigger copy from the SYNCH board. Table A.1 in Appendix A documents the TDC's operating settings.

Within this work, a dedicated EUDAQ producer has been implemented that interfaces this TDC to the main DAQ framework. This producer [39] configures the TDC and reads the data from the module's FIFO buffer at a a configurable polling rate. Empirically, the maximum readout rate is determined to be around 5 kHz.

In addition, the online data quality monitoring has been extended to both visualise the reconstructed beam occupancy (Fig. 6.14a) at each DWC as well as to present graphics illustrating synchronicity in the acquisition of the various DWCs' data (Fig. 6.14b) similar to how it is done for the MIMOSA26 at the DESY beam test site [12].

Microchannel Plate plus PMT Timing Detector (MCP)

MCPs are devices used for the detection of single particles via electron multiplication in secondary emissions. Time inference with resolutions around 50 ps for single particles and even less for electromagnetic showers are possible using MCPs [40]. Details on their construction and on their functionality can be found elsewhere, e.g. in Ref. [41].

Figure 6.13c shows the inside of one of the two MCPs lend by the CMS ECAL group for integration into the HGCAL beam test in October 2018. They were placed directly in front of the EE prototype calorimeter compartment. During operation, the inside is light shielded and a voltage of 2700 V is applied. Given the small size of the circular, sensitive area with a diameter of 1.8 cm, a smaller, namely $2{\times}2\,\text{cm}^2$, scintillator replaced the $4{\times}4\,\text{cm}^2$ when the MCP was included in the data taking. By this means, its acceptance matched better the trigger window and the yield of detected incident beam particles in the MCP could be increased.

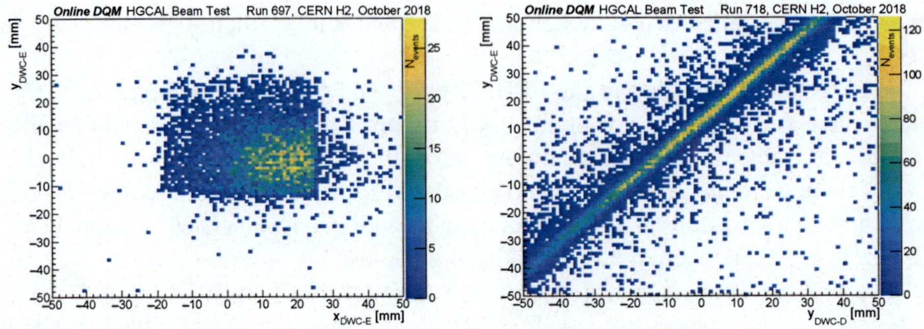

(a) (b)

Fig. 6.14 **a** Exemplary beam occupancy distribution measured by DWC E and **b** correlation in the y-coordinate determined by DWCs D and E as visualised by the online monitoring during the HGCAL prototype beam test in October 2018

Raw MCP signals were fed into and their waveform was digitised by a CAEN v1742 digitiser [15]. Like for the DWCs, a dedicated EUDAQ producer has been implemented [42] for its configuration and for polling its data from the module's FIFO. The DQM has been extended to show the recorded waveform of the last trigger, cf. Fig. 6.15a.

Due to the fast rise time of the expected MCP signal of around 1 ns and in order to sample the relevant rising edge with sufficient resolution, the CAEN v1742's sampling frequency, cf. Table A.2, was set to 5 GHz, i.e. its maximum frequency. At the same time, the number of samples per trigger buffered inside the module was limited to 1024 corresponding to a total readout window size of approximately 200 ns. Hence, delays in the (trigger) signal transmission in the DAQ system needed to be accounted for. For that reason, a delay line, as depicted in Fig. 6.12, had been interposed between the MCP and its readout which effectively delayed the MCP signal by 200 ns. Eventually, signal waveforms were recorded that are nicely centred in the digitiser's readout window.

Threshold Cherenkov Counter (XCET)
Two XCETs are permanently installed in the H2 beam line whose signals are brought to the PPE172 control room. They are located roughly 80 m and 100 m, respectively, upstream the HGCAL prototype. In general, those threshold Cherenkov counters can be used for particle, in particular hadron, identification, e.g. see Ref. [43].

Each XCET is filled with gas attributed with a known refractive index n_0. Cherenkov radiation can only be emitted if a particle with mass m and momentum p traverses the gas for which the applied gas pressure P exceeds a threshold pressure P_T. The latter is determined by the particle's kinematics and the gas' refractive index n_0 at a reference pressure P_0. Equation 6.5 states the predicted relationship.

$$P_T = P_0 \cdot \left(\frac{m \cdot c}{p}\right)^2 \cdot \frac{1}{n_0^2 - 1} \qquad (6.5)$$

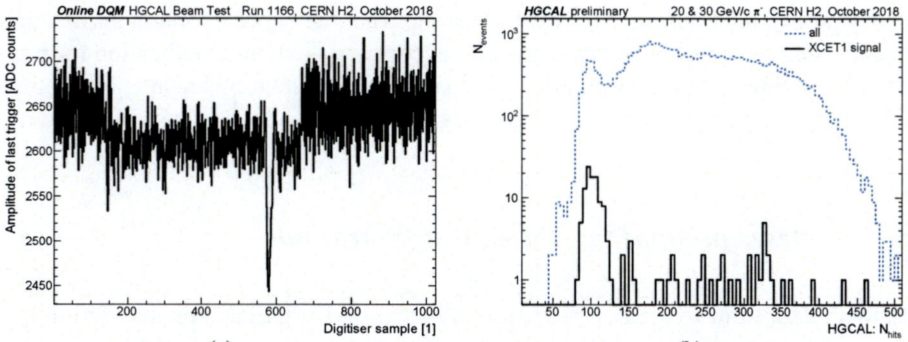

(a) (b)

Fig. 6.15 **a** Online monitoring of the last signal in one of the two MCPs recorded by the digitiser. **b** Distribution of cell occupancies in the prototype calorimeter for pion beams at 20 and 30 GeV/c. The muon-like signatures are extracted when demanding presence of signal in XCET1, effectively purifying the beam with muons

In turn, the average number of detected photo electrons by the XCET (N_e) can be predicted according to Eq. 6.6. L represents the XCET's length and A is a quality factor subject to calibration, see Ref. [44] for a concrete example. The quoted formulas are explicitly derived in Appendix B, whereby an algebraic mistake in its original documentation [43] could be spotted and could be corrected for.

$$N_e \approx A \cdot L \cdot \frac{n_0^2 - 1}{P_0} \cdot (P - P_T) \tag{6.6}$$

In this permanent installation, Cherenkov light-induced signals were discriminated before being transferred to the PPE172 area's control room. From there, the signals were routed down into the experimental area and were injected into a second CAEN v1290 TDC whose operating settings were equal to the ones shown in Table A.1. The 1.6 μs readout window is found to be sufficiently long to register eventual XCET signals.

In the October 2018 beam test, both XCETs were operated with Helium gas to maximise the distance between the pressure thresholds for different hadron types in the beam. Unfortunately, the XCET's length amounts to less than 2 m each. Together with the rather low quality factor for these two devices, creation of less than one photo electron could be expected on average for traversing hadrons in the SPS momentum range. Therefore, the efficiency for particle detection, cf. Fig. B.1a, at gas pressures slightly above the threshold pressure is found to be too small to yield sufficient data sample sizes for sophisticated analyses.

Nonetheless, the proof of principle of the successful incorporation of the XCETs into the HGCAL beam tests is demonstrated as follows. For a few runs with 20 and 30 GeV/c pion beam, the Helium pressure inside XCET1 was set to the pressure threshold of pions. Requiring a signal from XCET1 should be equivalent to the beam purification with particles that are lighter than pions, i.e. muons (e.g. from in-flight-decay pions). And indeed, Fig. 6.15b illustrates that the expected muon purification occurs adequately: Low values in the cell occupancy—a typical indication for non-showering particles (such as muons) traversing the calorimeter—are mostly obtained when imposing the presence of XCET1 signals that can only be caused by muons.

6.4.4 *Description of the Tested Configurations*

In total, data from four calorimeter prototypes tested at CERN are analysed in this thesis. The configuration highly depended on the number of available prototype modules at the time of the tests. A summary of the tested configurations is given in Table 6.3 and discussed in more detail in the following. In June 2018, 28 prototype modules had been delivered to CERN with which a fully equipped electromagnetic (EE) compartment with equal number of layers was constructed. By placing it on a hydraulic table, the setup could be moved through the beam for full illumination of the module area with muons.

By the following October, 94 prototype modules were at hand for beam tests at CERN. In total three prototype configurations were built and exposed to beam successively within two weeks in that campaign.

The first configuration again consisted of a fully equipped EE compartment with 28 sensitive layers, equal to the structure investigated in June. The remaining 66 modules formed the hadronic (FH) compartment prototype. It comprised nine layers equipped with seven modules followed by three single module layers, see Fig. 6.16a.

Table 6.3 List of HGCAL prototype calorimeter configurations and the corresponding number of installed prototype modules tested with beam at CERN in 2018. Also the availability of beam characterising detectors and the possibility of moving the setup through the beam is reported. The XCETs are included although their data is not used in any analysis in this thesis

Setup title	Working (defective) modules	EE	FH	Beam detectors	Movability
June	28 (0)	28 layers: 1 module/layer	–	4×DWC	Hydraulic table
October Config. 1	93 red (1)	28 layers: 1 module/layer	12 layers: 9×7 modules/layer 3×1 module/layer	4×DWC	Stationary
October Config. 2	93 (0)	28 layers: 1 module/layer	11 layers: 2×1 module/layer 9×7 modules/layer	4×DWC 2×MCP 2×XCET	Stationary
October Config. 3	91 (1)	8 layers: 1 module/layer	12 layers: 7 modules/layer	–	Stationary

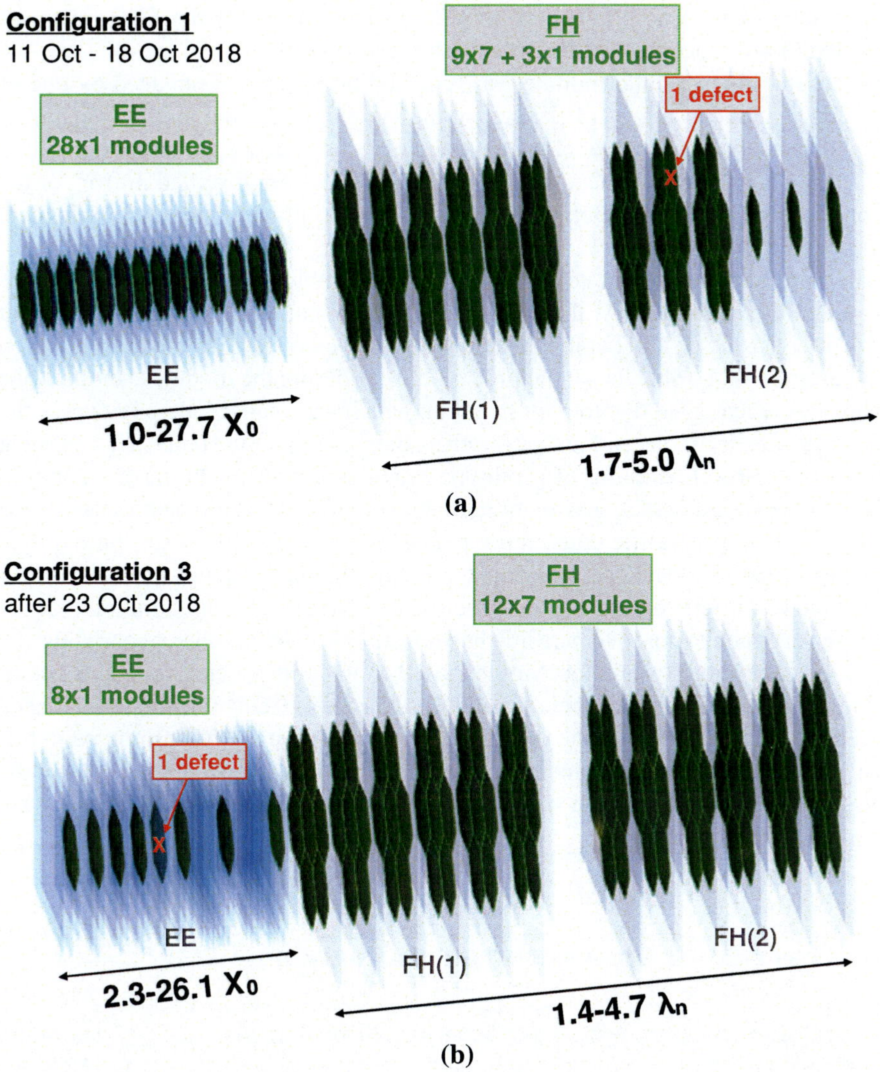

Fig. 6.16 Sketches of the HGCAL prototype calorimeter setups for **a** the configuration 1 and **b** configuration 3 during the October 2018 beam test in the PPE172 area. The sensitive prototype modules are drawn in green while passive material is represented as grey (iron) and blue (lead) volumes

During data taking, one non-central module located in one of the FH layers was identified to give corrupt data. It could not be recovered.

As result, this particular module was removed later and was replaced by one of the modules in the last three FH layers. At the same time, the remaining two single-module layers in the FH were relocated to the FH's front defining configuration 2 in the October beam test. By then, the MCPs and XCETs were installed in front of the EE or had been filled with Helium gas, respectively, and were included in the data taking.

The third and with it last configuration was assembled shortly before the nominal end of the beam test. At the cost of a sparse EE compartment with only eight layers, it consisted of a FH compartment with twelve layers each made from seven modules, see Fig. 6.16b. During the process of re-ordering the modules, another two modules were damaged due to poor handling which was realised already during the intervention itself for one and just later during data taking for the other. This incident stresses that careful handling of prototype modules is critical. From this configuration, only the data which was recorded with so-called parasitic beam is analysed in this thesis. The parasitic data taking period refers to the time after the nominal beam test. There the HGCAL calorimeter setup remained inside the PPE172 area and was now located downstream a test setup from the HERD collaboration [45] for one week and then exposed to parasitic beam from NA61 for another week. Thus, the prototype calorimeter was effectively exposed to undefined beam remnants representing a pile-up like environment. In addition, the $10\times10\,\text{cm}^2$ scintillator had been exchanged by a much larger one ($> 40\times40\,\text{cm}^2$) to expose the full prototype to this undefined beam. Again, the coincidence with the scintillator between the FH and the AHCAL prototype formed the trigger signal. Beam characterising detectors were not operated at this time.

6.5 Overview on the Measurement Programme

Several distinct measurements have been carried out throughout 2017-18 whose data serve as the basis for qualification and performance validation of the HGCAL beam test prototypes. The recorded datasets neither cover all aspects simultaneously nor does the chronology in their recording reflect the methodology applied in this work. This circumstance is related to the evolving availability of functional prototype modules, to the adaption to experimental conditions at the beam test sites, and to changing data taking priorities at the time of their collection. Instead, the different datasets serve distinct purposes.

The ones analysed in this thesis are defined below, whereby data accumulated with prototype modules are referred to as *events*. The hereby introduced abbreviations for the datasets, or equivalently programmes, are used throughout the next chapters of this thesis. Table 6.4 gives an overview on their usage in the respective studies.

Table 6.4 List of the measurement programmes and the usage of the therein collected datasets in the different studies of this thesis

Dataset/Section	IV	CV	TOM	M1	M2	JUN	O1M	O1E	O1H	O2	O3
8.1.1	✓										
8.1.2		✓									
8.2			✓								
8.3.1					✓	✓	✓				✓
8.3.2				✓		✓					
8.4.1						✓					
8.4.2				✓	✓	✓	✓				
8.4.3				✓							
9.1				✓		✓	✓				✓
9.2								✓	✓		
9.3										✓	
10.2								✓			
10.3									✓		
10.4								✓	✓		
11.6								✓			

- **Leakage Currents (IV)**: Per-diode leakage current measurements for 23 6″ prototype sensors with applied voltages ranging from 25 V to 1000 V. Data were recorded in spring 2017 at CERN with the probe station discussed in Sect. 6.1.
- **Capacitances (CV)**: Per-diode capacitance measurements for two 6″ prototype sensors with applied voltages ranging from 25 V to 450 V. Data were recorded in April 2018 at CERN.
- **Tomography (TOM)**: DATURA telescope data including $440 \cdot 10^6$ readouts with 3.0 GeV/c electron beam for configuration 1, cf. Fig. 6.9a, taken during the beam test at DESY in March 2018.
- **March 1 (M1)**: $0.5 \cdot 10^6$ events taken with 2.4–6.0 GeV/c electrons incident on one module in configuration 2, cf. Fig. 6.9b, recorded during the beam test at DESY in March 2018. The central module was operated at bias voltages ranging from 125 V to 200 V. The dataset is complemented by data from the DATURA telescope.
- **March 2 (M2)**: $2.5 \cdot 10^6$ events taken with 1.6 and 3.0 GeV/c electrons incident on different modules placed in configuration 2, cf. Fig. 6.9b, recorded during the beam test at DESY in March 2018. The bias voltage of the central module was kept constant at 200 V. The dataset is complemented by data from the DATURA telescope.
- **June (JUN)**: $0.5 \cdot 10^6$ events taken with 120 GeV/c muons incident on a movable and fully equipped EE calorimeter prototype recorded during the beam test at CERN in June 2018. Only modules equipped with silicon sensors with active thicknesses of 300 μm were used. The dataset is complemented by data from the DWCs.

- **October, Config. 1, Muons (O1M)**: $0.2 \cdot 10^6$ events taken with 200 GeV/c muons incident on the stationary calorimeter prototype configuration 1, cf. Fig. 6.16a, recorded during the beam test at CERN in October 2018. The dataset is complemented by data from the DWCs.
- **October, Config. 1, Electrons (O1E)**: $1.2 \cdot 10^6$ events taken with 20–300 GeV/c positrons incident on the stationary calorimeter prototype configuration 1 recorded during the beam test at CERN in October 2018. The dataset is complemented by data from the DWCs.
- **October, Config. 1, Hadrons (O1H)**: $0.8 \cdot 10^6$ events taken with 20–300 GeV/c negative pions incident on the stationary calorimeter prototype configuration 1 recorded during the beam test at CERN in October 2018. The dataset is complemented by data from the DWCs.
- **October, Config. 2 (O2)**: $1.0 \cdot 10^6$ events taken with 20–300 GeV/c positrons and negative pions incident on the stationary calorimeter prototype configuration 2 recorded during the beam test at CERN in October 2018. The dataset is complemented by data from the DWCs, MCPs and XCETs.
- **October, Config. 3 (O3)**: $1.3 \cdot 10^6$ events taken with parasitic beam incident on the stationary calorimeter prototype configuration 3, cf. Fig. 6.16b, recorded after the nominal beam test at CERN in October 2018.

References

1. Brondolin E, et al. (2019) ARRAY: an open source, modular and probe-card based system with integrated switching matrix for characterisation of large area silicon pad sensors. Nucl Inst Methods Phys Res, A 940:168–173. https://doi.org/10.1016/j.nima.2019.06.007
2. CMS Collaboration (2017) The phase-2 upgrade of the CMS endcap calorimeter. CERN-LHCC-2017-023 (CMS-TDR-019). https://cds.cern.ch/record/2293646
3. Keysight Technologies (2019) Impedance measurement handbook - a guide to measurement technology and techniques, 6th edn. https://literature.cdn.keysight.com/litweb/pdf/5950-3000.pdf. Accessed 31 July 2019
4. EP-LCD, CERN, "HexDAQ." https://gitlab.cern.ch/CLICdp/HGCAL/HGCAL_sensor_tests/releases. Tag: v1.3.22. Accessed 16 July 2019
5. Borg J, et al. (2017) SKIROC2_CMS an ASIC for testing CMS HGCAL. JINST 12:C02019. https://doi.org/10.1088/1748-0221/12/02/c02019
6. Rubinov P (2016) Development of flexible, scalable, low cost readout for beam tests of the high granularity calorimeter for the CMS endcap. https://doi.org/10.1109/NSSMIC.2016.8069667
7. Akchurin N, et al. (2018) First beam tests of prototype silicon modules for the CMS high granularity endcap calorimeter. JINST 13:P10023. https://doi.org/10.1088/1748-0221/13/10/p10023
8. EUDAQ Development Team (2016) EUDAQ user manual (v1.7). http://eudaq.github.io/manual/EUDAQUserManual_v1.pdf. Accessed 18 Nov 2019
9. Liu Y (2017) EUDAQ2 user manual. https://cds.cern.ch/record/2314266
10. Acar B, et al. (2021) The DAQ system of the 12,000 channel CMS high granularity calorimeter prototype. JINST 16:T04001. https://doi.org/10.1088/1748-0221/16/04/T04001
11. CMS Collaboration (2017) The CMS trigger system. JINST 12:P01020. https://doi.org/10.1088/1748-0221/12/01/P01020

12. Spannagel S (2016) Test beam measurements for the upgrade of the CMS pixel detector and measurement of the top quark mass from differential cross sections. PhD thesis, Fakultät für Mathematik, Informatik und Naturwissenschaften der Universität Hamburg, Germany. http://dx.doi.org/10.3204/DESY-THESIS-2016-010
13. Ahlburg P, et al. (2020) EUDAQ - A data acquisition software framework for common beam telescopes. JINST 15:P01038. https://doi.org/10.1088/1748-0221/15/01/p01038
14. CAEN (2010) Technical information manual, MOD. V1290 A/N, revision n. 11
15. CAEN (2011) Technical information manual, MOD. V1742, revision n. 0
16. Behnke T, et al. (2007) Test beams at DESY. https://www.eudet.org/e26/e28/e182/e283/eudet-memo-2007-11.pdf
17. Diener R, et al. (2019) The DESY II test beam facility. Nucl Instrum Meth A 922:265–286. https://doi.org/10.1016/j.nima.2018.11.133
18. Jansen H, et al. (2016) Performance of the EUDET-type beam telescopes. EPJ Tech Instrum 3:7. https://doi.org/10.1140/epjti/s40485-016-0033-2
19. Linac II and PIA (2019) https://min.desy.de/linac_ii__pia/. Accessed 19 July 2019
20. PETRA III (2019) http://petra3.desy.de. Accessed 18 July 2019
21. Matthews JL, Owens RO (1973) Accurate formulae for the calculation of high energy electron bremsstrahlung spectra. Nucl Instrum Meth 111:157–168. https://doi.org/10.1016/0029-554X(73)90105-5
22. Schütz A (2015) Simulation of particle fluxes at the DESY-II test beam facility. Master's thesis, Karlsruhe Institute of Technology. http://dx.doi.org/10.3204/DESY-THESIS-2015-017
23. Baudot J, et al. (2009) First test results of MIMOSA-26, a fast CMOS sensor with integrated zero suppression and digitized output. IEEE Nucl Sci Symp Conf Rec, pp 1169–1173. https://doi.org/10.1109/NSSMIC.2009.5402399
24. User manual (2019) https://telescopes.desy.de/User_manual. Accessed 20 July 2019
25. Cussans D (2009) Description of the JRA1 trigger logic unit (TLU), v0.2c. https://www.eudet.org/e26/e28/e42441/e57298/EUDET-MEMO-2009-04.pdf
26. CMS Collaboration (2018) Observation of $t\bar{t}H$ production. Phys Rev Lett 120:231801. https://doi.org/10.1103/PhysRevLett.120.231801
27. CMS Collaboration (2018) Search for $t\bar{t}H$ production in the $H \to b\bar{b}$ decay channel with leptonic $t\bar{t}$ decays in proton-proton collisions at $\sqrt{s}$=13 TeV with the CMS detector. CMS PAS HIG-17-026. https://cds.cern.ch/record/2308267
28. CERN (2019) The super proton synchrotron. https://home.cern/science/accelerators/super-proton-synchrotron. Accessed 20 July 2019
29. Abdullin S, et al. (2008) Design, performance, and calibration of CMS hadron-barrel calorimeter wedges. Eur Phys J C 55:159–171. https://doi.org/10.1140/epjc/s10052-008-0573-y
30. Charitonidis N (2019) H2 beam line, CERN secondary beam areas. https://sba.web.cern.ch/sba/BeamsAndAreas/H2/H2_presentation.html. Accessed 20 July 2019
31. Atherton HW, et al. (1980) Precise measurements of particle production by 400 GeV/c protons on beryllium targets. CERN 80-07. https://doi.org/10.5170/CERN-1980-007
32. SPY Collaboration (1998) Pion yield from 450 GeV/c protons on beryllium. Phys Lett B 425:208–214. https://doi.org/10.1016/S0370-2693(98)00237-8
33. Marchionni A (1999) Measurement of charged particle production from 450 GeV/c protons on beryllium. Nucl Phys B - Proc Suppl 75:194–196. https://doi.org/10.1016/S0920-5632(99)00240-6
34. Abgrall N, et al. (2014) NA61/SHINE facility at the CERN SPS: beams and detector system. JINST 9:P06005. https://doi.org/10.1088/1748-0221/9/06/p06005
35. Sefkow F, Simon F (2018) A highly granular SiPM-on-tile calorimeter prototype. J Phys: Confer Ser 1162. https://doi.org/10.1088/1742-6596/1162/1/012012
36. Charpak G, et al. (1968) The use of multiwire proportional counters to select and localize charged particles. Nucl Instrum Meth 62:262–268. https://doi.org/10.1016/0029-554X(68)90371-6
37. Spanggaard J (1998) Delay wire chambers - a users guide. http://cds.cern.ch/record/702443

38. Manarin A, Vismara G (1985) The delay wire chamber (DWC) description. http://cds.cern.ch/record/97367. LEP/BI-TA/NOTE 85-3
39. Quast T (2019) https://github.com/HGCDAQ/eudaq/tree/tb2018-October/user/cmshgcal_dwc. Accessed 18 Nov 2019
40. Brianza L, et al. (2018) Response of microchannel plates to single particles and to electromagnetic showers. Nucl Instrum Meth A 797:216–221. https://doi.org/10.1016/j.nima.2015.06.057
41. Barnyakov AY, et al. (2007) Investigation and development of microchannel plate phototubes. Nucl Instrum Meth A 572:404–407. https://doi.org/10.1016/j.nima.2006.10.276
42. Quast T (2019) https://github.com/HGCDAQ/eudaq/tree/tb2018-October/user/cmshgcal_mcp. Accessed 18 Nov 2019
43. Dannheim D, et al. (2013) Particle identification with cherenkov detectors in the 2011 CALICE tungsten analog hadronic calorimeter test beam at the CERN SPS. https://cds.cern.ch/record/1545809
44. Charitonidis N, Karyotakis Y, Gatignon L (2017) Estimation of the R134a gas refractive index for use as a Cherenkov radiator, using a high energy charged particle beam. Nucl Instrum Meth B 410:134–138. https://doi.org/10.1016/j.nimb.2017.08.020
45. HERD The High Energy cosmic Radiation Detection facility. http://herd.ihep.ac.cn/. Accessed 24 July 2019

Chapter 7
Data Reconstruction Algorithms

Raw data collected in the beam tests of the HGCAL prototype calorimeter require reconstruction into physics quantities before their analysis. These quantities are 5-dimensional calorimeter hits, impact positions and kink angles of particle trajectories measured with the DATURA beam telescope, impact positions of trajectories derived from the delay wire chambers and time of arrival information using the recorded MCP waveforms. As part of this thesis, algorithms have been designed, implemented and applied for inferring these quantities. This chapter deals with these algorithms as they are the basis for the characterisation and validation studies in the next chapters.

The overall reconstruction procedure is regarded as a workflow consisting of multiple steps which depend on each other and rely on the presence of configuration- and meta data. Bookkeeping of the interdependencies and of reliances to intermediate files is essential. Section 7.1 gives an overview on the beam test data reconstruction emphasising on the workflow solution realised in this thesis. A detailed explanation is given in Sect. 7.2 on the computation of calorimeter hits starting from the binary SKIROC2-CMS data. Illustrative examples of each method are provided therein. In addition, current limitations and possible further improvements are discussed. The algorithms' performances are validated as part of the later chapters.

The remaining sections in this chapter present the reconstruction of the data from the beam characterising detectors. This is accompanied by performance studies yielding an impression on the achieved measurement precision of each detector. Inference of particle trajectories from the DATURA beam telescope data in the presence of significant passive material in between both telescope arms is presented in Sect. 7.3. The developed procedure may be re-applied for future beam tests that comprise similar setups to the ones tested in March 2018 at DESY, cf. Fig. 6.9. Section 7.4 documents the reconstruction of particle trajectories using the delay wire chambers in the PPE172 area at CERN. The deduction of a beam particle's time of arrival at the setup from the recorded MCP waveforms is described in Sect. 7.5.

T. Quast, *Beam Test Calorimeter Prototypes for the CMS Calorimeter Endcap Upgrade*, Springer Theses, https://doi.org/10.1007/978-3-030-90202-5_7

7.1 Reconstruction Workflow

Reconstruction of the beam test data starts with unpacking of the binary SKIROC2-CMS data stored inside the EUDAQ format and ends with writing the derived calorimeter hits to TTrees in the ROOT framework [1], in the following referred to as *ntuples*. Similarly, information from the beam characterising detectors are reconstructed and synchronised. Ultimately, the ntuple provides all necessary inputs to the calibration and analysis of the beam test data. Reconstructed ntuples are shared with and used by the members of the CMS collaboration who are actively assessing the HGCAL prototype's performance with beam test data.

A series of reconstruction algorithms, bundled into certain steps, are applied. Figure 7.1 illustrates an example progression of these steps that are executed sequentially for the joint reconstruction of HGCAL prototype data in combination with data from the delay wire chambers. Each arrow hereby represents one step in the sequence. The invoked algorithms are mainly written in C++. The code is compiled either to executables or to libraries which in turn are run directly or interfaced via additional steering files written in Python. The algorithms are discussed in detail later in this chapter. Here, it should be noted that the steps are not autonomous because they require the presence of the output from their predecessors in the sequence. Furthermore, meta files such as setup-reliant calibrations, geometry definitions or electronics mappings need to be entered to the workflow at various stages.

These circumstances motivate the usage of a lucid workflow management system. In general, a workflow's main responsibility is the organisation of steps and the bookkeeping of all dependencies between them. Any workflow can in principle be controlled manually, for instance by executing customised shell commands one after another in their correct order. But this effort often results in a challenging, typically time consuming and error prone endeavour. In order to circumvent such inconveniences, the Python pipelining tool LUIGI [2–4][1] has been used as a workflow solution as part of this work. In the context of HGCAL beam test data reconstruction, it is now being deployed successfully to accomplish a semi-automated processing of terabytes of data taken with various HGCAL prototypes.

In this solution, steps are implemented as Python objects that inherit from LUIGI's "Task" class. Run numbers or other dataset indicators are passed as parameters to the objects' constructors for two reasons: First, it allows for dataset-independent implementation of such steps, rendering them generic templates. Furthermore, the availability of dataset indicators inside the steps is essential for their correct association to configuration files from a global database during reconstruction.

LUIGI's core feature is that it constitutes a target-oriented paradigm. As such, each step must at least specify its output file location and usually defines a runtime behaviour that creates this output. When a file in the foreseen output location exists, the given step is considered complete. In order to properly map the aforementioned dependencies between the distinct steps, a task should also state its requirements. By this means, initiation of a certain task first toggles the execution of the preceding steps

[1] invented by Spotify.

if their respective output does not yet exist. Similarly, shared requirements between steps are resolved automatically. A process is run only as often as necessary, resulting in an efficient usage of computing resources.

This overall GNU “make-like” [5] system can basically steer complex workflows fully automatically. It is straightforward to set it up if the management and execution of steps on one single machine is intended. However, taking 300 GeV/c positron data from 93 prototype modules as an example, the full workflow for reconstruction of one data taking run with 10^4 events lasts up to 2 h/CPU. Processing all the O* datasets ($\approx 4.5 \cdot 10^6$ events) would take more than nine days on a computer with four CPU cores that is not limited by I/O bandwidth in reading and writing out the intermediate reconstructed data files. Hence, steps containing computing-intensive procedures do not run on the workflow hosting machine itself but are submitted to the HTCCondor batch system at CERN with approximately 190,000 CPU cores [6]. Doing so, the entire beam test datasets can be reconstructed overnight.

Although remote execution of LUIGI tasks is not supported by default, the job submission benefits from the developed workflow solution as well. A dedicated task gathers and memorises the core runtime behaviour of each step that would normally be run locally. Then, the same task writes all accumulated commands to a bash script used for subsequent submission to the batch system. The submission itself is either done by the very same task again or can be done manually for a group of data taking runs. Unlike more sophisticated workflow solutions based on LUIGI, e.g. in Ref. [5], status monitoring and potential resubmissions of jobs are not implemented in the current reconstruction workflow system which leaves room for further improvement in the future.

7.2 Signal Reconstruction of HGCAL Prototype Data

Reconstruction of the binary HGCAL prototype data commences with their unpacking from the EUDAQ format. For each SKIROC2-CMS readout channel, the digitised Time-over-Threshold (*TOT*), Time-of-Arrival (*TOA*) as well as the high- and low gain samples from 13 switched capacitor array *SCA* memory cells are unpacked. The information on the trigger time is stored both in the headers of the RDOUT board ORMs and for each ROC individually. The former assures event synchronisation among all ORMs and eventually enables offline event synchronisation with other data streams, cf. Sect. 6.2.4. The latter is used for the assignment of the SCAs to time-ordered samples.

Reconstruction of HGCAL prototype data implies the conversion of this raw data into calorimeter hits. In its simplest form, a calorimeter hit can be represented by a five-entry tuple, namely the associated hit energy (E_i), its time of deposit (T_i) and its location given in Cartesian coordinates (x_i, y_i, z_i).

$$\text{HGCAL-reconstruction}: \text{ raw data} \rightarrow \{(x_i, y_i, z_i, E_i, T_i),\ i \in [1, 2, \ldots, N_{\text{hits}}]\}\ \forall \text{ events}$$

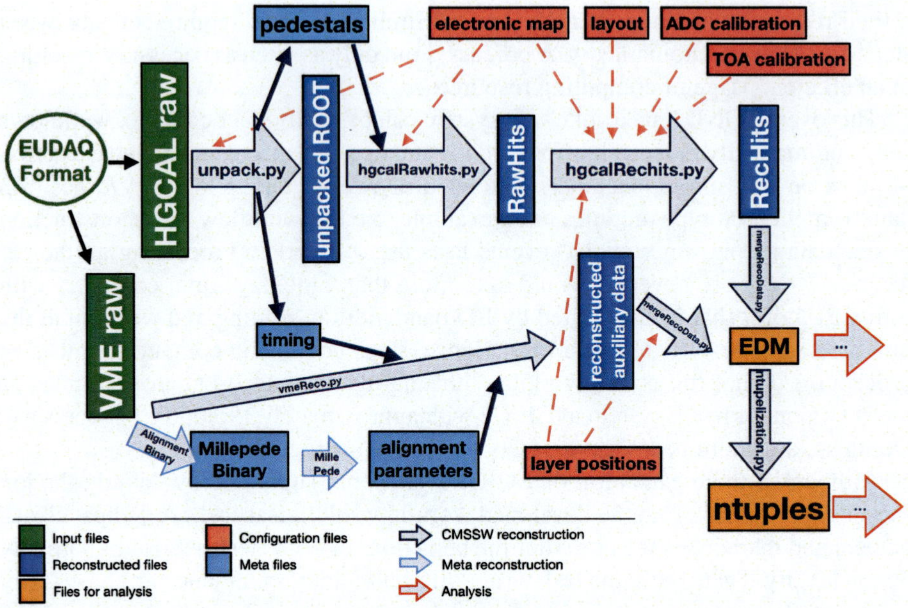

Fig. 7.1 Sequence of steps and their dependencies for the reconstruction of the HGCAL prototype's raw data into calorimeter hits (*RecHits*). This data is merged with likewise reconstructed information from beam characterising detectors. Here, the case for the delay wire chambers is illustrated. Reconstructed data are stored in the EDM file format for analysis with CMSSW, or as ntuples for analysis with ROOT. This workflow solution is similarly applied for the DATURA beam telescope data whose reconstruction also proceeds in numerous, interdependent steps

The coordinates are retrieved from hard-coded maps of the ROC ID's to their locations inside the different beam test prototypes ("layout" and "layer positions" files in Fig. 7.1) as well as from the mapping of the connected channels to the diodes on the sensors ("electronic map" in Fig. 7.1). This mapping requires precise documentation of the modules placements and orientations inside the calorimeter stack. Furthermore, the distinct diode-to-channel associations for the different versions of the prototype PCBs must be minded. Consequently, dedicated analyses are applied to aid with the a posteriori validation of the geometry mapping, cf. Sect. 8.4.1.

The following sections document the methods applied for the inference of the deposited energy and its time of deposition inside a silicon diode. Computation of the electronic baseline, called pedestals, is addressed in Sect. 7.2.1. The procedure for subtraction of common mode noise is motivated and explained in Sect. 7.2.2. Pulse fitting to the time samples, like it is described in Sect. 7.2.3, and the subsequent choice of the appropriate gain, described in Sect. 7.2.4, conclude the hit energy reconstruction. In addition to the raw TOA, the hit energy is required for the inference of the hit time due to timewalk effects which is elaborated in Sect. 7.2.5.

All calorimeter hit reconstruction procedures are implemented as C++ producer plugins within the CMSSW (9.3.0) framework [7] hereby benefitting from already existing data types and event flow infrastructure. Calorimeter hits are stored as HGCalTBRechit objects (*RecHits*) that contain reconstructed information supplementary to the simple 5D tuple. Among this information are the module, chip and channel identifiers, indicators on the corresponding sensor diode geometry, and the reconstructed ADC in each gain. This makes them suitable as input to calibration and sophisticated analysis procedures. The HGCAL beam test software repository is hosted on GitLab and has undergone numerous revisions since 2018. In this thesis, the reconstruction of all data is analogous to the framework's *v13-patch1* release [8].

7.2.1 Pedestal Subtraction

Even in the absence of any signal inside a diode, the readout chips (*ROCs*) are configured such that the digitised signal inside a SCA reaches positive values. The signal baseline in each SCA is called *pedestal*. Pedestals need to be estimated and subsequently subtracted, this way correcting for possible systematic baseline differences among different memory cells and accounting for their conceivable variations with time or with environmental parameters.

The pedestal is a signal-free quantity. When inferred using beam test data like it was done in the October 2018 campaigns, its estimation procedure should not be biased by physical signal. By design of the SKIROC2-CMS configuration used during the 2018 HGCAL beam tests, the first time-ordered sample in each readout does not contain any signal. Therefore, it can be used to compute signal-free distributions of digitised ADC counts in each SCA. Moreover, only runs with small expected signal yield in the HGCAL prototype are used as reference for pedestal subtraction circumventing the possibility of signal in the first time-ordered sample. As there are 13 memory cells, the probability for a given SCA to be assigned to this first sample is 1/13 in a given readout motivating the need for a sufficient number of events in order to obtain meaningful estimates of the raw ADC distributions for pedestal computation. Six example distributions from three SCAs connected to two channels on one prototype module including their pedestals are presented in Fig. 7.2a.

After having constructed such distributions for all SCAs, their medians are assessed and are defined as proxies for the pedestals $P_{k,\text{sca}}$[2]:

$$P_{k,\text{sca}} := \text{med}\left(\{\text{digitised signal}_{k,\text{sca}}^{e}\ ,\quad 1 \leq e \leq N_{\text{events}} | \text{sca} \leftrightarrow t = 0\}\right) \qquad (7.1)$$

In Eq. 7.1 and in the rest of this section, k is a multiindex for the readout channel, e denotes the event number, *sca* symbolises the memory cell and t is the time-ordered sample index. Alternative characterisation methods, e.g. fitting Gaussian curves to the raw ADC spectra, may yield systematic differences with respect to the evaluation using the median at the order of a few ADCs, which correspond to less than 10% of the signal induced by minimum ionising particles (*MIPs*), cf. Sect. 9.1.

[2] Gain specifications are not denoted explicitly to facilitate the notation.

Hence, a significant impact on the calorimetric performance from different pedestal subtraction schemes is not expected. The exact effect from exchange of the pedestal estimation scheme remains to be quantified. Furthermore, it should be remarked that additional modifications to the median-based pedestal assessment are required if the ADC underflows, e.g. in the presence of noise levels much higher than what was experienced in 2018.

Ultimately, a total of 3328 values (13 SCAs, 128 connected readout channels, 2 gains) covering all SCAs from all channels and respecting both gains are computed per module for a given data taking period. These pedestals are written to and are read from separate files within the reconstruction workflow, cf. Fig. 7.1.

Pedestals have been retrieved for each run allowing the study of hypothetical variations with time or their dependence to environmental parameters. The results of those studies are discussed in Ref. [9].

7.2.2 *Common Mode Noise Subtraction*

Recorded ADCs from channels that are physically connected to a silicon diode exhibit wider distributions than those from floating, i.e. unconnected, channels. This fact is depicted in Fig. 7.2a. The reason for this is the circumstance that the connected channels are affected by noise. In general, noise distorts the original signal from particles. While some contributions to the overall noise such as thermal and shot noise are fully random and hence untraceable, others are more systematic and can be corrected for. *Common mode noise* is such an example.

Common mode noise causes a simultaneous shift in the electronic baseline within (connected) channels on the same module. In the 2016 HGCAL beam test prototypes, common mode noise was primarily the result of ground loops in the DAQ chain and

Fig. 7.2 **a** Raw high gain signal for different SCAs of a randomly selected channel connected to the silicon sensor (="ch0") and one floating one (="ch1"). **b** Pedestal subtracted signal in high and low gain in the signal-free, first time-ordered sample (t = 0) before and after correction of the common mode noise (CM)

could be removed event-by-event during the offline reconstruction [10]. Signatures of common mode noise are also observed for the subsequent HGCAL beam tests using the SKIROC2-CMS ASIC and an algorithm for its subtraction is applied. The deployed algorithm has been adapted with respect to 2016 to deal with the signal sampling in the new ROC. It is described in the following whereby the procedure is identical for both gains. The algorithm starts with the median computation of pedestal subtracted signals in ADC counts sampled at the same time considering only channels connected to full hexagonal pads on the same module:

$$CM^{e}_{t,M} := \text{med}\left(\{\text{digitised signal}^{e}_{k,\text{sca}(t)} - P_{k,\text{sca}(t)}\,, \quad k \in H_M\}\right) \tag{7.2}$$

In Eq. 7.2, the per-event common mode noise estimate is introduced as $CM^{e}_{t,M}$ which will be used in the following, in addition to the definitions from the previous section. H_M stands for the set of channels connected to full hexagonal diodes on a given module M. Except for the untreated inner calibration pads, a scaling of $CM^{e}_{t,M}$ with respect to the sensor capacitance is applied for the few remaining channels that are not connected to full hexagonal diodes. Assuming proportionality between the capacitance and the planar diode area (a), the scaling reads:

$$CM^{e}_{t,M} \rightarrow CM^{e}_{t,M} \cdot \frac{a_k}{a_0} \quad (a_k = a_0 \quad \forall\, k \in H_M) \tag{7.3}$$

Afterwards, the pedestal and common mode noise corrected signal, referred to as *time sample* $TS^{e}_{k,t}$, is defined according to Eq. 7.4.

$$TS^{e}_{k,t} := \text{digitised signal}^{e}_{k,\text{sca}(t)} - P_{k,\text{sca}(t)} - CM^{e}_{t,M} \cdot \frac{a_k}{a_0} \tag{7.4}$$

As a result of this correction, the common mode noise is indeed largely removed like it is exemplified in Fig. 7.2b. However, the method relies on the majority of diodes in a module not being occupied with signal. As soon as more than half of the channels that are included into the median calculation in Eq. 7.2 record beam-induced signal in an event, the noise level is overcorrected and thus the actual physical signal is underestimated. As a solution to this deficit, a signal threshold, namely of 100 high gain (*HG*) ADC, in the sample with expected maximum signal (e.g. $t = 3$) has been introduced in this work. Channels exceeding that threshold are interpreted to contain signal and do not enter the median calculation:

$$H_M \rightarrow H^{e}_{M} := \{k \,|\, \text{digitised signal}^{e}_{k,\text{sca}(t=3)} - P_{k,\text{sca}(t=3)} \leq 100\,\text{HG ADC}\,, \quad k \in H_M\} \tag{7.5}$$

Pursuing studies on alternative common mode noise subtraction procedures is beyond the scope of this thesis.

7.2.3 Amplitude Reconstruction

The signal from the collected charge is shaped and sampled continuously at 40 MHz inside the SKIROC2-CMS. Before the signal saturates in a given gain setting, the amplitude of the resulting pulse is proportional to the true collected charge. Hence, extraction of the amplitude from the time sample sequence is a core ingredient to the energy reconstruction. In addition, knowledge on the signal's peaking time could potentially be of interest to support global timing measurements on the shower development with nanosecond precision.

Unlike detector operation at the LHC, the arrival of the particles in the beam test configuration is usually asynchronous to the detector's clock cycle. This is also true for the HGCAL beam tests. Therefore, the signal is sampled at random phases of the pulse rendering weight-based amplitude inference techniques, such as in Ref. [11], deprecated for application in the reconstruction of this beam test data. Instead, a digital representation is fitted to the time sample sequence for the extraction of the amplitude and the peaking time by minimisation of the χ^2 that is defined in Eq. 7.6.

$$\chi^2 \propto \sum_{t=0}^{t \leq N_t} (TS_t - A(25\,\text{ns} \cdot t))^2 \tag{7.6}$$

The used parameterisation for $A(t')$ is motivated by the work on signal reconstruction with the CMS ECAL crystals [12]. The quoted function therein is further extended to accommodate an undershoot by generalising the formula given by Eq. 4.69 in [13].

$$A(t') := \begin{cases} A_0 \cdot A_{\text{norm}} \cdot \left[\left(\frac{t'-t_0}{\tau}\right)^n - \frac{1}{n+1}\left(\frac{t'-t_0}{\tau}\right)^{n+1} \right] \cdot e^{-\alpha \cdot (t'-t_0)/\tau} & \text{, if } t' > t_0 \\ 0 & \text{, otherwise} \end{cases} \tag{7.7}$$

In Eq. 7.7, n, τ and α determine the signal pulse's shape and A_{norm} is a normalisation constant. These parameters are currently fixed and shared among all ROCs. Only t_0 and A_0 are subject to the fitting procedure. t_0 represents the start time of the signal pulse while A_0 is the signal amplitude to be related to the deposited charge in the silicon pads.

For the fitting, only the first five time samples ($N_t := 4$ in Eq. 7.6) enter the χ^2-minimisation to focus the fit on the rising part of the pulse. Running the χ^2-minimisation for all thousands of channels in two gains would take unreasonably long. In order to reduce the computing time, the signal fit is only performed for those channels that have recorded significant signal. The imposed criteria are based on the high gain information only and are listed in Table 7.1. Any channel fulfilling neither of the two criteria is discarded from the reconstruction, and in particular does not become a RecHit.

Figure 7.3 illustrates the pulse fitting for one exemplary readout channel. Signals from events are selected to which pulses with considerable amplitude could be fitted. The corresponding time samples are normalised by the reconstructed amplitude A_0

Table 7.1 Criteria imposed to the time samples to preselect readout channels with relevant signal prior to the pulse fit. t_m denotes the sample index with maximum signal after pedestal and common mode noise subtraction restricted to being either 2 or 3. c_S is a scaling parameter that is unity for modules equipped with 300 µm thick sensors and 2/3 for those with 200 µm thick sensors

Criterion 1	$TS_{t_m} > c_S \cdot 500$ [HG ADC counts]
Criterion 2	$TS_{t_m} > TS_{t_m+3}$
and	$TS_{t_m+1} > TS_{t_m+3}$
and	$TS_{t_m} > c_S \cdot 20$ [HG ADC counts]

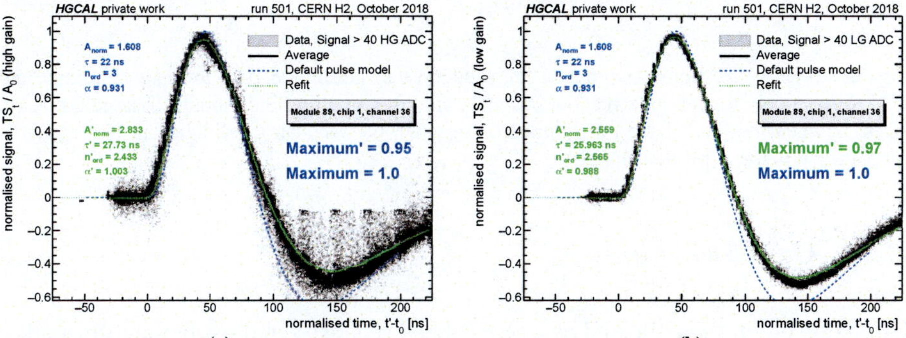

(a) (b)

Fig. 7.3 Illustration of the pulse fit model used for amplitude reconstruction in a readout channel for **a** the high- and **b** the low-gain ADC. Time samples are normalised by the fitted amplitude and start time for each event as obtained from the default pulse reconstruction fit using the parameterisation illustrated in red. The green curve represents a refit to the average, normalised samples where all parameters are allowed to vary. Its parameters seem to render a better description of the pulse shape incorporating even the undershoot

and by the start time t_0 in each event. The default pulse parameterisation corresponding to the *v13-patch1* release [8] of the reconstruction software is shown in red. All fixed parameters therein have previously been optimised by the HGCAL beam test group to yield an almost uniform distribution in t_0, that is consistent with the asynchronous beam with respect to the clock, and to result in a smooth continuity in the hit energy spectrum when the high gain saturates. The green curve shows a model for which all parameters float when refitted over the full range ($t \in [0,\ 10]$) of time samples. It reveals that further optimisations of the default settings might be possible because it describes the average undershoot and signal peak better than the default model in this example. As shown in Fig. 7.3, the difference in the fitted amplitudes from both parameterisations can amount to a few percent. But as long as both parameterisations result in reconstructed amplitudes that are proportional to the collected charge, any impact on the calorimetric performance by this systematic difference would be surprising. Intercalibration of the gains and subsequent energy calibration with MIPs should correct for such deficits in the signal amplitude inference.

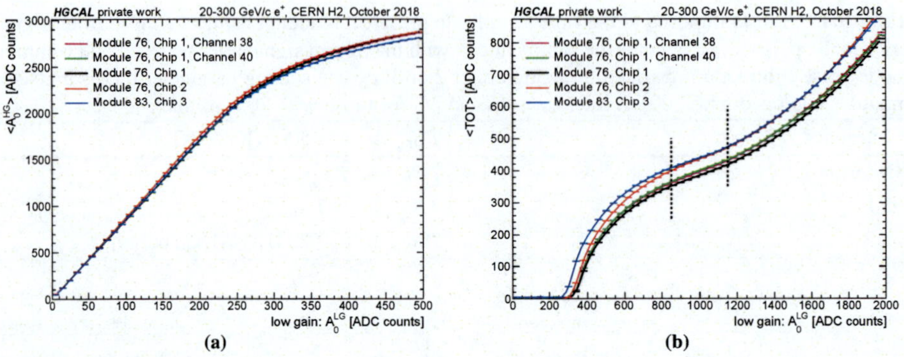

Fig. 7.4 Average relation between **a** the high and low gain amplitude, **b** the low gain amplitude and TOT, respectively for different readout channels on different chips and modules. The relationship is linear when the gains have not saturated, i.e. when the reconstructed signal in both gains is proportional to the deposited charge

7.2.4 *Gain Switching*

Up to this point, the collected charge has been reconstructed using the information from two gains resulting in the two amplitudes A_0^{HG} and A_0^{LG}, cf. Eq. 7.7. For high enough signals, typically for $A_0^{\mathrm{LG}} > 350$ ADC counts, the charge measurement is complemented by the digitised time-over-threshold (*TOT*). Determined by the signal size, the reconstruction selects the most appropriate information among these three for conversion into energy equivalents. The principal strategy is the choice of the gain that has not yet saturated for a given signal size. Subsequently, the signal is converted into the corresponding high gain equivalent, denoted as A'_{HG}.

Within the linear region, the reconstructed signal amplitude in a given gain is proportional to the deposited charge in the sensor. A strict cutoff at the turning point does not occur, but the proportionality slowly vanishes and the non-linearity increases. This behaviour is illustrated in Fig. 7.4a that shows the average reconstructed signal in high gain in bins of signal amplitudes in low gain. The relationship is linear for sufficiently small signals until the point where the high gain shaper saturates ($TP_{\mathrm{HG}} \approx$ 2000HG ADC counts). Similarly, the low gain and TOT establish a narrow region of common linearity as it is depicted in Fig. 7.4b.

Equation 7.8 states the gain switching decision logic. It includes calibration constants (TOT_{offset}, $m_{\mathrm{LG/TOT}}$, $m_{\mathrm{HG/LG}}$, TP_{LG}, TP_{HG}) whose meanings are implicitly defined by Eq. 7.8 and are elaborated later in Sect. 9.2. The constants are obtained by in situ calibration using beam test data and they are expected to mainly differ between the ROCs.

$$A'_{\mathrm{HG}} := \begin{cases} (TOT - TOT_{\mathrm{offset}}) \cdot m_{\mathrm{LG/TOT}} \cdot m_{\mathrm{HG/LG}} & \text{, if } A_0^{\mathrm{LG}} > TP_{\mathrm{LG}} \\ A_0^{\mathrm{LG}} \cdot m_{\mathrm{HG/LG}} & \text{, else if } A_0^{\mathrm{HG}} > TP_{\mathrm{HG}} \\ A_0^{\mathrm{HG}} & \text{, otherwise} \end{cases} \tag{7.8}$$

The conversion into reconstructed hit energies E is achieved by multiplication with a per-channel calibration constant which will be discussed in Sect. 9.1. Hereby, the energy scale is set by the amount of energy deposited by a MIP traversing the silicon pad. The electric response to MIPs is usually measured with the highest possible gain, referring to the high gain in the case of the SKIROC2-CMS, and individually for as many channels as possible.

$$E := M_{\mathrm{HG/MIP}}^{-1} \cdot A'_{\mathrm{HG}} \tag{7.9}$$

In total, an extensive database comprising six calibration constants (five in Eq. 7.8 and one in Eq. 7.9) per channel need to be derived and are injected into the reconstruction workflow ("ADC calibration" file, Fig. 7.1).

In Chap. 10 of this thesis, apart from the just explained default energy inference procedure, the calorimetric performance is also investigated after following an alternative recipe. It is intended to mimic the available gains of the HGCROC [14] in the final HGCAL design. The alternative hit energy reconstruction neglects the high gain information but makes use of the low gain and the TOT only. Equation 7.10 defines its logic replacing Eqs. 7.8 and 7.9.

$$E_{\mathrm{LG}} := M_{\mathrm{LG/MIP}}^{-1} \cdot \begin{cases} (TOT - TO!T_{\mathrm{offset}}) \cdot m_{\mathrm{LG/TOT}} & \text{, if } A_0^{\mathrm{LG}} > TP_{\mathrm{LG}} \\ A_0^{\mathrm{LG}} & \text{, otherwise} \end{cases} \tag{7.10}$$

7.2.5 Time-of-Arrival Reconstruction

The t_0 parameter introduced in Sect. 7.2.3 is obtained from the amplitude reconstruction and could serve as an estimate for the time of charge depositions in the silicon pad. But the associated $\mathcal{O}(1\,\mathrm{ns})$ timing resolution is considered too unreliable for detailed studies on the time evolution of hadronic showers. In order to achieve timing resolutions of less than 100 ps, the SKIROC2-CMS chip is designed to perform two precise measurements on a signal's time-of-arrival (TOA). For this, the charging of a capacitor is initiated as soon as the incoming signal exceeds a configurable threshold. The capacitor charges until the next rising edge of the global clock cycle (TOA_{rise}), or the next falling edge (TOA_{fall}), respectively [15]. The stored charge inside the capacitor is digitised and relates to the signal's arrival time within a clock cycle.

By construction, both definitions of the TOA should result in identical time measurements apart from an offset. Their equivalence is illustrated in Fig. 7.5a. Although the combination of the two would likely improve the timing resolution, it is sufficient to restrict the proof of principle demonstration in this thesis to the TOA_{rise} definition only.

Figure 7.5b shows representative averages of the TOA_{rise} as a function of the signal time estimates from the amplitude reconstruction. Hereby, it visualises the two principal challenges of the time measurement: First, the relationship between the true and measured time-of-arrival exhibits a non-linearity for large capacitor

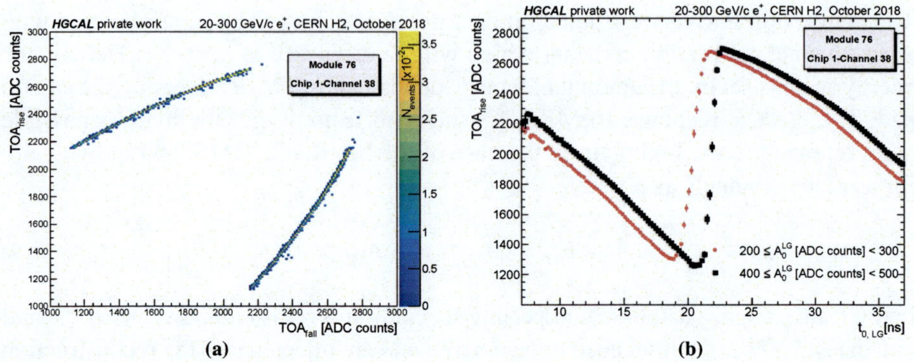

Fig. 7.5 **a** Equivalence of the TOA_{rise} and TOA_{fall}. **b** Relation of the TOA_{rise} to the time from the amplitude fit in the low gain for different ranges of the signal amplitude

charges. And, due to the fixed threshold in the *TOA* acquisition, the measured time depends on the signal size (*timewalk*). Consequently, the generic conversion of the *TOA* and the hit energy E to hit times T contains multiple, non-trivial terms as given in Eq. 7.11.

$$T = T(TOA_{\text{rise}}, E) = f_{\text{TOA}}\left(\frac{TOA_{\text{rise}} - TOA_{\text{rise}}^{\min}}{TOA_{\text{rise}}^{\max} - TOA_{\text{rise}}^{\min}}\right) + f_{\text{TW}}(E) + \Delta l_0 \quad (7.11)$$

In the argument of the function f_{TOA}, $TOA_{\text{rise}}^{\min}$ and $TOA_{\text{rise}}^{\max}$ denote the minimum and maximum values of the TOA taken by a given channel in a given dataset. This normalisation of the TOA_{rise} to $[0,\ 1]$ is done for convenience and facilitates the comparison of calibration constants among different channels. Evaluation of f_{TOA} returns the time when a recorded signal exceeds the configured *TOA* acquisition threshold. The function f_{TW} yields the timestamp correction due to timewalk. The parameterisations for f_{TOA} and f_{TW} are taken from Ref. [16] and are given by Eqs. 7.12 and 7.13. They are designed to yield a good description of the linear part of the calibration curve.

$$f_{\text{TOA}}(x) := a_{\text{TOA}} \cdot x + b_{\text{TOA}} + \frac{c_{\text{TOA}}}{x - d_{\text{TOA}}} \quad (7.12)$$

$$f_{\text{TW}}(E) := a_{\text{TW}} \cdot E + b_{\text{TW}} + \frac{c_{\text{TW}}}{E - d_{\text{TW}}} \quad (7.13)$$

The additional parameter Δl_0 in Eq. 7.11 symbolises a remaining offset, e.g. from the delay owing to different path lengths serving the clock signal to the readout channels.

In summary, nine additional calibration constants, besides the maximum and minimum TOA_{rise} values, are necessary for the derivation of physical hit timestamps. Here, the proof of principle makes use of the functional representations quoted in this section. Nevertheless, it must be noted that not all details of a common calibration strategy in the HGCAL beam test group have been sorted out at the time of writing this thesis. The parameterisations might be specified further in the future, in order to incorporate certain boundary conditions.

7.3 Tracking with the DATURA Beam Telescope

A raw DATURA beam telescope event is equivalent to a list of active pixels on the MIMOSA26 ASICs of the six telescope planes, cf. Fig. 7.6a. Given this set of pixels for each plane, the main objective of the track reconstruction is the construction of all particle trajectories that have traversed the planes within a readout. In particular, one is interested in the particles' impact positions and the kinks of their trajectories at the device-under-test.

Altogether, three aspects are covered by the reconstruction procedure which was developed in the work for this thesis:

- The pixel indexes are mapped to physical coordinates in a common reference frame. This mapping relies on comprehensive documentation and correct implementation of the tracking plane's geometry and their location in the experimental setup.
- Noise and the possibility of having more than one particle traversing the MIMOSA26 planes within the 115.2 μs integration window result in numerous active pixels per readout and per plane. Identification of the relevant ones for the track measurement is a core ingredient to the reconstruction procedure. This is achieved by masking of noisy pixels, clustering of neighbouring entities and their combination to sequences that are consistent with a particle trajectory.
- By default, the accuracy of the tracking planes' placement is limited by mechanical precision. Therefore, alignment corrections must be inferred and be applied in order to achieve the optimal track pointing resolution.

Both the EUTelescope [17] and the currently evolving CORRYVRECKAN [18] software frameworks offer the necessary infrastructure to accomplish all these tasks with more or less effort by the user. Owing to the latter framework's simplicity and modular design, its *v0.6* tag has been chosen as basis for reconstruction of the DATURA telescope data in this thesis. Functionality for unpacking the raw data, masking of noisy pixels and the clustering is already provided. Hence, these steps are summarised only briefly in Sect. 7.3.1. Section 7.3.2 describes the generic combinatorial algorithm for construction of straight lines that was developed as part of this thesis. The principal strategy hereby is that data from different arms (three planes per arm, cf. Fig. 6.9) are first treated separately. Afterwards, they are combined depending on similar orientations and impact points of bisected tracks at a reference plane. By this means, it can cope with significant scattering material located in between both arms. Moreover, this method does not require full alignment corrections of the experimental setup for its application. By the contrary, this algorithm can even be deployed for the derivation of the alignment corrections which is demonstrated in Sect. 7.3.3.

Section 7.3.4 describes the obtained tracking performances in the HGCAL beam test at DESY in March 2018.

7.3.1 *Noise Rejection and Clustering*

Using the EUDAQ framework in their recording, MIMOSA26 data are stored as "StandardPlane" objects. CORRYVRECKAN offers an EUDAQ event loader that reads these objects and converts them into naive "Pixel" objects as input to the reconstruction procedure.

Noisy pixels are identified by evaluation of the mean global hit rate of pixels on a plane. If a pixel exceeds this rate by a factor of ten, it is considered noisy and rejected from further analysis. In practice, this scheme applies well to the cases where the full active area is exposed to the beam. By contrast, it would fail if the exposed area is limited to a few pixels only. In the dataset used here, approximately 400–600 pixels in each of the six planes, corresponding to approximately 1 ‱ of the active area, are found to be noisy and are masked consequently. Afterwards, the reconstruction proceeds with the combination of pixels to clusters like it is schematically illustrated in Fig. 7.6b. A cluster is made from merged entities that are directly adjacent to each other. A centre of gravity is computed determining the clusters' positions. With the data analysed in this thesis, approximately three pixels are combined on average to form one cluster. The average cluster multiplicity amounts to 7–10 for each plane and reaches a maximum of 20 for plane 5.

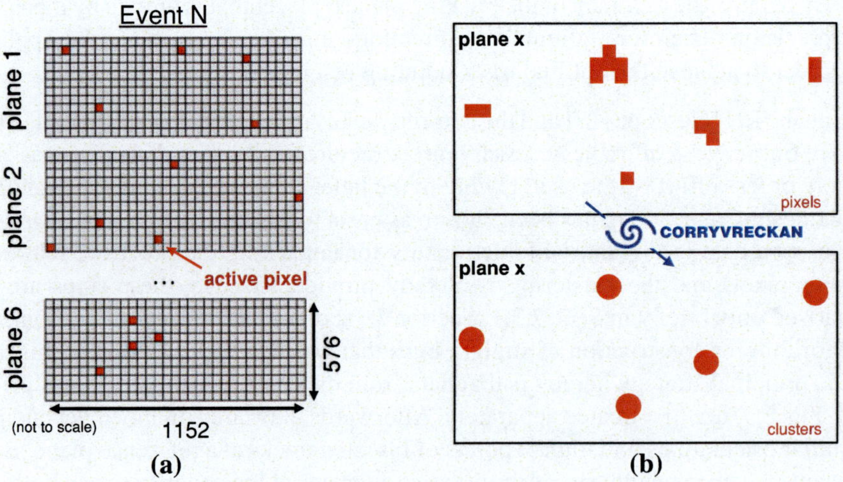

Fig. 7.6 **a** Schematic illustration of the DATURA beam telescope's raw data and **b** the pixel clustering provided by CORRYVRECKAN

Subsequently, the list of clusters is input to the track finding algorithm described in the next section.

7.3.2 *Ntlet-Tracking Algorithm*

As stated in the introduction of Sect. 7.3, the principal strategy for identification of interesting clusters in a given readout is based on their combinations that are consistent with the trajectory of relativistic electrons in the beam. In most tracking applications, such a trajectory is well approximated by straight lines going through all six planes. The reconstruction modules offered in CORRYVRECKAN with its $v0.6$ tag rely on that approximation. However, HGCAL prototype modules with substantial thicknesses were placed in between the telescope arms. Given the momentum realm at the DESY test beam line, the material causes considerable multiple scattering of the incident electrons. In consequence, the straight line model through all planes is an inaccurate model for reconstructing the trajectories. In fact, when using the provided CORRYVRECKAN modules for basic tracking, reasonable results could not be obtained.

The challenge of dealing with significantly distorted trajectories in between the telescope arms is overcome by deploying an alternative track pattern recognition scheme. This method initially treats both telescope arms independently hereby identifying straight lines in the set of clusters on either arm (*tracklets*). Then, two tracklets are merged to *full tracks*, meaning those with clusters from all planes, if the distance between their extrapolated impact points at a common reference plane and their kink angle are sufficiently small. With it, the algorithm is insensitive to the multiple scattering at the material in between both arms. Figure 7.7 explains the steps for the relevant experimental situation, namely with each three tracking planes up- and downstream of the material. The algorithm is written generically and can be applied for the reconstruction with at least three but otherwise arbitrary number (N) of tracking planes in arm 1 ($N_1 \geq 3$) and in arm 2 ($N_2 - N_1 \geq 3$). Hence, it is named "Ntlet-Tracking".

Table C.1 in Appendix C documents the parameter settings chosen for the DATURA beam telescope data reconstruction in this thesis. Running this algorithm on an Intel© Xeon© CPU E5-1620 for the reconstruction of a representative dataset collected with 3 GeV/c electrons and a 1 mm thick aluminum plate placed in the middle takes less than 0.5 ms per readout which is only three times longer than the preceding pixel clustering step. Systematic tuning of the parameters for optimising the computation time, tracking efficiencies or for reducing track multiplicities is left open for the future.

Finally, the trajectory parameterisation of the full tracks is specific to the targeted application in the analysis. Straight lines through all six planes are used for the derivation of alignment corrections. Separate parameterisations from matched tracklets are maintained for the kink angle measurements in the context of the PCB tomography study that is presented in Sect. 8.2. The General Broken Line formalism

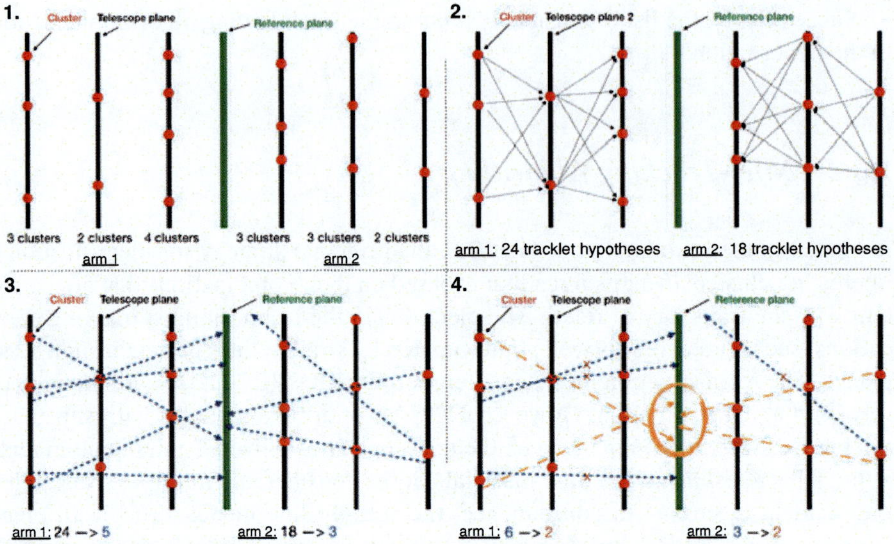

Fig. 7.7 Illustration of the Ntlet-Tracking algorithm (implementation: [19]) for the setup with two telescope arms and three planes each. The main scattering material in the middle is defined as the reference plane.
(1.) Starting situation with the computed clusters subsequent to the rejection of noisy pixels.
(2.) Combination of all clusters in planes 1–3 and 4–6 separately making tracklet hypotheses. By construction, each tracklet hypothesis contains only one cluster per plane. Each hypothesis is attributed a χ^2/ndf that quantifies the correspondence to a straight line.
(3.) Discard of tracklet hypotheses that are inconsistent with straight lines. Only those with a χ^2/ndf below a configurable, arm-dependent threshold (χ_i^2/ndf) are kept. Limitation of the combinatorial count and with it the computation time is achieved by selecting only the first $N^{\mathrm{max}}_{\mathrm{tracks,\,arm}}$ tracklets in each arm for further analysis.
(4.) Combination of tracklets from both arms. A pair of tracklets is merged to full tracks only if two conditions are met: Their absolute distance when extrapolated to the reference plane is below a configurable limit $\Delta D_{z_{\mathrm{ref}}}$. Also, their kink angle must not exceed a configurable value Δk. By convention, the merging is not unique. A given tracklet can be input to form multiple full tracks

(*GBL*) [20] includes the effect of multiple scattering by the incident electrons into the track model. It accounts for the kinks introduced at each scattering material such as the MIMOSA26 planes and the prototype modules. Using the method derived in Appendix D, the GBL formalism can even incorporate the scattering effect from air in between the planes. GBL is used in order to maximise the track pointing resolution at the prototype module for the studies presented in Sect. 8.4.

7.3.3 Telescope Alignment

By default, each MIMOSA26 plane constitutes its own coordinate system in which pixels and clusters are localised. Their physical position and orientation inside the laboratory frame is determined mechanically and the accuracy in their placement is thus limited. Uncorrected displacements and misorientations with respect to the naive experiment geometry can drastically worsen the tracking performance, see e.g. Ref. [21]. Therefore, an essential aspect of the DATURA track reconstruction is the alignment of all planes, i.e. the assessment and correction of relative displacements and misorientations between them offline. After successful alignment, all cluster positions are in a common coordinate system and the tracking resolution is improved.

Specifically for the DATURA beam telescope geometry, each plane may in principle be displaced and misoriented. With one plane defining the common coordinate system, 30 independent alignment parameters covering all possible degrees of freedom would need to be inferred. Established expert-level alignment tools like MILLEPEDE [22, 23] would be suitable for this task. Instead, intuitive alignment corrections based on the information from the Ntlet-Tracking algorithm are used in this thesis. These could serve as pre-alignment corrections for studies that rely on more sophisticated alignment corrections.

The applied alignment procedure consists of four steps that can be categorised twofold. In the first part, the telescope arms are aligned with respect to each other making use of the tracklets. For the second part, the plane-to-plane alignment corrections are determined from full tracks. All correction steps are explained hereafter. Their effects are illustrated for the alignment of DESY configuration 1 analysing a dedicated run where no scattering material was placed in between the arms. The steps and their effects are analogous for configuration 2. Again, the Ntlet-Tracking's parameter settings for each step can be found in Table C.1 in Appendix C.

Step 1: Arm-to-Arm Orientation, cf. Fig. 7.8a

This alignment step ensures an identical orientation of both telescope arms by comparison to the true beam axis. For this purpose, the mean angle of tracklets with respect to the nominal z-axis is computed. Afterwards, the two arms are rotated accordingly such that the mean beam angle measured by each arm is zero.

Step 2: Arm-to-Arm Placement, cf. Fig. 7.8b

Particle impact positions evaluated at the reference plane are compared between tracklets from either arm. Subsequently, all planes in the second arm are translated identically by the resulting average difference.

By this means, the relative displacement of both telescope arms is corrected for.

Step 3: Plane-to-Plane Placements, cf. Fig. 7.9a

After tracklets have been merged to full tracks with parameterisations $\mathbf{f}_{\text{Track}}(z)$, biased residuals $(\Delta x(y)^b_P)$ are computed for every plane P at position z_P:

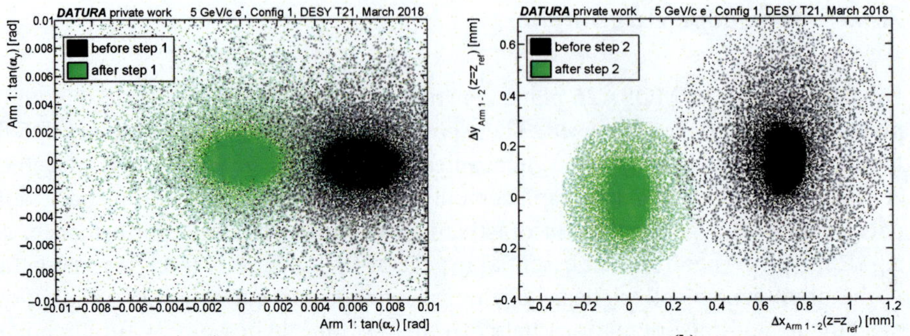

(a) (b)

Fig. 7.8 Illustration of the telescope arm-to-arm alignment procedures: **a** Effect of alignment step 1 correcting for their different orientations, here shown for arm 1. The tracklet angle is denoted as $\alpha_{x(y)}$. **b** Alignment step 2 in which the second arm is translated to amend its relative displacement, $\Delta x_{\text{Arm 1-2}}$ (y analogous), with respect to the first arm

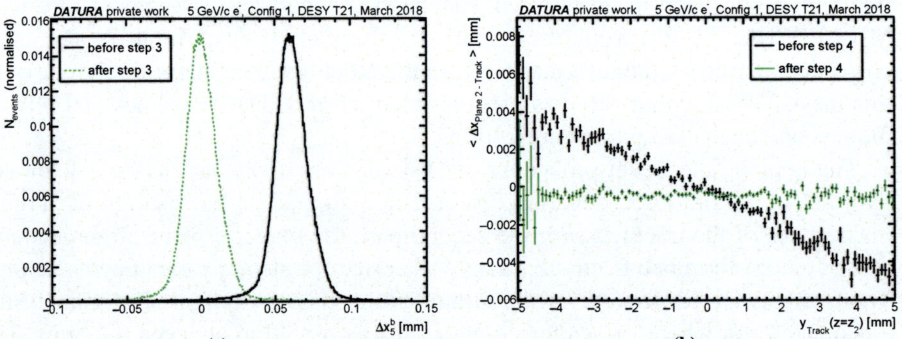

(a) (b)

Fig. 7.9 Illustration of the telescope plane-to-plane alignment procedures: **a** Effect of alignment step 3 amending relative displacements between the planes. **b** Alignment step 4 correcting the planes' relative rotations along the z-axis

$$\Delta x_P^b := x_P - \mathbf{f}_{\text{Track}}(z_P) \cdot \mathbf{e}_x \ , \quad y \text{ analogous} \tag{7.14}$$

The residuals are called *biased* because the point $x_P(y_P)$ is used for computation of the reference trajectory $\mathbf{f}_{\text{Track}}$. The fact that their distributions are not centred at zero is caused by the relative displacements of the planes. This is corrected for after translating all planes by the negative of their respective mean residual.

Step 4: Coplanarity Orientation, cf. Fig. 7.9b

In the last step, the orientations of the x-y coplanarities are unified by computing their rotations α_P around the plane's z-axis. The corrections can be measured by correlating the mean biased residual in x(y) to a full track's impact position in y(x) at a given plane, cf. Eqs. 7.15 and 7.16.

$$< \Delta x_{\text{Plane P-Track}} > \;=\; + \sin \alpha_{P,1} \cdot \mathbf{f}_{\text{Track}}(z_P) \cdot \mathbf{e}_y + \text{offset} \tag{7.15}$$

$$< \Delta y_{\text{Plane P-Track}} > \;=\; - \sin \alpha_{P,2} \cdot \mathbf{f}_{\text{Track}}(z_P) \cdot \mathbf{e}_x + \text{offset} \tag{7.16}$$

The negative average of the two, ideally comparable, angles determines the corrective rotation along the z-axis for each plane:

$$\alpha_P := -\frac{\alpha_{P,1} + \alpha_{P,2}}{2} \tag{7.17}$$

In summary, the presented alignment procedure includes independent translations of each plane transverse to the beam direction, individual plane rotations along the beam axis as well as common translations and rotations of the planes on the same arm. For the remaining degrees of freedom, CORRYVRECKAN provides a variational approach in which the χ^2 of fitted six-point straight-line tracks is minimised by sequential modification of the geometry parameters. Empirically, this algorithm is found appropriate for the material-free configuration used for the alignment of configuration 1. By contrast, the same method yields unphysically high rotational corrections up to 12° when applied to configuration 2 where substantial scattering material is placed in between both arms. A possible reason could be the oversimplification of particle trajectories traversing the scattering material to straight lines in this approach. In consequence, this additional correction method is not applied for further alignment of configuration 2. The following section demonstrates that the resulting pointing resolution, after following the alignment procedure as elaborated in this section, is already satisfactory for the purpose of the studies in this thesis.

7.3.4 Tracking Performance

Two tracking performance aspects are especially interesting in the context of the studies with the HGCAL prototype modules.

- Without any additional information from other detectors, the multiplicity of reconstructed tracks in a readout determines the ultimate event yield for further analysis of the combined HGCAL and DATURA beam telescope data.
- The positioning resolution at the centrally placed module should allow for studies of the region between the silicon pads on the sensor.

The following performance assessment focusses on these aspects. It is based on the data taken with DESY configuration 2.

This experimental setup differs from the underlying configurations used in the published work on the DATURA's performance [24] not only because of the presence of significant scattering material but also due to a different arrangement of the telescope arms. Anyways, an exact imitation of the methodology that is applied in this publication is not intended.

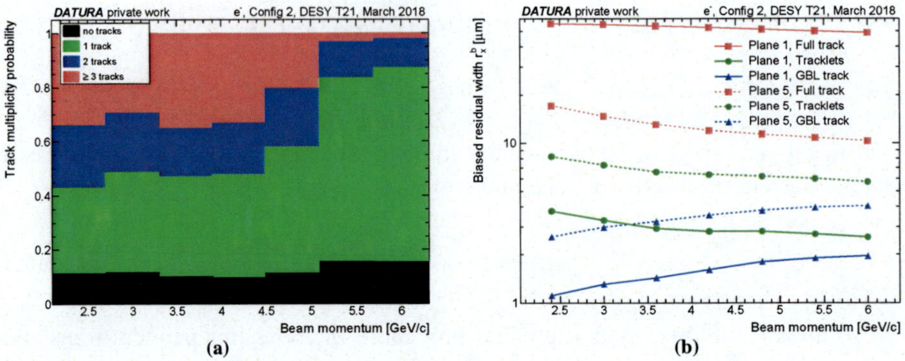

Fig. 7.10 **a** Multiplicity of reconstructed tracks with the DATURA beam telescope during the HGCAL beam test at DESY in March 2018. **b** RMS of biased residual distributions for tracklets, full and GBL tracks at planes 1 and 5

Therefore, the results presented hereafter do not disagree with the findings of Ref. [24]. The MIMOSA26's integration time exceeds 100 µs and consequently the probability of recording multiple particles in one readout is considerable. Figure 7.10a depicts the multiplicity of reconstructed tracks per event. The probability of obtaining exactly one reconstructed track in one readout window for 2.4 GeV/c electrons reaches 30% while it converges towards 70% for increasing momenta. This observation is consistent with the decrease of the particle rate provided by the DESY beam test facility. The integration time of the SKIROC2-CMS in the HGCAL prototypes is much faster than the MIMOSA26. A recorded HGCAL event in test beam at DESY contains practically always one particle shower exclusively. Hence, only events with exactly one reconstructed particle track from the telescope can be used for joint analysis with the telescope data. Independent of the particle momentum, the probability of reconstructing no track at all is approximately 10–15% hinting at the possibility of optimisation of the tracking parameters in Table C.1.

As a measure for the agreement between the clusters and the fitted trajectory, biased residuals, cf. Eq. 7.14, are computed for various parameterisations, planes and electron momenta. The widths $r^b_{x(y)}$ of the respective distributions are shown representatively for planes 1 and 5 in Fig. 7.10b. Full tracks exhibit the largest widths among all track models because the material in between the arms distorts the flight path from the straight line trajectory. The tracklets' residuals are not affected by the multiple scattering due to the material in between the arms and thus the width in their distribution is significantly smaller. Incorporating the deflections due to multiple scattering, the physically most accurate trajectory model is indeed the GBL track. It yields residual distribution widths comparable to the tracklets' despite including clusters from all planes up- and downstream the thick reference plane simultaneously. The slight dependence on the electron momentum and the imbalance between inner and outer tracking planes is qualitatively explained in Ref. [24].

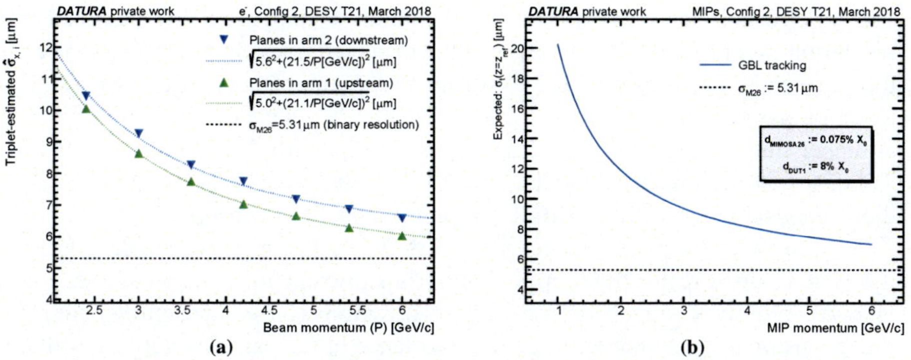

Fig. 7.11 **a** Intrinsic pointing resolution estimates from doublet analysis for each telescope arm. **b** Expected track pointing resolution at the reference plane using GBL tracks and all six telescope planes. The used tool can be found in Ref. [25]. Positive effects from charge sharing between the clusters are neglected and the intrinsic sensor resolution is assumed to be $18.4\,\mu\text{m}/\sqrt{12} \approx 5.31\,\mu\text{m}$

Cluster points at each arm are used for the estimation of the intrinsic pointing resolution by each sensor. In this connection, doublet tracks constructed from clusters from two planes (k, j) on the same arm determine the impact position on the remaining plane i. The difference between the inter- or extrapolation to the cluster defines an unbiased residual whose distribution exhibits a width $r^u_{x(y)}$. Assuming identical pointing resolution $\hat{\sigma}_{x(y)}$ for each plane on the same arm and neglecting multiple scattering in the doublet tracks, the former can be approximated using Eq. 7.18, where $z_{j(k)}$ are the cluster coordinates from the aligned telescope geometry.

$$\hat{\sigma}_{x(y),i} \;:=\; < 1 + \left(1 - \frac{z_i - z_j}{z_k - z_j}\right)^2 + \left(\frac{z_i - z_j}{z_k - z_j}\right)^2 >^{-0.5} \cdot r^u_{x(y)}, \quad z_i : \text{plane position} \tag{7.18}$$

The hereby obtained values for planes on the same arm are practically identical, $\hat{\sigma}_{x(y),1} \approx \hat{\sigma}_{x(y),2} \approx \hat{\sigma}_{x(y),3}$ and $\hat{\sigma}_{x(y),4} \approx \hat{\sigma}_{x(y),5} \approx \hat{\sigma}_{x(y),6}$. It is interesting that the pointing resolution estimates shown in Fig. 7.11a for the downstream planes is systematically higher than for the upstream arm which might be related to energy loss in the central material. Since the effects due to multiple scattering are not included in this analysis, these estimates decrease with increasing electron momentum. More importantly, they barely reach the MIMOSA26's naive, binary resolution of 5.31 µm even for the highest momenta. While it is obvious that this deficit is attributed to significant contribution from multiple scattering, the extrapolation to highest momenta does not suggest that the claimed geometry-independent intrinsic sensor resolution of 3.24 µm is actually achieved in this data. In fact, the average cluster size of less than 3 obtained here, and with it the positive effect from charge sharing, is less than what is reported elsewhere [24, 26].

As a consequence, a pessimistic, namely binary, intrinsic sensor resolution is assumed for the purpose of this assessment. This is equivalent to ignoring charge sharing effects in the cluster position calculation. Deploying the GBL-based track resolution tool [25], Fig. 7.11b presents the expected tracking resolution at the prototype module for configuration 2 for various beam momenta. Due to the pessimistic assumption on the intrinsic sensor resolution, the shown curve can be understood as an upper limit of its true value. It amounts to 12 μm for 2 GeV/c and improves to 7 μm for 6 GeV/c (electron-) MIPs which is less than half of the minimal extent of diode gaps on the investigated HGCAL prototype silicon sensors.

In summary, the performance of the DATURA beam telescope deploying the newly developed reconstruction procedure is sufficient for detailed studies on the HGCAL prototype silicon sensors' intergap regions. Of course, further tuning of the reconstruction parameters, e.g. in dependence of the particle energy, is still conceivable. For future studies with electrons as MIPs at the DESY beam test facility, running with higher momenta is recommended in order to maximise the event yield for combined analysis of HGCAL prototype and DATURA beam telescope data.

7.4 Impact Position Measurement with Delay Wire Chambers

The inference of particle trajectories in the PPE172 area at the CERN SPS H2 beam line starts with the timestamps of signals reaching either end of the delay line which are recorded as hits in the TDC, cf. Sect. 6.4.3. The average hit multiplicity per TDC readout and per channel is found to be consistent with one and a combinatorial hits analysis is not necessary. The first hit timestamp in a channel's readout window is considered for evaluation of Equation 6.4. By this construction, not more than one impact position on each DWC is obtained.

Without specification of the track model, correlation analysis like the one exemplified in Fig. 6.14b both validates the assignment of TDC-to-DWC readout channels and helps constructing a common orientation of all local coordinate systems already during the data taking. The deviation from a straight line trajectory due to multiple scattering of a 200 GeV/c muon travelling through 10 m of air corresponds to less than 0.1 mm which is less than the claimed DWC's intrinsic pointing resolution of 0.2 mm [27]. Hence, multiple scattering is neglected for tracking at this energy regime and straight lines are adopted for beam particle trajectories instead of GBL. Biased residuals are computed for translation alignment in the plane transverse to the beam line using MILLEPEDE [22, 23]. All other degrees of freedom, in particular any relative rotations, are not corrected for. During the alignment procedure, the coordinates of the two outermost chambers, i.e. DWCs ext and E, are fixed and define the common coordinate system.

Unbiased residuals (Δx^u, Δy^u), i.e. residuals in which the measured point is not input to the track fit to which it is compared to, are computed in order to estimate the intrinsic DWC pointing resolution. A representative distribution of unbiased residuals on DWC D is shown in Fig. 7.12a. The core of the distribution can be

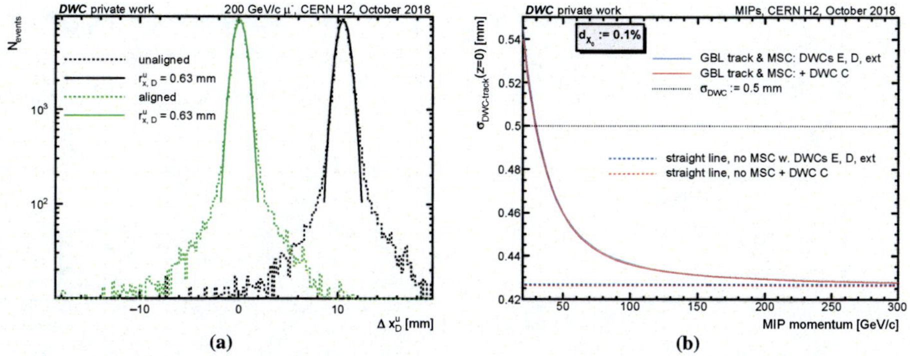

Fig. 7.12 **a** Distribution of unbiased, horizontal residuals on DWC D before and after translation alignment. The core can be approximated by a Gaussian whose width amounts to 0.63 mm. **b** Expected tracking resolution at the beginning of the HGCAL prototype calorimeter for different track models and momenta of the MIP-like particles. The thickness of each DWC is assumed to be 0.1% X_0 (30 m of air corresponds to approximately 1% X_0) and the intrinsic DWC pointing resolution is set to 0.5 mm. The used tool can be found in Ref. [25]. In this work, delay wire chamber tracks are reconstructed as straight line trajectories

approximated by a Gaussian whose width is interpreted as the biased residual width, e.g. $r^u_{x,D} \approx 0.6$ mm. In the limit of true straight line particle trajectories, the unbiased residual width $r^u_{x(y),i}$ is linked to the intrinsic DWC (index i) pointing resolution $\hat{\sigma}_{x,i}$. The relationship is stated in Eq. 7.19.

$$\begin{aligned}\left(r^u_{x,i}\right)^2 &= \hat{\sigma}^2_{x,i} + \sigma^2_{\text{track},i} \\ \sigma^2_{\text{track},i} &= \frac{1}{\Delta^2} \cdot \sum_{j \neq i} \left(S_{zz} - z_j \cdot S_z + z_i \cdot z_j \cdot S - z_i \cdot S_z\right)^2 \cdot \hat{\sigma}^2_{x,j} \\ S := \sum_{j \neq i} 1\,,\quad S_z &:= \sum_{j \neq i} z_j\,,\quad S_{zz} := \sum_{j \neq i} z_j^2\,,\quad \Delta := S \cdot S_{zz} - S_z \cdot S_z\end{aligned} \tag{7.19}$$

For numerical evaluation of the latter equation, the resolution of DWCs ext, D and E are assumed to be identical. The result is $\hat{\sigma}_{x(y),ext(D,E)} \approx 500\,\mu$m which is more than twice the reported value in the DWC manual [27].

By contrast, the y-residual width, and with it the estimated intrinsic pointing resolution, of DWC C is found to be significantly higher. Although the exact cause for this behaviour has not been established, it is speculated that it might be related to a defective cathode in this particular DWC or a malfunctioning NIM discriminator processing its signals. In any case, data from this DWC is excluded from the particle track reconstruction in the following. This can be done without significant loss of track pointing resolution at the calorimeter prototype which is demonstrated in Fig. 7.12b. It depicts the expected pointing resolution at the beginning of the HGCAL prototype's EE compartment. The values are idealised estimates because they are based on MIP-like trajectories, whereas all test particles apart from muons shower in the

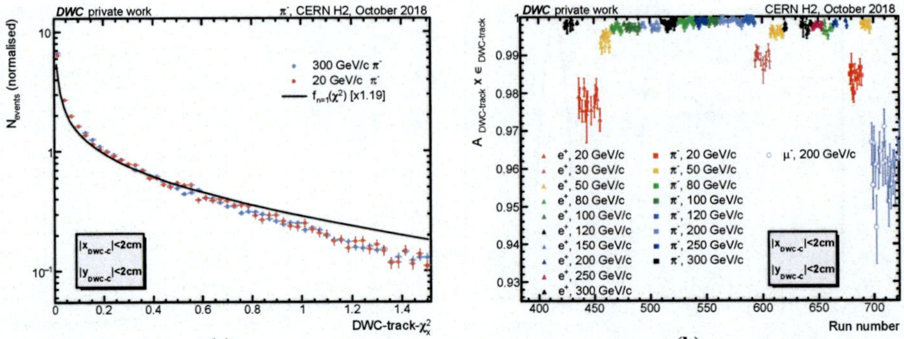

(a) (b)

Fig. 7.13 **a** χ^2-distribution of straight line trajectories fitted to x-coordinate measurements by three DWCs (ext, D, E) for 20 and 300 GeV/c pions. The theoretically expected distribution in case of Gaussian uncertainties in the DWC measurements is drawn as a black line. **b** Probability to obtain measurements from all three DWCs in an event when a central impact on DWC C is reconstructed for different runs in the *O1** datasets. This quantity can be understood as the combined efficiency and acceptance. Its values are consistent with the claimed DWC particle detection efficiency above 99%

calorimeter. Any intrinsic deviation of the shower axis to the initial trajectory is not considered. Furthermore, the track resolution amounts to approximately 0.43 mm which is less than the intrinsic pointing resolution of each DWC, profiting from the large distances between the DWCs. When excluding DWC C in the trajectory reconstruction, the resolution is practically unchanged. Qualitatively, the track resolution at the HGCAL prototype is driven by the two DWCs closest to it. Hence, the precision is only marginally enhanced when including a (nominally performing) DWC at position C. This statement even holds true when incorporating the effects of multiple scattering into the trajectory model deploying GBL. The convergence of the GBL track-based towards the straight line trajectory-based resolution curves at high particle momenta implies equivalence of both models at this regime where multiple scattering can be neglected.

Figure 7.13a illustrates the χ_x^2-distributions used for the fit of straight line trajectories for 20 and 300 GeV/c pions. They exhibit basic agreement to the theoretical χ^2-function with a degree of freedom equal to one. Deviations are attributed to non-Gaussian effects which will be incorporated for the studies in Sect. 10.2.3. Moreover, the presented distributions from both datasets with different positron momenta are consistent with each other[3] although the theoretical deflections due to multiple scattering are increased by a factor of 15 for 20 GeV/c with respect to 300 GeV/c pions. This finding supports the neglection of this effect in the tracking model.

The probability of reconstructing trajectories given all three impact points from DWCs E, D and ext in an event using information from DWC C as a tag is presented in Fig. 7.13b. This quantity can be understood as an upper limit for the product $\epsilon_{\text{DWC-track}}$ of individual DWC detection efficiencies $\epsilon_{\text{E(D,ext)}}$ multiplied by some acceptance

[3] p-value = 23% using an implementation of the method described in Ref. [28].

$A_{\text{DWC-track}}$ with values between zero and one.

$$\epsilon_{\text{DWC-track}} = \epsilon_E \cdot \epsilon_D \cdot \epsilon_{\text{ext}} \leq 1 \text{ , uncorrelated to } \epsilon_C \quad (7.20)$$

For runs with limited transverse extent of the beam, see Appendix E for an overview of reconstructed beam profiles from the *O1** datasets, the acceptance $A_{\text{DWC-track}}$ approaches values close to unity. In those cases, $\epsilon_{\text{DWC-track}}$ exceeds 99% which, in fact, is in agreement with the claimed detection efficiency in the DWC manual [27].

7.5 Time Reference Inference with MCP Waveforms

For validation of the TOA measurement principle implemented in the SKIROC2-CMS ASIC, knowledge of a given incident particle's time of arrival at the setup relative to the provided clock is essential. This timing information is derived from MCP signal waveforms that were digitised with 1024 12-bit samples using a CAEN v1742 digitiser [29], cf. Sect. 6.4.3.

The MCP signals were recorded as part of the *O2* dataset. Figure 7.14a shows an example of the raw waveforms of both MCPs and the digitised clock signal. The rising part of a typical MCP signal is 1 ns long and the full amplitude of the signal is contained within 5 ns.

Reconstruction of the reference timestamp proceeds in two conceptual steps. In the first part, each MCP signal waveform is characterised through constant fraction discrimination. Thereafter, the clock phase inside the readout window ΔT_ϕ is computed in order to indicate the absolute peaking time of the signal waveform T_{MCP} with respect to the previous falling edge of the clock cycle ΔT_{MCP}. The periodicity of the 40 MHz clock is taken into account.

$$\Delta T_{\text{MCP}} = \left(T_{\text{MCP}} - \Delta T_\phi\right) \bmod 25\,\text{ns} \quad (7.21)$$

MCP Waveform Characterisation (T_{MCP})

The baseline of the recorded MCP waveform is estimated and subtracted using the average from samples 5–49 which are outside the signal window. Common baseline shifts across all digitiser channels are computed and corrected for by using averages from simultaneous samples from four unconnected channels. The RMS from the noise corrected values in the baseline serves as a measure for the remaining noise in the waveform. After sign inversion, the maximum sample value is identified. Quality criteria on the MCP waveforms are applied. The minimum signal is required to be at least 100 ADC counts and the signal must exceed three times the estimated noise. Imposing these conditions, the signal efficiency can be evaluated as a function of the reconstructed impact position by the closest delay wire chamber. As shown for one of the two MCPs in Fig. 7.14b, the acceptance of the MCPs' sensitive area is hereby

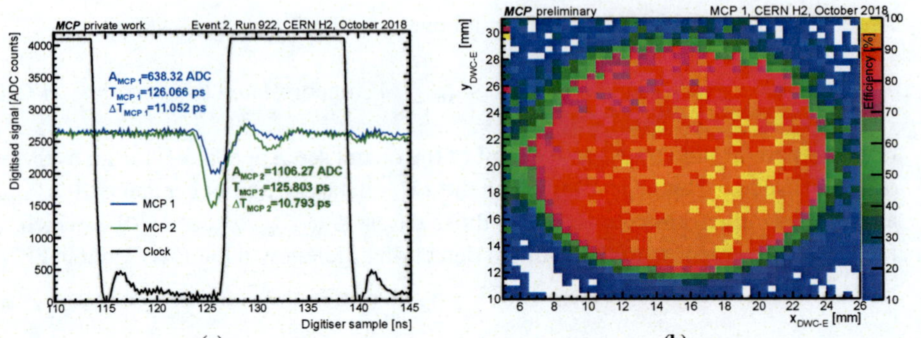

(a) (b)

Fig. 7.14 **a** Example of the digitised waveforms of both MCPs together with the 40 MHz clock signal. **b** Signal reconstruction efficiency for MCP 1 for different beam impact positions as measured by DWC E

reproduced. Tighter criteria would improve the timing performance of the MCP but would also reduce the efficiency of the acceptance region.

Subsequently, the amplitude A is inferred by fitting a second degree polynomial within a symmetric 2 ns wide window around the waveform's maximum. With this information, a linear fit is applied to the five samples preceding the maximum corresponding to the rising part of the MCP signal. The reconstructed timestamp is defined as the time when this linear model of the rising part exceeds 50% of the fitted amplitude. Figure 7.15 illustrates the result of a tag-and-probe analysis of the MCP's timing capabilities whereby events with similar MCP signals are selected and both MCPs are assumed to perform identically. The resolution depends on the reconstructed signal amplitude. It is empirically found to scale according to Eq. 7.22 reaching values below 20 ps for high enough signal amplitudes.

$$\frac{\sigma^2_{T,\mathrm{MCP\ 1\text{-}2}}}{2} = c^2 + \frac{n^2}{A^2_{\mathrm{MCP\ 1}}} \tag{7.22}$$

Clock Phase Reconstruction (ΔT_ϕ)

Reconstruction of the clock phase is based on the identification of falling clock edges. Neither pedestals nor common baseline shifts are corrected for in this procedure because the detection of falling edges makes explicit use of the ADC saturation inside the digitiser. Namely, a falling edge is detected if a given sample amounts to $2^{12} - 1 = 4095$ while the following two samples fall below that value. In that case, the four subsequent samples are subject to a straight line fit. The time of the falling edge is computed from the point where the value of this linear fit is equal to 3200 ADC counts. This procedure is iterated for all identified falling edges in the acquired waveform. ΔT_ϕ is defined as the average clock edge timestamp after subtracting the 25 ns clock period for each entry. The RMS of all reconstructed falling edge

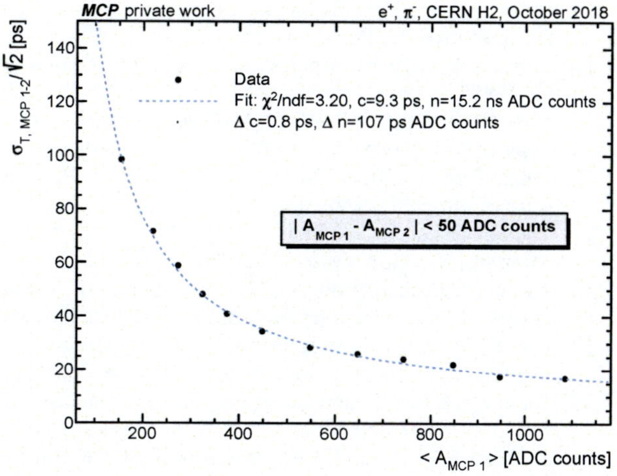

Fig. 7.15 Timing resolution estimates obtained through a tag-and-probe analysis for MCP 1 as function of the reconstructed signal amplitude using MCP 2 as reference. Events with similar MCP signal amplitudes are selected and both MCPs are assumed to have identical timing resolutions

timestamps in the dataset amounts to 2.7 ps on average and thus it is negligible when put in contrast to the MCP timing resolution.

References

1. Brun R, et al. (2018) ROOT data analysis framework - user's guide
2. The Luigi Authors, Luigi. https://github.com/spotify/luigi. Accessed 18 Nov 2019
3. Rieger M (2019) Search for higgs boson production in association with top quarks and decaying into bottom quarks using deep learning techniques with the CMS experiment. PhD thesis, RWTH Aachen, Germany. http://doi.org/10.18154/RWTH-2019-06415
4. Quast T (2016) Search for top quark pair associated higgs boson production decaying into bottom quarks at $\sqrt{s} = 13$ with the cms detector. Master's thesis, RWTH Aachen, Germany.
5. Erdmann M, et al. (2017) Design and execution of make-like, distributed analyses based on spotify's pipelining package luigi. J Phys: Confer Ser 898. https://doi.org/10.1088/1742-6596/898/7/072047
6. CERN Information Technology Department, Batch service. http://information-technology.web.cern.ch/services/batch. Accessed 01 2019
7. CMS Collaboration (2021) CMS offline software. https://github.com/cms-sw/cmssw. Accessed 23 June 2021
8. CMS HGCAL Beam Test Group, Test beam CMSSW framework. https://gitlab.cern.ch/cms-hgcal-tb/TestBeam/releases. Tag: v13_patch1. Accessed 18 2019
9. Acar B et al (2021) Construction and commissioning of CMS CE prototype silicon modules. JINST 16:T04002. https://doi.org/10.1088/1748-0221/16/04/T04002
10. Akchurin N et al (2018) First beam tests of prototype silicon modules for the CMS high granularity endcap calorimeter. JINST 13:P10023. https://doi.org/10.1088/1748-0221/13/10/p10023
11. CMS Collaboration (2006) Reconstruction of the signal amplitude of the CMS electromagnetic calorimeter. Eur Phys J C 46:23–35. https://doi.org/10.1140/epjcd/s2006-02-002-x
12. Paganini P, Van Vulpen I (2004) Pulse amplitude reconstruction in the CMS ECAL using the weights method 025. http://cds.cern.ch/record/801816

13. Helmuth S (2012) Semiconductor detector systems. Semiconductor Science and Technology. Oxford University Press, Oxford
14. Lobanov A (2019) Electronics and triggering challenges for the CMS high granularity calorimeter. JINST 13. https://doi.org/10.1088/1748-0221/13/02/c02056
15. Borg J et al (2017) SKIROC2_CMS an ASIC for testing CMS HGCAL. JINST 12:C02019. https://doi.org/10.1088/1748-0221/12/02/c02019
16. Pitters F, et al. (2019) Time and energy calibration of timepix3 assemblies with thin silicon sensors. http://cds.cern.ch/record/2649493
17. The EUTelescope Authors (2019) EUTelescope Wiki. https://github.com/eutelescope/eutelescope/wiki. Accessed 22 2019
18. The Corryvreckan Authors (2019) Corryvreckan - the maelstrom for your test beam data. https://gitlab.cern.ch/corryvreckan/corryvreckan. Accessed 18 2019
19. Quast T (2019) NtletTracking. https://gitlab.cern.ch/tquast/corryvreckan/tree/master/src/modules/NtletTracking. Tag: HGCAL-2018. Accessed 18 Nov 2019
20. Kleinwort C (2012) General broken lines as advanced track fitting method. Nucl Instrum Meth A 673:107–110. https://doi.org/10.1016/j.nima.2012.01.024
21. Schleper P, et al. (2008) Alignment of the CMS silicon tracker using Millepede II. J Phys: Confer Ser 119. https://doi.org/10.1088/1742-6596/119/3/032040
22. Blobel V, Kleinwort C (2002) A new method for the high-precision alignment of track etectors. arXiv:hep-ex/0208021v1 [hep-ex]
23. Blobel V (2019) Millepede II - draft manual. http://www.desy.de/~kleinwrt/MP2/doc/html/draftman_page.html. Accessed 03 2019
24. Jansen H et al (2016) Performance of the EUDET-type beam telescopes. EPJ Tech Instrum 3:7. https://doi.org/10.1140/epjti/s40485-016-0033-2
25. Spannagel S, Jansen H, GBL track resolution calculator v2.0. http://dx.doi.org/10.5281/zenodo.48795
26. Jansen H (2016) Resolution studies with the DATURA beam telescope. arXiv:1610.02505 [physics.ins-det]
27. Spanggaard J (1998) Delay wire chambers - a users guide. http://cds.cern.ch/record/702443
28. Gagunashvili ND (2006) Comparison of weighted and unweighted histograms. arXiv:physics/0605123 [physics.data-an]
29. CAEN (2011) Technical information manual, MOD. V1742, revision n. 0

Chapter 8
Silicon Sensor and Module Qualification

The calorimetric performance can be impaired in case of poor sensor quality, malfunctioning readout chips (*ROCs*) or inadequate assembly of the prototype modules. In addition, it can be altered if the modules contribute substantial passive material to the total calorimeter depth. In the following, these aspects are investigated closer with four rather independent studies. Theoretical motivations and explanations of the applied methods as well as practical suggestions for their future application are given when appropriate.

The structure of this chapter is analogous to the procedural steps in the module assembly. Measurements on the electrical properties of the bare sensors are presented in Sect. 8.1 investigating if leakage currents and capacitances are according to expectation. The printed circuit board (*PCB*) may principally contribute passive material to the calorimeter depth. Its material budget is assessed and the expected impact on the energy response is estimated in Sect. 8.2. The pedestals and with it data integrity of the HGCAL prototype ROC for beam tests, the SKIROC2-CMS, are shown, also the observed electronic noise level is described in Sect. 8.3. In the last Sect. 8.4, the efficiencies to detect the smallest relevant energy depositions by minimum ionising particles are studied.

Altogether, it will be established that the prototype modules, apart from only a few low-efficiency cells, are mostly operational. Therefore, the modules are suitable for the following tests with showering particles.

8.1 Electrical Properties of 6″ Silicon Sensors

More than hundred 6″, 300 µm thick prototype HGCAL silicon sensors, cf. Table 4.1, have been produced and delivered by Hamamatsu Photonics K.K. (HPK) for beam test module assembly (and sensor design optimisation studies). Randomly selected

T. Quast, *Beam Test Calorimeter Prototypes for the CMS Calorimeter Endcap Upgrade*, Springer Theses, https://doi.org/10.1007/978-3-030-90202-5_8

samples were electrically characterised prior to their assembly to prototype modules corresponding to the *IV* (current vs. voltage) and *CV* (capacitance vs. voltage) datasets.

In this section, the results of their analyses are presented. Hereby, it is shown that the tested sensors are of excellent quality in terms of leakage currents and that the pad capacitances as well as the depletion voltages behave as expected. At the same time, the positive results also establish that the newly developed switch- and probe-card system, cf. Sect. 6.1, is in fact suitable for effective and time efficient electrical characterisation of HGCAL silicon wafers during mass production.

More detailed simulation and experimental studies of HGCAL prototype sensor properties have been reported elsewhere, e.g. in Ref. [1].

8.1.1 Leakage Current

For each of the 23 sensors for which IV measurements were performed, summary graphics such as the ones in Fig. 8.1 are produced for different bias voltages. These graphics are called *HexPlots* and they visualise the measured leakage current in each pad at their physical location on the wafer. In order to allow for a unbiased comparison between sensors measured in different environmental conditions, the currents are subsequently normalised to the same temperature, namely $T_{24} = 24°\text{C}$, as stated in Eq. 8.1.

$$I_{24} = I_T \cdot \left(\frac{T_{24}}{T}\right)^2 \cdot \exp\left(\frac{E_g}{2 \cdot k_b} \cdot \left(\frac{1}{T} - \frac{1}{T_{24}}\right)\right) \tag{8.1}$$

Relative current corrections per temperature changes by one degree for temperatures around 24°C amount to $8 - 9\%$.

As can be seen in Fig. 8.2a, the pad leakage current is "remarkably low and homogeneous" [2], mostly 1 nA or less at the typical beam test operation voltage of 200 V. This result is consistent with the 7-needle based measurements of prototype sensors with different p-stop geometries [1]. The leakage current at 1000 V significantly exceeds 10 nA only for two sensor pads in this sample, cf. Fig. 8.2b. Beginning of breakthrough of one single pad on each of these two sensors is observed, cf. Fig. 8.1b, d. It occurs at voltages greater than 600 V for one of the pads and at voltages greater than 800 V for the other one, respectively, cf. Fig. 8.3.

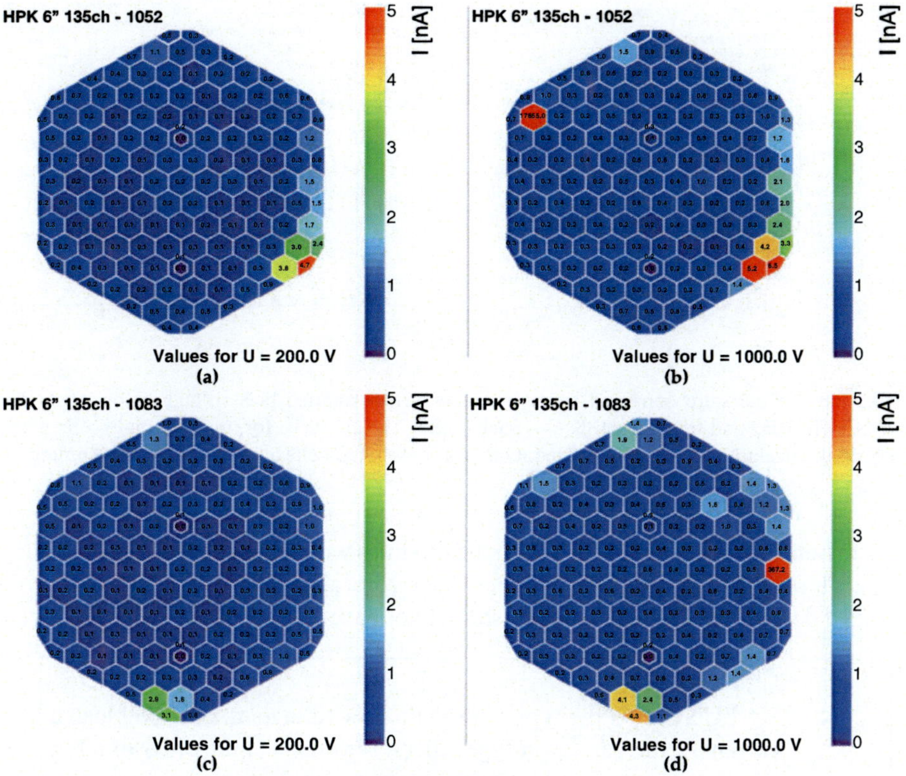

Fig. 8.1 Hexplots showing the measured leakage currents of sensor HPK 6″ 135ch–1052 **a** at an applied bias voltage of 200 V, **b** at 1000 V and **c** for sensor HPK 6″ 135ch–1083 at 200 V and **d** at 1000 V. Cells with currents above the indicated scale are coloured red. The temperature during the measurements ranged from 21.5–24.5°C. No temperature correction is applied in these graphics

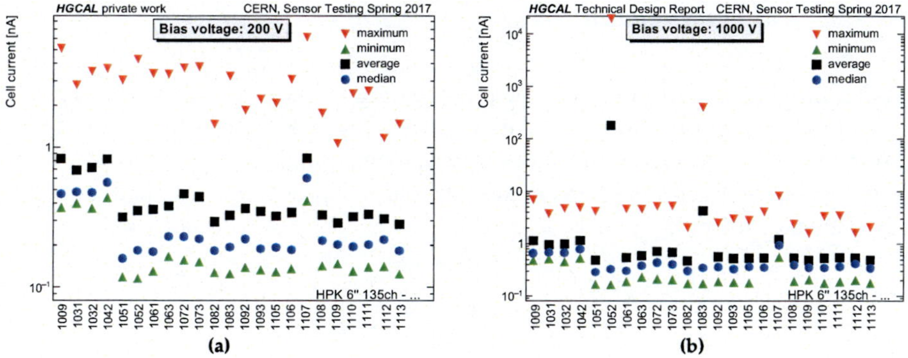

Fig. 8.2 Summary of the temperature corrected leakage currents **a** at a bias voltage of 200 V and **b** at 1000 V for all characterised silicon prototype sensors in the *IV* dataset. The summary includes the leakage currents from cells with full hexagonal geometries only

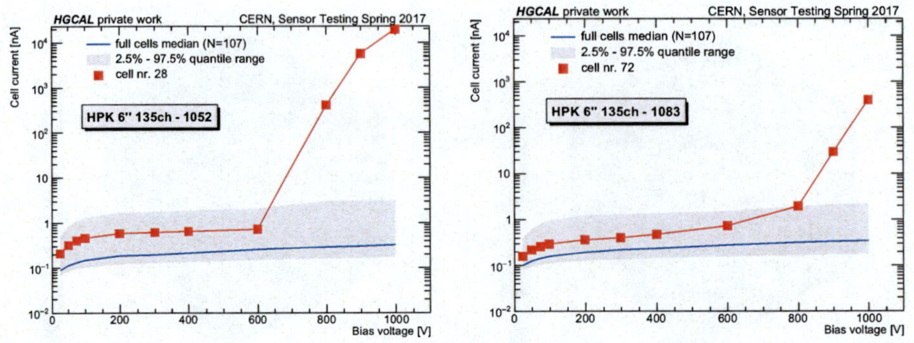

(a) (b)

Fig. 8.3 Temperature corrected leakage current versus applied bias voltage **a** for sensor HPK 6″ 135ch–1052 and **b** sensor HPK 6″ 135ch–1083. The IV curve for the two single cells with an observed breakthrough are highlighted. Only full hexagonal cells are included in this summary

In both cases, the regime at which the modules were typically operated during the beam tests is well below these voltages. Therefore, abnormally high leakage currents due to defects on the silicon sensor should not have been an issue in the beam tests.

✓ **MQ-FUNC.1** At the typical module bias voltage, the per-cell leakage current of the bare prototype silicon sensors is at the order of a few nanoamperes.

8.1.2 *Capacitance and Depletion Voltage*

After modification of the switch- and probe-card system's circuitry, CV curves for two prototype sensors were re-taken in 2018 at a LCR frequency of $f = 10\,\text{kHz}$. Applying an open correction [3] of the measured (complex) impedances Z_x through appropriate subtraction of the system's impedance Z_s, cf. Eq. 8.2, the pad capacitance is calculated using Eq. 8.3.

$$Z_c = \left(\frac{1}{Z_x} - \frac{1}{Z_s}\right)^{-1} \tag{8.2}$$

$$C = \frac{-1}{2\pi \cdot f \cdot \text{Im}(Z_c)}, \quad f = 10\,\text{kHz} \tag{8.3}$$

Figure 8.4 shows the Hexplot for the derived pad capacitances for one of the two sensors at a bias voltage of 200 V. The capacitance amounts to roughly 43-45 pF for

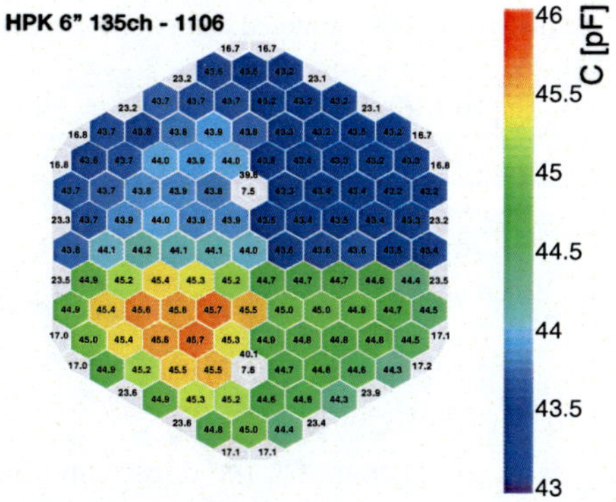

Fig. 8.4 Hexplot of sensor HPK 6″ 135ch–1106 showing the open-corrected pad capacitances at an applied bias voltage of 200 V. It illustrates the four quadrants with different interpad distances affecting the measured pad capacitance

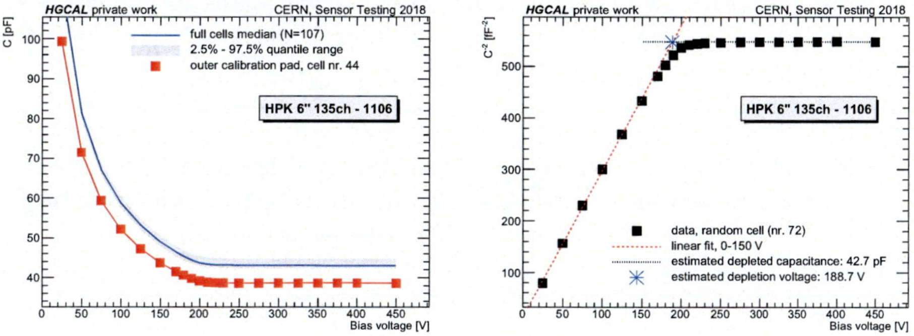

(a) **(b)**

Fig. 8.5 **a** Summary of measured CV curves for sensor HPK 6″ 135ch–1106 after application of the open correction. Only full hexagonal pads are included in this summary. One outer calibration cell (nr. 44) with 8/9 of the full hexagonal cell area is shown in addition. **b** Example estimation of the depletion voltage for a full cell (nr. 72) on sensor HPK 6″ 135ch–1106

full hexagonal geometries. Astonishingly, the measurement precision with this setup allows for distinction of wafer quadrants with different interpad distances that are known to affect the pad capacitances.

Figure 8.5a presents the average CV curve of full hexagonal pads and the one for cell number 44 which is an outer calibration pad with approximately $\frac{8}{9}$ of the nominal full hexagonal pad area. Cell number 44 exemplifies the fact that the capacitance, in first approximation, scales with the cell area.

The CV curves are the basis for estimation of the depletion voltages. For this purpose, the quadratic inverse of the open-corrected capacitance is plotted against the applied bias voltage.

The relation is linear until full depletion is reached, cf. Eq. 4.2. The intersection of two linear fits, one to the rising part at low voltages and the other to the constant part

Table 8.1 Summary of the distribution of estimated depletion voltages. Only full hexagonal pads are included in this summary

	HPK 6″ 135ch–1105 (V)	HPK 6″ 135ch–1106 (V)
Average	194.2	191.2
95%-quantile	200.0	195.7
Median	193.3	191.3
5%-quantile	190.1	186.1

at higher voltages, serves as an estimate for the depletion voltage of the investigated pad. An example of this analysis is shown in Fig. 8.5b.

As it is summarised in Table 8.1, the depletion voltages are determined to be a little less than 200 V for the pads characterised as part of the *CV* dataset. During the beam tests in October 2018, the bias voltage was set to approximately 250 V whereas it was set to 200 V during the tests in June and in March 2018. If this voltage was indeed fully transferred to the sensor, the latter should have been fully depleted during those tests.

✓ **MQ-FUNC.2** The silicon sensors are fully depleted around the expected depletion voltage and the per-cell capacitance decreases with increasing bias voltage until full depletion is reached.

8.2 Tomography of the Prototype PCB

In general, the inclusion of passive absorber material in a sampling calorimeter may influence its energy response to particle showers. For instance, the tracker contributes passive material equivalent to 0.4–2.0 radiation lengths (X_0) to the currently installed ECAL in the CMS detector. This leads to systematic shifts in the calorimeter response which need to be compensated for. In this particular case, the relative corrections amount to 6–10 % [4]. Although the additional tracker material in this example is placed in front of the calorimeter, it is intuitive that similar effects can be obtained when placing additional absorber inside. The expected response may differ systematically from the measurement if some passive material is not accounted for accurately in the simulation model.

The question arises on the required level of detail in the implementation of the HGCAL prototype calorimeter's geometry for a trustworthy comparison of beam test data to simulation. In this section, the most heterogeneous passive element in the HGCAL prototype calorimeter configuration, namely the printed circuit board (*PCB*), is investigated in this regard. The PCB houses bonding holes, the SKIROC2-

CMS ASICs, connectors and other electronic elements. Both its thickness and the resulting impact on the energy response are assessed in the following. With that, it is contemplated if the PCB can be approximated safely as a homogeneous element in the simulation.

The effect of multiple scattering of high-energy electrons is exploited to infer the PCB's material budget. The feasibility of this tomography principle has already been demonstrated in Ref. [5]. In general, the Highland formula [6, 7], cf. Eq. 8.4, relates the standard variation σ_θ of the kink angle between the trajectories of incoming particles (with charge z, travelling at momentum p and velocity β) and the outgoing trajectories to the material's thickness ϵ in between.

$$\sigma_\theta = \frac{13.6\,\text{MeV}}{\beta \cdot p \cdot c} \cdot z \cdot \sqrt{\epsilon} \cdot (1 + 0.038 \cdot \ln \epsilon) \tag{8.4}$$

Using the *TOM* dataset, tracklets of electrons impinging onto the PCB are reconstructed up- and downstream the PCB, cf. Sect. 7.3.2. Then, the kink angle, denoted as k, is calculated from the scalar product of both tracklets whose orientation vector is denoted as $\mathbf{v}_{up(down)stream}$. After defining a suitable coordinate system, Eq. 8.5 states the relationship between k and the 2D kink angles $\theta_{x(y)}$ from the tracklet parameterisations.

$$\begin{aligned} \mathbf{v}_{\text{upstream}} &\propto \mathbf{e_z} \\ \mathbf{v}_{\text{downstream}} &\propto \tan\theta_x \cdot \mathbf{e_x} + \tan\theta_y \cdot \mathbf{e_y} + \mathbf{e_z} \\ \cos^2 k &:= \frac{1}{\tan^2\theta_x + \tan^2\theta_y + 1} \end{aligned} \tag{8.5}$$

Note that for small angles $\theta_{x(y)} \ll 1$, the scalar product angle is well approximated by the quadratic sum in Eq. 8.6.

$$\begin{aligned} 1 - k^2 &\overbrace{\approx}^{\theta_{x(y)} \ll 1} 1 - \theta_x^2 - \theta_y^2 \\ \rightarrow k &\approx \sqrt{\theta_x^2 + \theta_y^2} \end{aligned} \tag{8.6}$$

The exact relationship between the kink angle k and the material budget ϵ is calibrated in Sect. 8.2.1. Afterwards, the material budget measurements for dedicated regions on the prototype PCB are presented in Sect. 8.2.2. Finally, the results are interpreted in Sect. 8.2.3 in terms of the expected impact on the calorimeter response.

For outreach and education purposes, the applied methodology in this part was transformed into a Python-based analysis tutorial [8], where a simplified version of the *TOM* dataset is used.

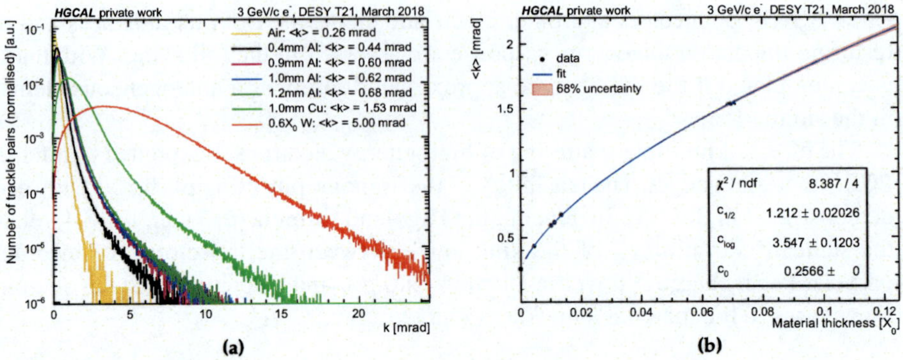

Fig. 8.6 **a** Distributions of tracklet kink angles for different reference materials placed in between the telescope arms. The mean electron momentum is 3 GeV/c. (b) Calibrated relation between the material thickness in radiation length equivalents (X_0) and the mean kink angle. The point corresponding to the 0.6 X_0 tungsten plate is not shown in this graph. The data point for air is not directly included in the fitting, but it defines c_0 in Eq. 8.9

8.2.1 Kink Angle Calibration

Figure 8.6a presents the distribution of measured kink angles for different materials, namely various aluminum plates as well as copper and tungsten plates, with a priori known thicknesses. They are used for calibration of the tomography procedure. As can be seen, the mean kink increases with larger material thicknesses. Due to its robustness to acceptance related cutoffs [5] and due to the feasibility to evaluate it independently from any underlying probability density, the mean kink angle is chosen as observable. Extending the original feasibility study, deriving this quantity's expected relation to the material budget proceeds as follows: Starting from Eq. 8.6, the kink angle is interpreted as a random number formed from the squared sum of two Gaussian-distributed numbers. An equivalent statement, if these two Gaussian distributions N share the same parameters, is that it is computed from the square root of numbers drawn from the gamma distribution, cf. Eq. 8.7.

$$\begin{aligned} k &\in \sqrt{\left(N(0,\,\sigma_{\theta_x})^2 + N(0,\,\sigma_{\theta_y})^2\right)} \\ &\overbrace{\equiv}^{\sigma_\theta=\sigma_{\theta_x}=\sigma_{\theta_y}} \quad k \in \sqrt{\Gamma(1,\,2\cdot\sigma_\theta)} \end{aligned} \tag{8.7}$$

A number constructed by the square root of a gamma-distributed number, follows a Nakagami distribution [9] whose mean value can be computed analytically from shape parameters of the underlying gamma-distribution. Finally, the mean kink angle parameterisation according to Eq. 8.8 is obtained, whereby σ_θ is taken from Eq. 8.4.

$$\langle k\rangle \;=\; \sqrt{2}\cdot\frac{\gamma(1.5)}{\gamma(1.0)}\cdot\sigma_\theta \;\approx\; 1.25331\cdot\sigma_\theta \tag{8.8}$$

However, for the purpose of calibration, additional degrees of freedom are allowed in the theoretical parameterisation of σ_θ, cf. Eq. 8.9.

$$\langle k \rangle := \sqrt{2} \cdot \frac{\gamma(1.5) \cdot 0.0136}{\gamma(1.0) \cdot 3} \cdot \sqrt{\epsilon} \cdot \left(c_{1/2} + c_{\log} \cdot 0.038 \cdot \ln(\epsilon)\right) + c_0 \tag{8.9}$$

Especially, the factors $c_{1/2}$ and $c_{\log}$ therein are subject to optimisation. A fixed offset c_0 is included to account for the finite tracking resolution and the non-zero contribution from the (neglected) multiple scattering in air in the tracklet model.

Statistical errors on the mean kink angles as well as systematic uncertainties on the material thickness are incorporated into the fit. More precisely, a 1% uncertainty on the thickness of the aluminum plates is assumed while the uncertainty for the copper and tungsten plates is set to 2% and 5%, respectively. The latter two choices are motivated by unknown purities of the used materials which would affect the respective proportionality to the radiation length equivalents. Ultimately, the fitted model exhibits a χ^2 per degree of freedom close to 2 that is dominated by the data point taken with the 0.6 X_0 tungsten plate. Except for that particular data point, the calibrated relationship between the mean kink angle and the material describes the data well, cf. Fig. 8.6b.

8.2.2 Results

The evaluation of the mean kink angle as a function of the tracklets' impact positions at the DUT yields tomography images with an astonishing level of detail like the ones in Fig. 8.7a–c. Each of these images is computed from $\mathcal{O}\left(10^8\right)$ tracklet pairs from 3 GeV/c electrons equivalent to approximately one day of data taking with the DATURA beam telescope at the DESY (Hamburg) beam test facility.

Figure 8.7d illustrates the distribution of kink angles for four distinct areas on the PCB: the air gap, the bare PCB's plastic without any surrounding electronic elements, the SKIROC2-CMS ASIC on the PCB and the capacitor that is depicted in the bottom left in Fig. 8.7a. Table 8.2 supplements the mean kink angles with the derived equivalences in radiation length.

8.2.3 Impact on the Simulated Calorimeter Response

As listed in Table 8.2, the thickness of the SKIROC2-CMS contributes an additional 1% X_0 to the bare PCB. Since this ASIC's area is significantly larger compared to all other central, on-board electronics, only its particular impact on the energy response is studied next. For this purpose, an independent implementation of an EE prototype-like geometry with 28 sensitive modules in 14 cassettes (configuration "102" in Ref. [10]) using GEANT4 (version 10.4) is examined.

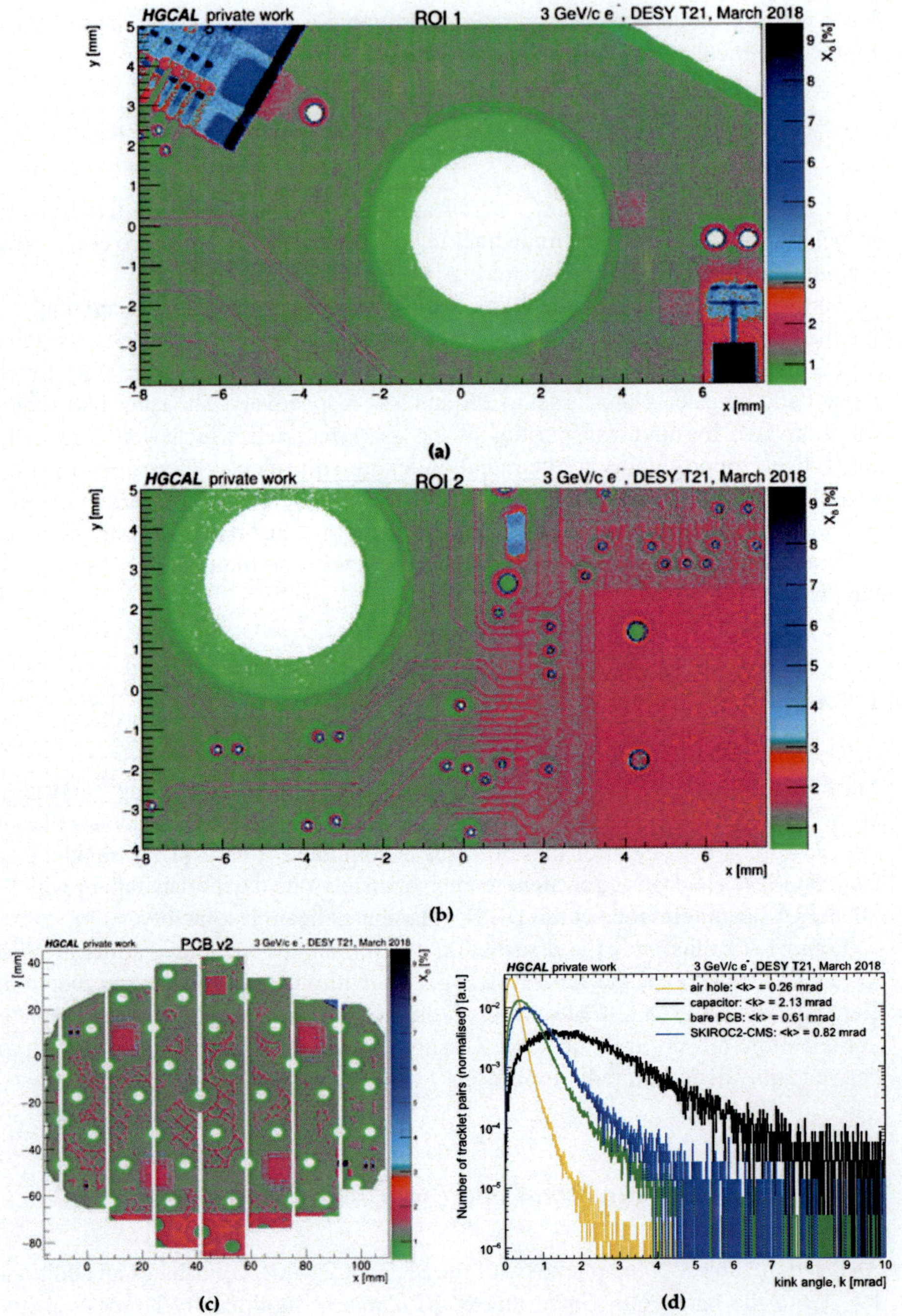

Fig. 8.7 **a**, **b** Tomography images of two dedicated regions on the HGCAL prototype PCB (v2). Underflow values in the images are drawn white. **c** Full tomography image of a PCB. The holder of the PCB in the beam corresponds to the pink area at the bottom. **d** Distribution of kink angles for electron tracklets impinging onto defined areas on the HGCAL prototype PCB

Table 8.2 Measured mean kink angles and material thicknesses for four distinct regions on the prototype PCB. The mean kink angle measured within the air hole defines c_0 in Eq. 8.9. Systematic errors correspond to the uncertainties in the calibration

Region	$\langle k \rangle \pm$ stat. [mrad]	Thickness $\pm$ stat. $\pm$ sys. [% X_0]
Air hole	0.2593 ± 0.0001	$:= 0$
Bare PCB	0.6087 ± 0.0006	$1.066 \pm 0.002^{+0.009}_{-0.011}$
SKIROC2-CMS	0.822 ± 0.002	$2.076 \pm 0.008^{+0.024}_{-0.021}$
Capacitor	2.135 ± 0.005	$12.56 \pm 0.05^{+0.39}_{-0.36}$

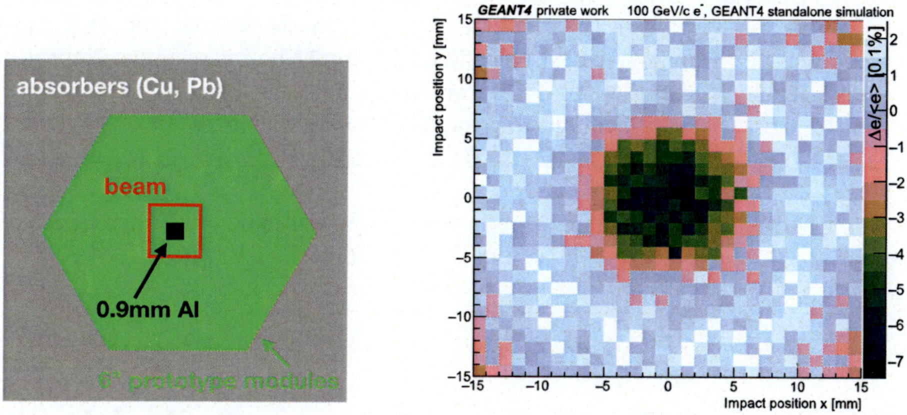

(a) (b)

Fig. 8.8 **a** Sketch of the simulated geometry and the beam profile for assessment of the impact by the SKIROC2-CMS's material budget on the calorimeter response. 1 cm × 1 cm and 0.9 mm thick aluminum plates are used as proxies for the SKIROC2-CMS. **b** Relative difference of the simulated response ($\Delta e/e$) of an EE-like calorimeter to 100 GeV/c electrons

The situation is sketched in Fig. 8.8a. Models of the HGCAL prototype modules are extended by centrally mounted 0.9 mm thick and 1×1 cm large aluminum plates mimicking the contribution by the SKIROC2-CMS. By this means, the air gap between the cassettes and the subsequent passive absorber is reduced from nominally 1.5 mm to 0.6 mm at the location of these additional plates. Furthermore, all 14 cassettes are placed such that all SKIROC2-CMS proxy plates are perfectly aligned along the beam axis.

Based on a calorimeter simulation with the FTFP_BERT physics list, Fig. 8.8b shows the relative change in the energy response e to 100 GeV/c electrons. The difference amounts to 0.9% whereby the response is visibly reduced for electrons impinging on the proxy plates for the SKIROC2-CMS. A possible explanation for this effect could be linked to the reduced air gap. Smaller air gaps inhibit the transverse shower spread in the air volume between the absorbers and cassettes and in turn reduce the sampled energy.

In summary, the HGCAL prototype PCB is indeed heterogeneous also in its material budget. Some capacitors on this prototype are found to contribute more than 12% X_0, whereas the bare PCB amounts to 1% X_0. The largest additional contribution over a considerable area is made by the SKIROC2-CMS ASIC. A specific assessment of its impact on the energy response exhibits a systematic shift of less than 1%. By comparison, exchange of the GEANT4 physics list in the simulation can impact the response by $\mathcal{O}(5\%)$. Moreover, unlike in this study and also in the final HGCAL design, the SKIROC2-CMS ASICs were not perfectly aligned in the beam test configurations. Therefore, the obtained value of 1% for the relative difference should even be understood as an upper limit. In conclusion, the exact details of the prototype PCBs do not need to be entered into the prototype simulation for the comparisons to beam test results later in this thesis. Instead, the PCB can be approximated as homogeneous. Nonetheless, it might be interesting to repeat this study with the HGCROC on the PCB foreseen for the final design. This could justify similar geometry simplifications in future HGCAL simulations.

✓ **MQ-PCB** It is validated that the PCB may be assumed as a homogeneous element in the simulation model.

8.3 Electronic Characterisation of the Beam Test Modules

In this section, the electronic baseline of the digitised signals, referred to as *pedestals*, and the observed noise level of the HGCAL prototype modules in the beam tests are presented. This electronic characterisation can be seen as the verification of the SKIROC2-CMS's proper functionality when operated after installation on such a module. Simultaneously, it serves as partial validation of the data reconstruction procedure.

Section 8.3.1 deals with the pedestals, hereby investigating if the recipe as it is described in Sect. 7.2.1 yields reasonable pedestal values, meaning if the recorded data are consistent at all. Furthermore, the pedestal variation among different switched capacitor array (*SCA*) memory cells and their stability with time is discussed.

Thereafter, the common mode noise subtraction procedure is demonstrated and the intrinsic noise is evaluated. In particular, the following two topics are addressed in Sect. 8.3.2: Reduction of the noise in the modules via common mode noise estimation and its subtraction as outlined in Sect. 7.2.2, as well as the level of intrinsic noise, i.e. the noise after subtraction of the common mode.

Most of the results that are presented next were obtained originally within the work for this thesis. Otherwise, selected findings from Ref. [11] are quoted as indicated in the text.

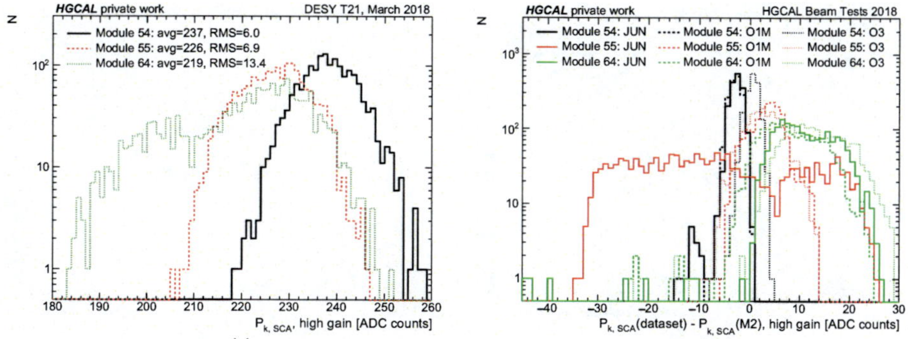

(a) (b)

Fig. 8.9 **a** Distribution of high gain pedestal values for connected channels on three modules tested at the DESY beam test facility in March 2018 corresponding to the *M2* dataset. **b** Distribution of pedestal differences with respect to the *M2* dataset derived from the other HGCAL beam test measurement programmes in 2018

8.3.1 Pedestal Uniformity and Stability

For the vast majority of the beam test modules, data integrity was given. Eventually, only one out of 94 modules was identified to send corrupt data and was excluded from further analysis. For all others, pedestal values in the expected range are retrieved and there are no signs that possibly indicate that the pedestal computation scheme does not work as intended.

Figure 8.9a shows the distribution of high gain (*HG*) pedestals independently for three modules that were qualified as part of the *M2* programme. For each distribution, the range of the entries is below 70 HG ADC whereas the RMS is below 14 HG ADC. Furthermore, the differences between pedestals in a given SCA memory cell taken from distinct setups is not consistent with zero, cf. Fig. 8.9b. Instead, pedestal instabilities, evident through their change between different data taking periods, up to 60 HG ADC are measured for some modules. Noteworthily, the observed variations and instabilities are not negligible with respect to the typical signal from minimum ionising particles which is approximately 40 HG ADC, cf. Sect. 9.1. By contrast, the pedestals are found to be rather stable within a fixed prototype calorimeter experimental setup, cf. Ref. [11]. A negligible alteration of the baseline by only 2–3 HG ADC counts across all runs with the same HGCAL prototype configuration during the most recent HGCAL beam test in October 2018 at CERN is reported therein. This value is much less than what is presented for the comparison between the diverse setups in Fig. 8.9b.

Amongst other things, differences in the operational bias voltages supplied to these modules and changes in environmental conditions such as temperature or humidity could influence the electronic baseline in the SKIROC2-CMS. A fair comparison between pedestals from different setups would require knowledge and synchronised incorporation of all these parameters which was not foreseen initially. Thus, the

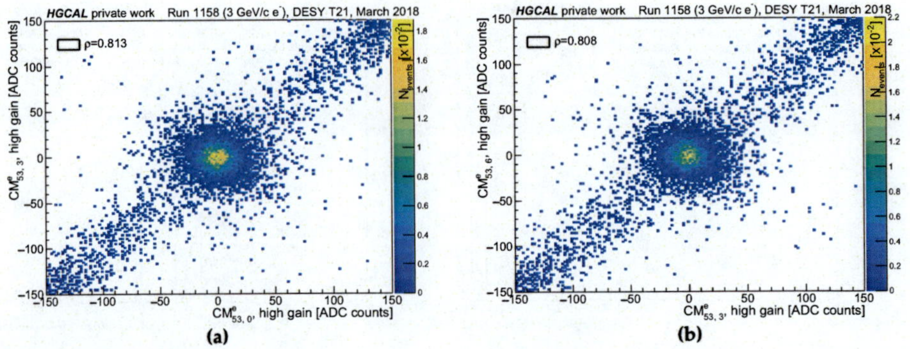

Fig. 8.10 Correlation of computed high gain common mode noise **a** in samples $t = 0$ versus $t = 3$, and **b** $t = 3$ versus $t = 6$, respectively, for the module tested as part of the *M1* programme. The correlation coefficient, indicating the equivalence of common mode noise for different samples t, is denoted as ρ

exact cause for the observed pedestal discrepancy between the setups has not been investigated further in this thesis.

As practical conclusion, the pedestal determination for each memory cell and data taking period separately, in order to cope with the considerable pedestal variation, is justified. By doing so, a systematic bias in the subsequent signal reconstruction is not anticipated.

✓ **MQ-FUNC.3** When operated on the fully assembled modules, the SKIROC2-CMS ASICs give consistent data that corresponds to the intended data format. The pedestal values are reasonable.

8.3.2 *Common Mode Noise*

Representative for all datasets, two-dimensional distributions of the calculated common mode noise in two different samples within the same event and the same module are displayed in Fig. 8.10. In these examples, the correlation coefficient for the common mode noise from samples that are separated by 75 ns amounts to 0.81. It is driven by a few events that are likely affected by low frequency noise. In most cases however, the estimated noise in a sample is uncorrelated to the preceding or consecutive sample which hints at predominant noise with frequencies higher than the 40 MHz sampling rate. It is known, e.g. exemplified in Fig. 7.2a, that in floating channels the level of total noise is significantly lower with respect to channels that are connected to the silicon sensor. Therefore, it is believed that one possible origin of the observed common mode noise could be "disturbances in the bias voltage lines" [11] that would propagate through the baseplate and through the sensor into the readout channel. A detailed assessment of the possible noise sources would compel additional measurements in the laboratory which is beyond the scope of this thesis. Here, Fig. 8.10 is solely intended to emphasise the need for sample-based instead of event-based common mode noise evaluation.

The resulting noise level reduction has been illustrated before in Fig. 7.2b and is established further by Fig. 8.11. The latter summarises the widths of signal-free ADC spectra before and after common mode noise subtraction for a larger set of prototype modules. The width of common mode noise corrected spectra can be interpreted as the intrinsic noise of a module, whereas the uncorrected ones correspond to the total noise. As can be seen, the intrinsic noise in both gains is significantly lower than the total noise for all shown modules. Similar trends and levels, that are stable with time, are obtained with the data taken within the $O*$ programmes [11]. Moreover, Fig. 8.11 illustrates that the application of the RMS is in agreement (well within 1 ADC count) with the width evaluation using the standard deviation σ from a Gaussian fit to the ADC spectrum implying the presence of Gaussian (intrinsic) noise in the system.

The right-most modules in the Fig. 8.11 were assembled using older versions of the HGCAL prototype PCB. They exhibit a larger total noise compared to the modules comprising an updated PCB version with improved routing of digital signals. In addition, noise levels are systematically lower for modules equipped with two layers of Kapton© or for those comprising a PCB as a baseplate, while they are unaffected by the sensor's active thickness [11].

Altogether, common mode noise subtraction effectively reduces the noise towards the intrinsic noise. Its corrected, median level for the hereby investigated modules equipped with 300 μm sensors, that enters the following signal amplitude reconstruction, does not exceed 9 HG ADC (2 LG ADC) counts, cf. Fig. 8.11a (Fig. 8.11b), which corresponds to a fraction of the expected MIP signal that amounts to ∼ 40 HG ADC (∼ 5 LG ADC) counts, cf. Sect. 9.1.

✓ **MQ-FUNC.4** The noise level of the modules is manageable and can be corrected for. The intrinsic noise level does not exceed a fraction of the signal induced by MIPs in both the high and the low gain.

8.4 Module Qualification with Minimum Ionising Particles

One important ingredient to the qualification of the HGCAL prototype modules is the verification that the smallest, physically interesting energy depositions in the sensors, namely due to minimum ionising particles (*MIPs*), can be measured. This strategy is pursued in this section.

The reconstruction of MIP signals in this section is limited to the high gain samples only. Indirect qualification of the SKIROC2-CMS's low gain and TOT is demonstrated later in Sect. 9.2 on the intercalibration of the gains. Its timing capabilities is demonstrated as part of the TOA calibration in Sect. 9.3. As a reminder of the corresponding requirements listed in Sect. 5.1, the HGCAL prototype modules are considered operational in this thesis if the following conditions are met.

- The smallest energy depositions in a particle shower, i.e. induced by MIPs, can be reconstructed efficiently for most of the module's area. The fraction of dead cells is below a few percent.
- Apart from inefficient cells, the integrated MIP detection efficiency is close to 100% for a given module.
- Fake rates, i.e. the probability of reconstructing a calorimeter hit despite the respective cell not being traversed by particles, are less than 1%.
- Diodes on the sensors are isolated units for energy measurements at the MIP scale. This means that small energy depositions are only detected by the channel connected to the corresponding diode that is traversed by a particle. The induced charge due to MIPs is not significantly shared with neighbouring diodes.

This section is structured as follows. MIP efficiency maps for modules that were fully exposed to muon beam are presented in Sect. 8.4.1. Inefficient areas and dead channels are visually highlighted and hence are intuitively identified through such an analysis. Quantification of the efficiency as well as the fake rate for representative modules are provided in the following Sect. 8.4.2. The studies on the MIP efficiency in the region between neighbouring diodes and the assessment of hypothetical charge sharing between them are discussed in Sect. 8.4.3.

Although the subsequently explained qualification procedure was not iterated for every module due to time constraints and due to their limited availability during the beam tests, the reported findings will generally confirm the requirements listed above. With it, the HGCAL prototype module design is functional and suitable for energy measurements. The documentation of the underlying methodology should facilitate an iteration of this procedure in beam tests with future HGCAL prototypes.

8.4.1 Search for Inefficient Areas

The construction of the efficiency maps is based on particle trajectories reconstructed from the upstream delay wire chambers (*DWCs*) with assigned tracking $\chi^2_{x,y}$ below 10. Impact points at each layer of the calorimeter are evaluated in a common coordinate frame between the DWCs and the prototype modules' local coordinate systems. For this purpose, translation alignment corrections are first inferred from residual distributions for each layer, cf. Fig. 8.12, and are then applied to the calorimeter hit positions. The efficiency to detect MIPs is measured in $0.2\,\text{cm}\times 0.2\,\text{cm}$ area segments. The efficiency is defined as the quotient of events with at least one calorimeter hit

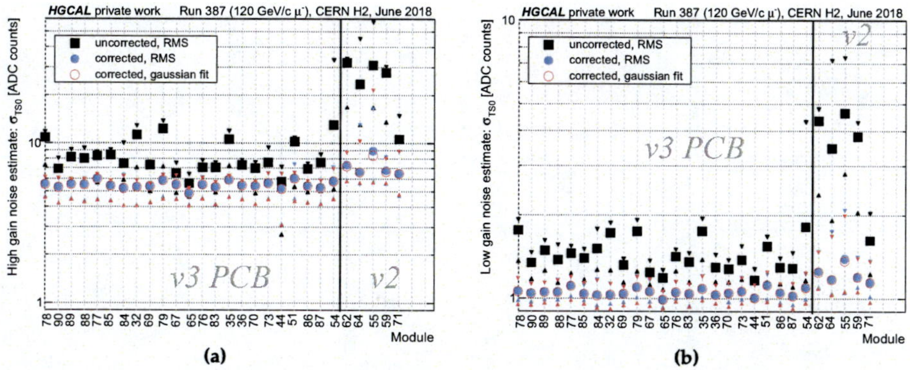

Fig. 8.11 Measured noise levels derived from the width of per-channel pedestal distributions before and after subtraction of common mode noise for all modules tested within the *JUN* dataset **a** in high and **b** low gain. Boxes and circles represent the median values taken over all channels connected to the sensor. Up- and down markers depict the 95% and 5% quantiles, respectively. The red markers represent Gaussian fits as alternative estimates of the noise levels whereas all others are inferred from computation of the RMS

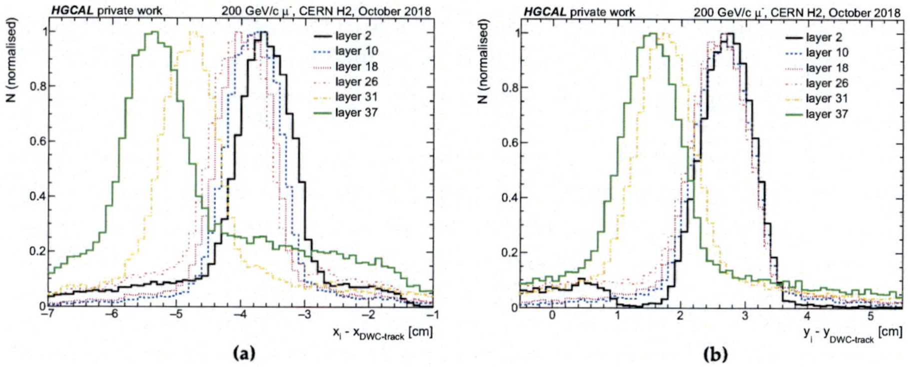

Fig. 8.12 Illustration of the HGCAL-DWC alignment using minimum ionising particles that is performed for both the *JUN* and the *O1M* datasets. **a** Exemplary distribution of hit positions with respect to the extrapolated DWC-track (*residuals*) in x and **b** y using the *O1M* dataset. The peak position determines the translation alignment corrections applied to the calorimeter hits

in some vicinity of the DWC-track divided by the number of good reconstructed DWC-tracks pointing to that segment, cf. Eq. 8.10.

$$\text{MIP efficiency} := \frac{\#\,(\text{good tracks pointing to area} \wedge \text{at least one hit in vicinity of impact})}{\#\text{good tracks pointing to area}} \tag{8.10}$$

Only those hits are considered whose reconstructed signal amplitude in high gain is above 20 ADC counts (only modules with 300 μm sensors in the considered *JUN* dataset) and if their distance to the reconstructed track impact does not exceed 1 cm

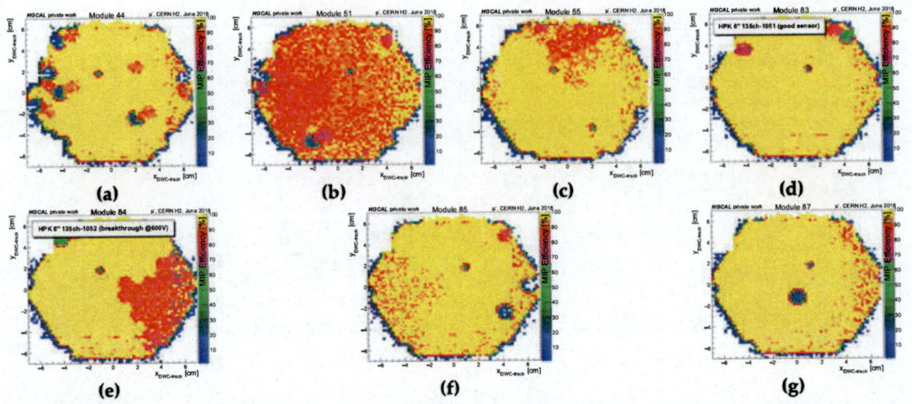

Fig. 8.13 MIP detection efficiency maps for 7 of the 28 modules tested within the *JUN* programme that show a higher degree of inefficiencies compared to the others. The probability to detect a MIP-like hit is colour-coded and is shown in dependency on the impact position inferred from delay wire chamber measurements. Darker colours depict areas with MIP efficiency below 100%

in each direction. Note that the efficiency defined by Eq. 8.10 is sensitive to incorrect mappings of readout channels to their position inside the calorimeter. In fact, the provided mappings could be corrected for some modules while performing this measurement.

Muons are preferred as test particles because they do not tend to initiate particle showers inside the calorimeter. This way, they are suitable for characterisation of various prototype modules simultaneously that may already be installed within a complete calorimeter prototype and that are placed within the beam acceptance. It is remarked that this strategy can similarly be applied to all other MIP-like particles, if muons are not available. A corresponding analysis would have to identify the depth of the shower starting primary interaction for each event, cf. Sect. 10.3.1, and to reject the downstream hits. Besides, the test particle's momentum should be chosen such that the effect of multiple scattering does not deteriorate the tracking resolution. Worsening of the tracking resolution due to multiple scattering is not relevant for high enough beam momenta. As an example, the initial trajectory of the 120 GeV muons is expected to be smeared by only 0.2 mm at a depth of 26 X_0 and 50 cm which is less than the intrinsic DWC tracking resolution in the setups tested at CERN.

After these considerations, this analysis is applied to the muon dataset recorded within the *JUN* programme. The decisive feature of this particular dataset is that the prototype calorimeter was installed on a hydraulic table and could be moved through the muon beam. By this means, a large fraction of the module areas could be characterised with MIPs.

Figure 8.13 shows the outcome for six of the 28 modules with comparatively non-optimal MIP efficiencies with respect to the others. Evidently, six diodes in module 44, cf. Fig. 8.13a, yield efficiencies close to zero. The same observation is made for two diodes on module 85, cf. Fig. 8.13f, and for one each on modules 51 and 87, cf. Fig. 8.13b, g. In none of the cases[1] for which results on the electrical characterisation are also available, cf. Sect. 8.1.1, the inefficiency could be linked to high leakage currents of the bare sensors: First, the diode showing early breakdown on sensor 1052, cf. Fig. 8.1b, is not pronounced in the efficiency map of the associated module 84. Secondly, the rather inefficient diode on module 83, cf. Fig. 8.13d, is located on sensor 1051 which did not show any breakthrough behaviour during the IV measurements, cf. Fig. 8.2b. This circumstance leads to the interpretation of these inefficiencies as the result from other causes, such as bad wire bonding of the ROC to the diodes.

The triangular areas of lower efficiencies in modules 55 and 84, cf. Fig. 8.13c, e, correspond to the area served by one SKIROC2-CMS with systematically lower response to MIP-induced charges. An instructive example for lowered MIP efficiency for a full module due to high leakage current is given by module 51, cf. Fig. 8.13b. The total leakage current of this module was measured after its assembly. It exceeded 400 μA at 190 V which should have effectively lowered the bias voltage on the sensor during operation. With it, the depleted, i.e. sensitive, sensor thickness decreased which explains why the sensor's response to MIPs was reduced.

Apart from these modules, all remaining ones exhibit a remarkably good efficiency in the detection of MIPs. The efficiency maps of these modules are shown in Fig. 8.14. Areas with lower efficiencies correspond to the diodes close to the module's data and communication port and to the unconnected calibration pads, cf. Fig. 4.11.

In summary, MIP efficiency maps are advantageous for the identification of inefficient areas on the module either due to a malfunctioning readout or due to a reduced response. In this connection, it is found that the wire bonding to only ten out of more than 3000 exposed diodes in the *JUN* beam test programme appears to be inadequate. Moreover, the MIP efficiency for the diodes close to the data and communication tends to be lowered systematically. It should be noted for future beam test iterations that these maps have proven helpful in practical aspects, too. They were essential both in the debugging of the assignment of readout channels to the connected diodes and in the adjustment of the calorimeter hit selection thresholds in Table 7.1.

✓ **MQ-FUNC.6** Only a few isolated cells exhibit low MIP efficiency.

[1] By chance, only three of the previously characterised sensors were assembled to modules that were used in the *JUN* programme. The correspondence of the sensors assembled to the modules is documented in Ref. [12].

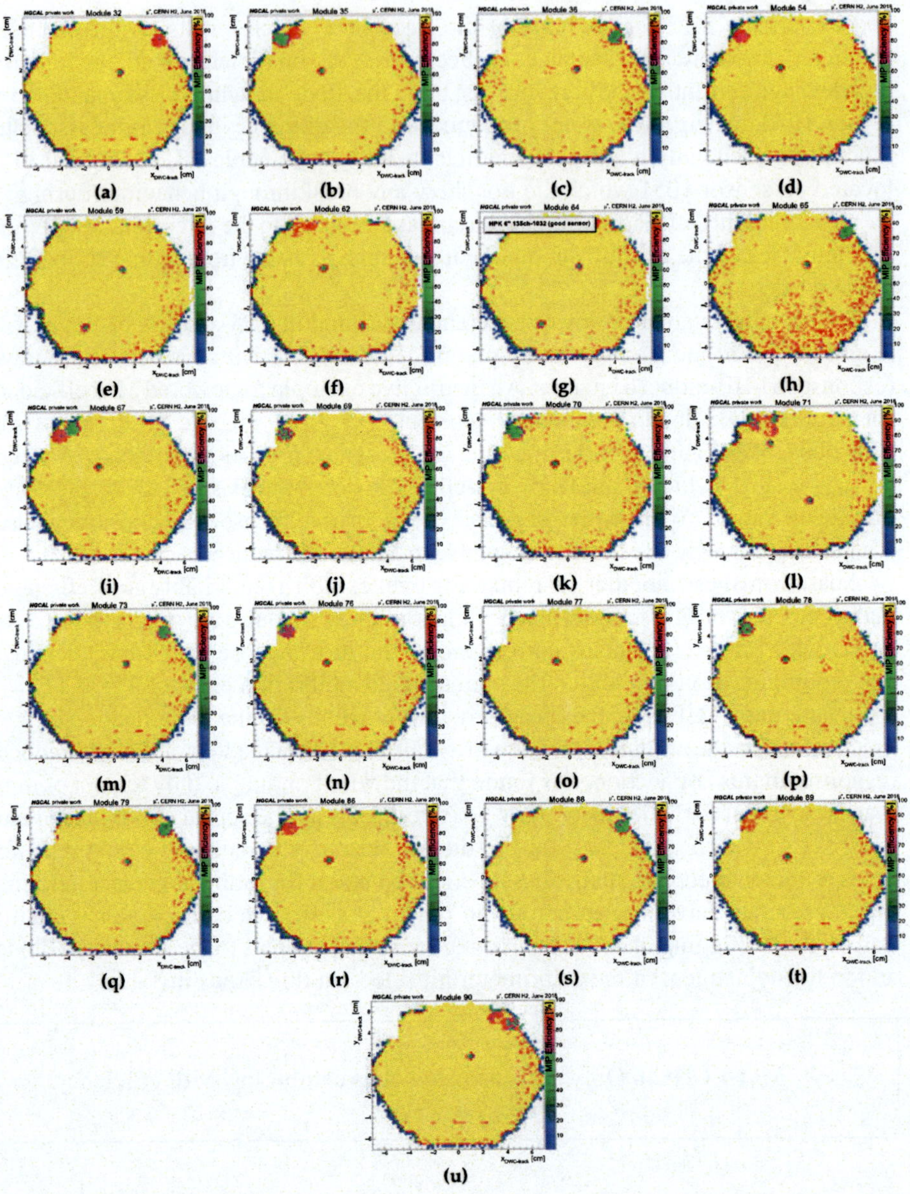

Fig. 8.14 MIP detection efficiency maps for the remaining 21 modules of the 28 tested within the *JUN* programme. The probability to detect a MIP-like hit is colour-coded and is shown in dependency on the impact position inferred from delay wire chamber measurements. The depicted modules have a different orientation owing to their different orientation in the prototype cassettes. Darker colours depict areas with MIP efficiency below 100%

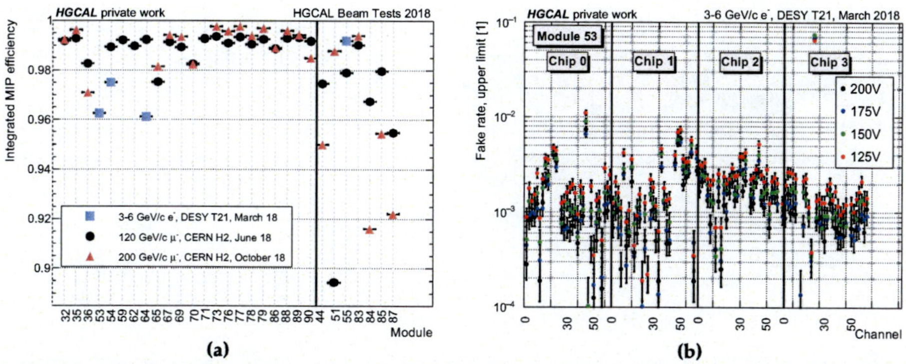

Fig. 8.15 **a** Module-integrated efficiency to detect minimum ionising particles for all modules tested in June 2018 (*JUN* dataset) and overlaid with the respective results from the other beam test campaigns. **b** Fake rates for different bias voltages and for all channels on module 53 that were outside the beam acceptance during the *M1* programme in March 2018 at DESY

8.4.2 Signal Efficiency and Fake Rate

Based on the efficiency maps, efficiency averages are computed over all area segments within a central 6 cm×6 cm window in the efficiency maps. Segments with at least 20 tracks are included in the sum. The bias due to non-uniform exposure of the modules to the beam has been studied and is found to be negligible. Moreover, the method is iterated for the *O1M* and the *M** datasets.

For the latter dataset recorded at DESY, electrons are considered as MIPs. Events with exactly one reconstructed trajectory from the DATURA beam telescope are selected because ambiguities in the assignment of calorimeter hits to multiple telescope tracks cannot be resolved otherwise. The hereby reported ultimate efficiency might underestimate the true module's MIP efficiency due to finite (but unknown) tracking inefficiencies. In the explicit example of having two physical trajectories measured within one readout, it may occur that the proper track for assignment to the calorimeter hit is not reconstructed and the other one is used instead. Then, the denominator in the efficiency calculation would be wrongfully incremented and the efficiency estimate would be lowered artificially.

Figure 8.15a presents the obtained efficiency numbers. Apart from the aforementioned "bad" modules, cf. Fig. 8.13, the average efficiency amounts to more than 96% for all modules, independent from the setup. For 16 modules, the measured efficiency even reaches 99% and more.

The indications of the signal efficiencies are now complemented by the corresponding fake rates. The latter are not defined per area but for each sensor diode. The fake rate is equivalent to the probability to reconstruct a calorimeter hit, cf. Table 7.1, in a readout when the incident particle traverses the module elsewhere. Its assessment under beam test conditions is straightforward. For the *M1* dataset, the fake rate is evaluated for all diodes outside the 2 cm×1 cm acceptance defined by

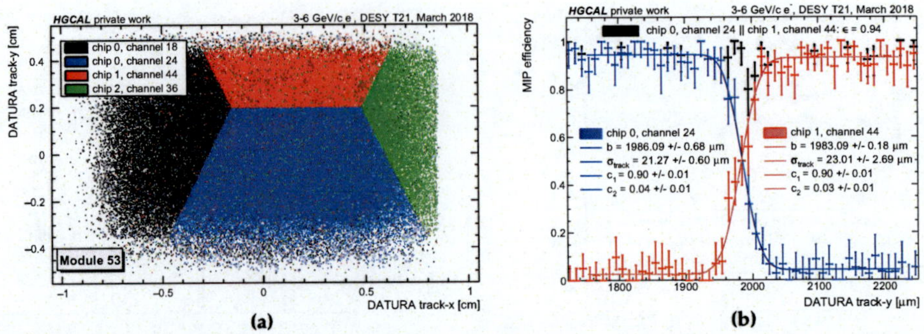

(a) (b)

Fig. 8.16 **a** DATURA track occupancy of events with one reconstructed trajectory. Entries are colour-coded depending on the simultaneously reconstructed hits. **b** MIP detection efficiency at the horizontal boundary region between channel 44 of chip 1 and channel 24 of chip 0. The physical gap between these diodes is designed 20 μm wide

the triggering scintillators. This strategy yields accurate results under the assumption that all signal causing particles in the tested prototype module are confined to the trigger acceptance. If this assumption is not fulfilled, the fake rate is generally overestimated.

Detection of additional particles outside the trigger acceptance was not feasible with the investigated setup. For this reason, the shown values in Fig. 8.15b should, in fact, be regarded as upper limits of the true fake rate. The measured fake rates amount to 0.1–1% for all channels except for the one connected to the diode close to the data and communication port that reaches 7%. It is observed that the fake rate tends to marginally increase when the bias voltage is reduced and with it the diode's capacitance and the expected intrinsic noise are increased.

In all cases, hit selection based on reconstructed signal amplitudes quickly suppresses these rates further. For instance, setting the selection threshold on the reconstructed signal amplitude to 40 high gain ADC, roughly corresponding to the signal of a MIP, the fake rate is reduced to values well below 1‰.

✓ **MQ-FUNC.5** The lowest relevant energy depositions due to MIPs are detectable with efficiencies close to 100%.
✓ **MQ-FUNC.7** The calorimeter hit fake rate is below 1%.

8.4.3 *Signal Confinement to Single Diodes*

Low fake rates as documented in the previous section are coherent with the statement that MIP-induced charge depositions are confined to single cells. As visual and intuitive support of this claim, Fig. 8.16a shows the occupancy of reconstructed electron trajectories of events with one reconstructed GBL track using the DATURA beam telescope. The entries are colour-coded according to the calorimeter hits that are reconstructed simultaneously. Hexagonal diode acceptances are clearly visible and they do not exhibit any holes.

The data on the occupancy along the border between the red- and blue-coloured cells is processed further to yield the efficiency curves in Fig. 8.16b. In agreement with the earlier findings, the per-cell MIP efficiency is close to 100% when the electron trajectory points to a diode and quickly falls to zero between the two cells. Besides, the turn-on does not extend more than 100 μm. Inside this region, the efficiency to detect particles in one or the other cell does not drop significantly. Consequently, the passive interpad region does not seem to introduce insensitive areas on the sensor.

Further quantification of the turn-on is attempted by fitting a rectangular step function Θ to the data. The step function is convoluted ($\circledast$) by a Gaussian in order to account for the finite tracking resolution with the DATURA beam telescope, cf. Eq. 8.11

$$\epsilon_{\text{blue}}^{\text{red}}(y) \;=\; [c_1 \cdot \Theta(\pm y \mp b) \;\circledast\; \text{Gaussian}(0,\; \sigma_{\text{track}})] + c_2 \qquad (8.11)$$

In this attempt, the derived σ_{track} parameter overestimates the expected tracking resolution, cf. Fig. 8.11b, by roughly a factor 2. It should be remarked however that the optimisation and precisions of the parameters b and σ_{track} are strongly affected by the chosen fitting range in this analysis.

Furthermore, the recorded and usable dataset after track multiplicity selection for this analysis contains only a few 100k events which limits the granularity of the binning. As a result, the sampling of the efficiency curve is still too coarse for reliable quantification of its turn-on part in the interpad region. For future improvement of the measurement's sensitivity to the tracking resolution and to potential sensor interpad effects, enough events should be accumulated where the beam remains focussed onto the region of interest. Under consideration of the multiplicities of reconstructed tracks, cf. Fig. 8.10a, and the experienced data acquisition rate, cf. Table 6.1, a rule-of-thumb calculation predicts a required data taking time of $\Delta T = $ 8–9 hours with 5.4 GeV/c electrons to obtain a 10% relative error on the efficiency measured in 0.5 cm × 10 mm bins.[2] Then, it would also be interesting to focus on other sensor areas, for example with increased interpad distances of up to 80 μm. Such data would be beneficial to the inference of rotational alignment corrections to the prototype modules with respect to the telescope planes. A conceptual shortcoming of the current, ad hoc approach is that rotational alignment could not be performed with the given data.

Nevertheless, it must be emphasised that detailed studies of the HGCAL sensor's interpad regions in beam tests together with a precise tracking telescope are possible. Returning to the original question at hand, the preliminary result presented by Fig. 8.16b reinforces the statement that the diodes are independent sensitive entities.

---

[2] Assuming a uniform beam profile: $\Delta T = \frac{1}{0.1^2} \cdot \frac{2\text{cm}}{0.5\text{cm} \cdot 10\,\mu\text{m} \cdot 0.65 \cdot 20\,\text{Hz}}$.

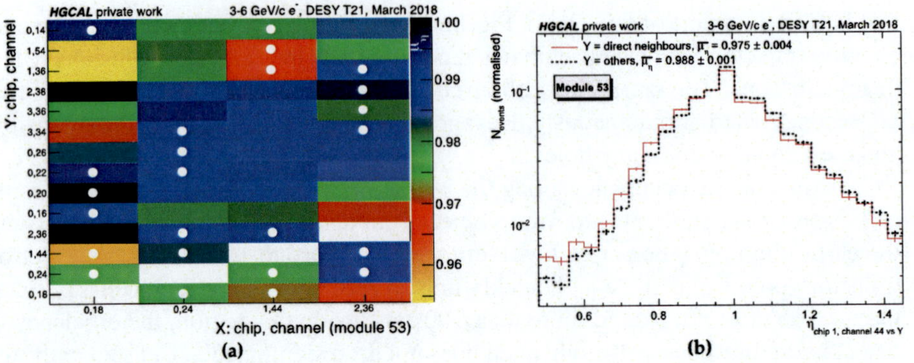

Fig. 8.17 **a** Summary of peak values μ_η of the η-distribution [13] computed for selected pairs of exposed channels to non-exposed ones. White dots symbolise pairs of directly neighbouring cells on this module. **b** Cumulative η-distribution computed for channel 44 on chip 1 with respect to its direct neighbours and to distant channels

Charge Sharing at the MIP Scale

A remaining possible concern is that the preselection criteria, cf. Table 7.1, might conceal subtle cross-talk effects due to capacitive coupling and charge sharing between the diodes. A complementary study deploying the η-algorithm [13] on pairs of diodes weakens this argument qualitatively. No preselection and no amplitude reconstruction are applied in this context. Instead, the third high gain sample is used as proxy for the deposited charge in a diode, cf. Eq. 8.12.

$$\eta_{X,Y} := \frac{TS_3^X}{TS_3^X + TS_3^Y} \tag{8.12}$$

In essence, the distribution of η peaks at 1 in absence of capacitive coupling. When capacitive coupling is present and with it a fraction K of the signal charge is shared between cells, the distribution is shifted to lower values. For small charge sharing, the fraction K can be related to the mean value (μ_η) of the η-distribution [13], cf. Eq. 8.13.

$$K \approx 1 - \mu_\eta \tag{8.13}$$

In order to rule out other causes, e.g. common mode noise, that could achieve a similar effect, the distributions are compared between direct neighbours (with hypothetical capacitive coupling) and distant neigbours with at least one diode in between (no capacitive coupling).

Figure 8.17a visualises the results. Although the peak positions μ_η are consistently smaller than 1, there is no systematic difference between them when computed between directly neighbouring and distant cells. Even for the signal cell with the most pronounced numerical deviation from 1, as it is shown in Fig. 8.17b, the difference in the combined mean $\bar{\mu}_\eta$ between direct and distant neighbours is not substantial.

While this result practically supports the absence of charge sharing at the MIP scale, it cannot be used to predict charge sharing for large signals of hundreds of

MIPs. Corresponding studies require dedicated test setups, e.g. with light or charge injection. Those were being developed for the HGCAL upgrade at the time of writing this thesis. Their description would go beyond the scope of this work.

✓ **MQ-FUNC.8** MIP signals are confined to single cells.

References

1. Pree E (2018) Development of large area silicon sensors for the high granularity calorimeter at CMS. PhD thesis, Technische Universität Wien Fakultät für Physik, Austria. http://katalog.ub.tuwien.ac.at/AC15192834
2. CMS Collaboration (2017) The phase-2 upgrade of the CMS endcap calorimeter. CERN-LHCC-2017-023 (CMS-TDR-019). https://cds.cern.ch/record/2293646
3. Keysight Technologies (2019) Impedance measurement handbook - a guide to measurement technology and techniques, 6th edn. https://literature.cdn.keysight.com/litweb/pdf/5950-3000.pdf. Accessed 31 2019
4. Benaglia A (2014) The CMS ECAL performance with examples. JINST 9:C02008. https://doi.org/10.1088/1748-0221/9/02/C02008
5. Jansen H, Schütze P (2018) Feasibility of track-based multiple scattering tomography. Appl Phys Lett 112. https://doi.org/10.1063/1.5005503
6. Highland VL (1975) Some practical remarks on multiple scattering. Nucl Instrum Meth 129:497–499. https://doi.org/10.1016/0029-554X(75)90743-0
7. Patrignani C, et al. (Particle Data Group) The review of particle physics. Chin Phys C 40:100001. https://doi.org/10.1088/1674-1137/40/10/100001
8. Quast T (2019) Electron beam tomography. https://gitlab.cern.ch/tquast/electron-beam-tomography. Tag: INFIERI2019. Accessed 18 Nov 2019
9. Nakagami M (1960) The m-distribution-a general formula of intensity distribution of rapid fading. Stat Methods Radio Wave Propag 3–6(6a):7–36. https://doi.org/10.1016/B978-0-08-009306-2.50005-4
10. Quast T (2019) Geant4 based event display & standalone simulation. https://github.com/ThorbenQuast/HGCal_TB_Geant4/. Tag: v1.0. Accessed 12 Aug 2019
11. Acar B et al (2021) Construction and commissioning of CMS CE prototype silicon modules. JINST 16:T04002. https://doi.org/10.1088/1748-0221/16/04/T04002
12. UCSB Physics - Sensors. http://hgcal.physics.ucsb.edu. Accessed 18 2019
13. Turchetta P (1993) Spatial resolution of silicon microstrip detectors. Nucl Instrum Meth A 335:44–58. https://doi.org/10.1016/0168-9002(93)90255-G

Chapter 9
In Situ Calibration of Prototype Modules

The principal result of the qualification studies in the previous chapter is that the HGCAL prototype modules provide reconstructable electronic signals for the measurement of deposited energies in the calorimeter cells. However, the conversion of such signals into a physical energy scale relies on prior calibration of the cells' energy responses as well as calibration of the gains involved in the readout. Referring to the energy derivation recipes formulated by Eqs. 7.8, 7.9 and 7.10 in Sect. 7.2.4, an equivalent statement is that all calibration constants for as many readout channels as possible remain to be derived. This task is discussed in this chapter. Here, the calibrations are not obtained from dedicated calibration experiments, e.g. from charge injection in the laboratory, but in situ from the beam test data directly using minimum ionising and showering particles.

Section 9.1 explains the response calibration and equalisation strategy with minimum ionising particles. This effort represents a major contribution to the CMS HGCAL's system group activity made as part of this work and the findings are partially to be published in Ref. [1]. The intercalibration of the gains was performed by other members of the HGCAL System Test group. Hence, its presentation in Sect. 9.2 is limited to a brief description of the underlying methodology and the presentation of the main results.

Apart from the energy measurement, the SKIROC2-CMS design features timing capabilities to measure the time of arrival of energy depositions (*TOA*) with unprecedented precision in calorimetry. The principle TOA functionality could be validated as part of this work. In particular, Sect. 9.3 demonstrates that the TOA can be calibrated incorporating external time reference measurements. The procedure requires calibrated hit energies and is also performed with beam test data. Finally, estimates for the achieved timing precision are reported.

T. Quast, *Beam Test Calorimeter Prototypes for the CMS Calorimeter Endcap Upgrade*, Springer Theses, https://doi.org/10.1007/978-3-030-90202-5_9

9.1 Energy Response to Minimum Ionising Particles

Remark: Italic passages in this section are substantively quoted from the author's original contribution to the publication draft (24 August 2019) of Ref. [1].

The frequency of deposited energy by minimum ionising particles (MIPs) in the silicon diode approximately follows a Landau distribution [2]. *Its maximum is expected around 86 keV for 300 μm (∼57 keV for 200 μm) thick sensors. Variations in the gains in the SKIROC2-CMS and differences in the energy response, e.g. due to differences in the depletion thickness, of the sensitive diodes are corrected for through the per-cell relation of the digitised signal to the energy deposited by MIPs.*

Due to their overall negligible energy loss compared to the initial momentum of multiple GeV/c, the amount of deposited energy per distance can be regarded independent of the calorimeter depth. In consequence, MIPs are suitable for the equalisation through calibration of the electric response throughout the detector.

The vast majority of recorded high-energetic muons in the O1M dataset can be visualised in event displays that look similar like the one shown in Fig. 9.1. *In general, muons at this momentum realm do not tend to initiate particle showers in the calorimeter and hence are particularly suitable candidates for the energy calibration with MIPs.*

Section 9.1.1 motivates the necessity of energy calibration with MIPs in this calorimeter system and describes the general procedure. Since the fake rate of reconstructing calorimeter hits is not zero, cf. Sect. 8.4.2, and since the muon beam often covered the full trigger acceptance in the beam tests, noise in the MIP energy spectra is not negligible. Hit selection based on the identification of MIP-like trajectories

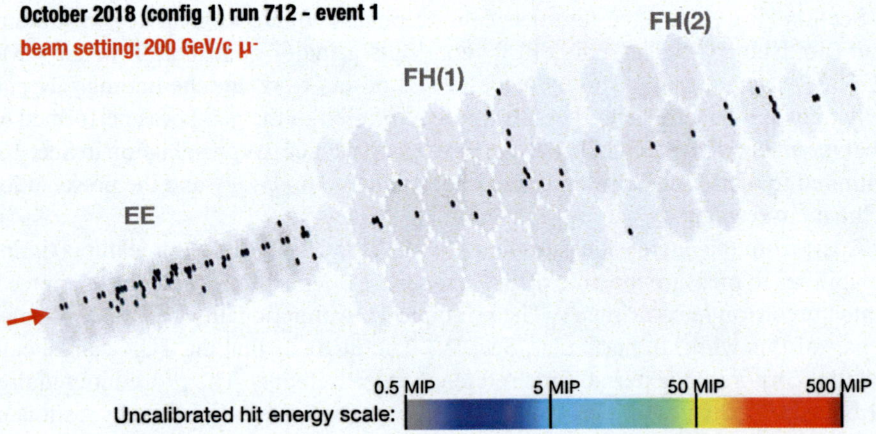

Fig. 9.1 Event display of a 200 GeV/c muon traversing the electromagnetic calorimeter (28 layers) and the hadronic calorimeter (12 layers) prototype compartments taken from the *O1M* dataset. The incoming muon enters the detector from the left-hand side. The HGCAL beam test prototype event display tool [3] was developed as part of this thesis

inside the HGCAL prototype calorimeter is essential for selecting energy depositions truly induced by MIPs. A similar identification strategy is then applied on the *O3* dataset taken with parasitic, i.e. undefined, particle beam in order to maximise the total number of calibrated channels. Section 9.1.2 explains the successful application of this HGCAL-based MIP tracking algorithm for energy calibration of ultimately more than ten thousand channels with MIPs in the so far largest HGCAL prototype calorimeter tested in October 2018. The principal findings from this effort are presented in Sect. 9.1.3.

9.1.1 Response Equalisation

As explained in Sect. 4.3, a MIP traversing the silicon sensor interacts with the atoms via ionisation and loses energy which is deposited in the material. The amount of the correspondingly induced charge carriers in the silicon that ultimately form the signal is determined by the sensor properties. For instance, a reduced depletion zone with respect to a fully depleted sensor is equivalent to less sensitive material from which detectable charge carriers can be collected. Although it is not relevant for the beam test prototypes, radiation effects in silicon may affect the charge collection efficiency [4]. Moreover, the same amount of induced charge in the sensors may be digitised differently by the different readout chips (*ROCs*). Such variations in the electronic response are typically more sizeable than variations due to differences in the silicon diodes to which they are attached. In account of both effects, the same amount of energy in a diode translates into different reconstructed signal amplitudes in the readout channels.

Thus, determination of the high (low) gain to MIP proportionality constants $M_{\mathrm{HG/MIP}}$ ($M_{\mathrm{LG/MIP}}$) in Eq. 7.9 (7.10) for each channel is crucial for uniformisation of the response. This calibration is commonly referred to as *MIP calibration*. It simultaneously defines a common energy scale in the experimental data and the simulation, namely the *MIP scale*, and thus enables comparisons of the measured energy depositions to the simulated counterpart. The sampled energy in a given sensor is indicated in units of energy equivalents that a MIP would deposit therein. However, it is important to note that a typical particle shower does not consist of a collection of MIPs. The statement that a certain number of MIPs is reconstructed does not imply that the shower comprises this number of particles, but represents a certain amount of deposited energy.

Continuous energy calibration with MIPs is foreseen for the final HGCAL operation at HL-LHC in order to cope with the anticipated degradation in the sensor response due to radiation effects. It has been studied that a 3% uncertainty in the MIP calibration constants would add 0.5% to the constant term in HGCAL's ultimate energy resolution [5]. Therefore, MIP calibration is conducted for all beam tests not only for comparison of the data to simulation but also to gain more insights in this regard.

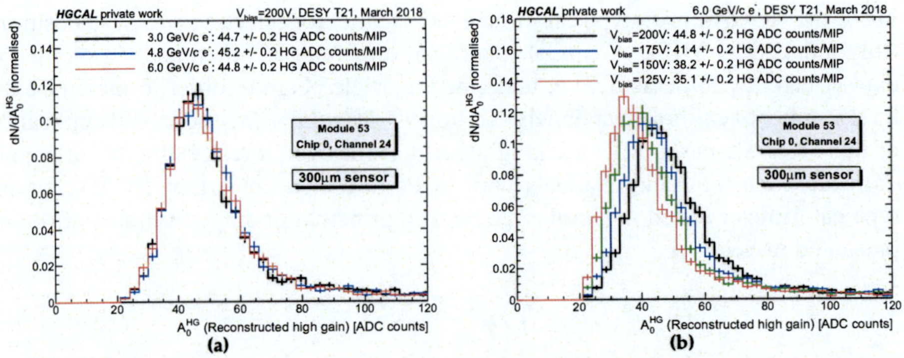

Fig. 9.2 **a** Distribution of the reconstructed high gain amplitudes of an exemplary readout channel due to incident electrons with different momenta and **b** due to 6 GeV/c electrons at different sensor bias voltages. The reported values in the legend correspond to the maximum value taken by the parameterisation, cf. Eq. 9.1, which is fitted to the spectra (not shown)

Distributions of reconstructed MIP-induced signal amplitudes are the fundamental objects in this procedure. Three illustrative examples of high gain amplitude spectra induced by non-showering electrons as MIPs at different momenta are shown in Fig. 9.2a. According to Eq. 33.12 in Ref. [6] or Fig. 14 in Ref. [7], the most probably energy loss due to MIPs reaches a constant plateau for $\beta \cdot \gamma > 100$. As consequence, the most probable values of the spectra in Fig. 9.2a do not show any dependence on the electron momentum because electrons at the considered momentum range of 1-6 GeV/c have $\beta \cdot \gamma \gg 100$. By contrast, the spectra of reconstructed signal amplitudes are substantially shifted when certain sensor characteristics are altered, such as the depletion zone thickness. In Fig. 9.2b, the latter circumstance is mimicked by lowering the applied bias voltage. For lower bias voltages, the response to MIPs is shifted to lower values in agreement with the intuition of having less sensitive material.

In principle, the reconstructed amplitudes would need to be normalised to the same effective MIP's path length in the sensor. The multiplicative correction is equal to $(\cos\alpha)^{-1}$, where α is the angle between the sensor's surface normal and the MIP's flight direction. A relative correction value of 0.5% would imply $\alpha \approx 100$ mrad. Note that in the beam test situation, the setups were always arranged such that MIPs impinge on the calorimeter perpendicularly. This expectation is confirmed a posteriori in Sect. 9.1.2 through construction of MIP trajectories within the prototype calorimeter's coordinate system. One obtains $\alpha \ll 100$ mrad, and thus path-length corrections can be neglected here.

For quantification of the spectra, the model given by Eq. 9.1 is fitted to the data. It mainly consists of a Landau function (L) convoluted ($\circledast$) by a Gaussian (G).

$$f\left(x \mid MP_L,\, \sigma_L,\, \sigma_n,\, \sigma_p,\, c_1,\, c_2,\, c_p\right) := c_1 \cdot \left(L(MP_L,\, \sigma_L) \circledast G(0,\, \sigma_n)\right)(x) + c_2 \cdot \left(L(2 \cdot MP_L,\, \sigma_L) \circledast G(0,\, \sigma_n)\right)(x) + c_p \cdot G(x, 0,\, \sigma_p) \tag{9.1}$$

All parameters therein (i.e. MP_L, σ_L, σ_n, σ_p, c_1, c_2, c_p) are subject to the fitting. The order of their arguments is defined as by the ROOT [8] software framework. This convolution term occurs twice in order to allow for conceivable cases in which multiple MIPs traverse a sensor in one readout. The last term $c_p \cdot G$ corresponds to a pedestal in the amplitude spectrum at low amplitudes to incorporate contributions from noise. Fits are performed from 15 to 120 high gain ADC counts (A_0^{HG}) and from 2 to 15 low gain ADC counts (A_0^{LG}), respectively, in the spectrum range. The impact due to different binnings has not been studied. In practice, a couple thousand entries in a spectrum is found to sufficient to obtain statistical uncertainties on the most probable value of the Landau function (MP_L) close to 0.5 %. By convention, the fitted curve's maximum value renders the sought calibration constant instead of MP_L. The search range for the maximum is chosen such that the pedestal term is excluded from the search window, cf. Eq. 9.2.

$$\begin{aligned} M_{\mathrm{HG/MIP}} &:= \max_{A_0^{\mathrm{HG}} \in [25,120]} f\left(A_0^{\mathrm{HG}} \mid MP_L^{\mathrm{HG}},\ \sigma_L^{\mathrm{HG}},\ \sigma_n^{\mathrm{HG}},\ \sigma_p^{\mathrm{HG}},\ c_1^{\mathrm{HG}},\ c_2^{\mathrm{HG}},\ c_p^{\mathrm{HG}}\right), \quad \forall\, \text{channel} \\ M_{\mathrm{LG/MIP}} &:= \max_{A_0^{\mathrm{LG}} \in [2,15]} f\left(A_0^{\mathrm{LG}} \mid MP_L^{\mathrm{LG}},\ \sigma_L^{\mathrm{LG}},\ \sigma_n^{\mathrm{LG}},\ \sigma_p^{\mathrm{LG}},\ c_1^{\mathrm{LG}},\ c_2^{\mathrm{LG}},\ c_p^{\mathrm{LG}}\right), \quad \forall\, \text{channel} \end{aligned} \tag{9.2}$$

With close to twelve thousand individual readout channels in the HGCAL calorimeter prototype tested in October 2018, the MIP calibration can be summarised as a "straightforward albeit laborious" [9] procedure. Proper selection of MIP-traversed cells in a given readout is beneficial for cleaning the spectra from pedestal contributions that otherwise may hide the actual signal. The developed strategy in this thesis for this task is discussed next.

9.1.2 *MIP Tracking with the HGCAL Prototype*

The first selection strategy of exposed cells is based on beam particle tracking upstream the calorimeter prototype using the delay wire chambers. This method has been realised successfully already in the HGCAL prototype beam tests in 2017 [10] and serves as reference to the approach that is presented in the following. *In the following study, the calorimeter itself is used as a MIP tracking device. In this context, MIP signatures are identified by combining reconstructed calorimeter hits that are consistent with one straight line trajectory per readout traversing the calorimeter.*

Tracking of Muons

For the purpose of tracking muons in HGCAL, a triplet-based tracking was proposed originally in the technical proposal for the CMS Phase 2 upgrades [11]. Under consideration of non-idealised assumptions on the calorimeter performance such as sizeable fake rates and MIP inefficiencies present in the beam tests, an alternative, however preliminary, algorithm was deployed for the MIP calibration of the 2018 beam test data. The approach has neither been studied on simulation nor it has been evaluated with Monte Carlo truth information.

Calorimeter hits with reconstructed high gain signal amplitudes between 25–200 high gain ADC counts are exclusively considered for the HGCAL-based "MIP Tracking" algorithm. The algorithm has been implemented as a module [12] using the CORRYVRECKAN framework (*v0.6*) [13] as part of the work for thesis. Its logic can be divided into seven steps:

1. A 40×40 grid of parameterisations of two-dimensional straight line track hypotheses is constructed. This is done for each dimension (x/y) separately at the beginning. The associated slopes $m_{x(y)}$ and offsets $b_{x(y)}$ are sampled uniformly in the intervals $m_{x(y)} \in [-0.2,\ 0.2]$ and $b_{x(y)} \in [-20\,\text{cm},\ 20\,\text{cm}]$.
2. Hits are associated to a two-dimensional track hypothesis if their distance to the track's impact is less than 1.2 cm. Tracks with associated hits from less than eight distinct layers are discarded, leaving n_x tracks remaining in $x - z$ and n_y in $y - z$.
3. All remaining n_x track hypotheses in x and and n_y hypotheses in y are combined to form $n_x \cdot n_y$ (three-dimensional) track candidates.
4. The number of distinct contributing layers is redetermined and the track candidates with the most are kept.
5. Track candidates with more than 1.5 associated hits per layer on average are discarded. This step aims at the rejection of particle showers.
6. The combinatorial analysis analogous to steps 2 and 3 of the Ntlet-Tracking in a telescope arm, cf. Fig. 7.7, is performed on all hits associated to a track candidate.
7. Among all combinations, the combination with best agreement to a straight line (i.e. minimal χ^2) defines the reconstructed MIP trajectory.

By construction, this algorithm reconstructs one trajectory per readout at most. The *O1M* dataset could be processed at a rate of approximately 70 Hz (Intel© Xeon© CPU E5-1620). MIP-like trajectories are reconstructed in more than 92% of all events in the *O1M* dataset. The efficiency of the MIP tracking with HGCAL decreases with an increasing minimum number of contributing layers. Especially at impact locations that correspond to locations of inefficient channels in the prototype calorimeter, cf. Fig. 9.3, the MIP tracking efficiency with HGCAL drops noticeably.

Figure 9.4a shows the occupancy distribution of reconstructed MIP trajectories illustrating the typical bias towards the hexagonal diode centres. As it is expected for muons that traverse the full calorimeter, all layers contribute rather equally to the reconstructed MIP trajectory, cf. Fig. 9.5a. The trajectories' orientations within the calorimeter coordinate system are determined and allow for inference of the beam angle with respect to the prototype calorimeter orientation in that particular setup. As it is exemplarily depicted in Fig. 9.5b, this angle amounts to ~10 mrad in the x-direction (~5 mrad for the y-direction) in support for the neglect of path-length corrections in the MIP calibration discussed in the previous section.

Finally, the initial signal size bias, that is explicitly imposed at the beginning for the construction of the MIP trajectory, is removed: For the computation of the MIP signal spectra, all hits with distances less than 1.2 cm to those trajectories are selected, independent of their reconstructed high gain amplitude. *Figure* 9.6 *displays*

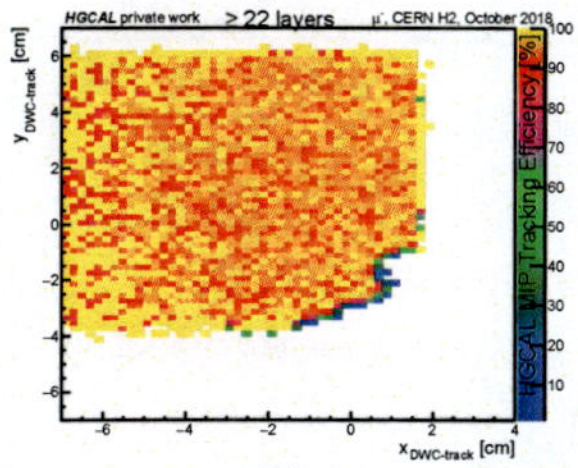

(a)

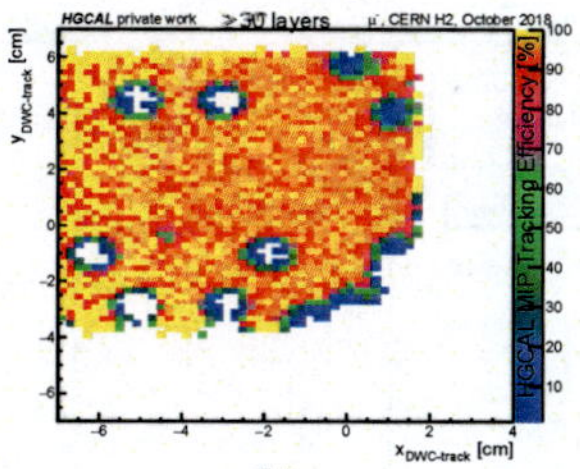

(b)

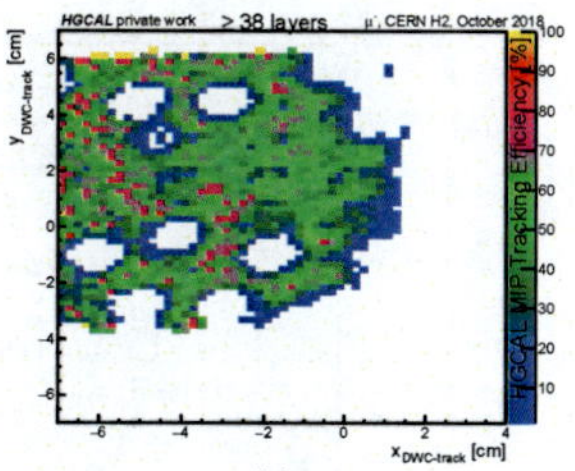

(c)

Fig. 9.3 Efficiency to reconstruct MIP-like trajectories with the HGCAL prototype as a function of the DWC-track extrapolation using the *O1M* dataset. Shown are the efficiency maps for trajectories with **a** at least 22, **b** 30 and **c** 38 contributing layers

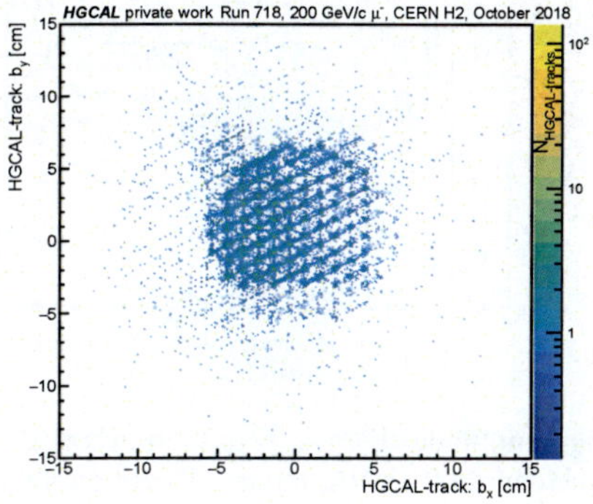

(a)

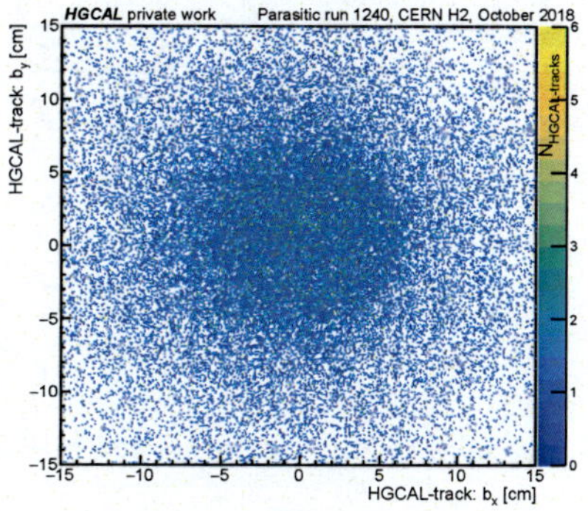

(b)

Fig. 9.4 HGCAL-based track occupancies of MIP-like trajectories for runs **a** with muon beam (*O1M* dataset) and **b** for those with undefined, parasitic beam (*O3* dataset): The trajectories are evaluated at the beginning of the prototype's EE compartment

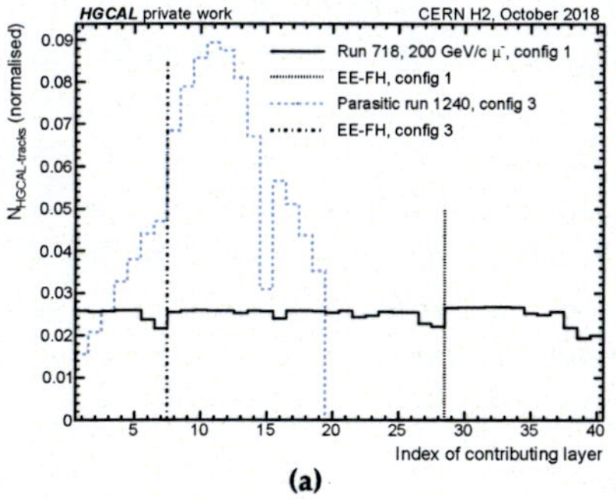

(a)

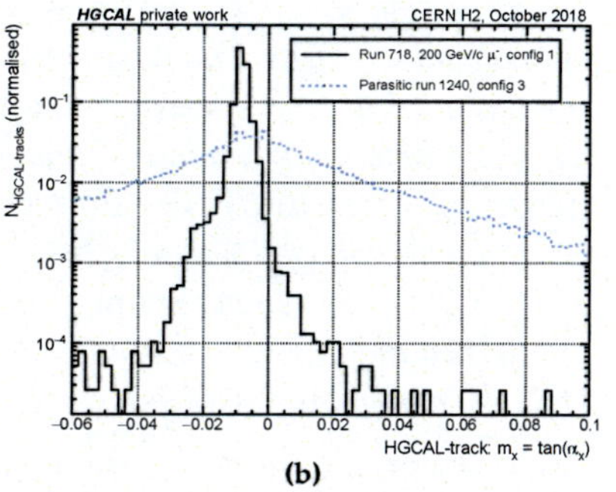

(b)

Fig. 9.5 **a** Distribution of the depth of hits contributing to the HGCAL-based MIP trajectories. **b** Distribution of the trajectory's horizontal direction of flight assessed in the prototype calorimeter's coordinate system. Results for runs with muon beam (*O1M* dataset) and for those with undefined, parasitic beam (*O3* dataset) are shown

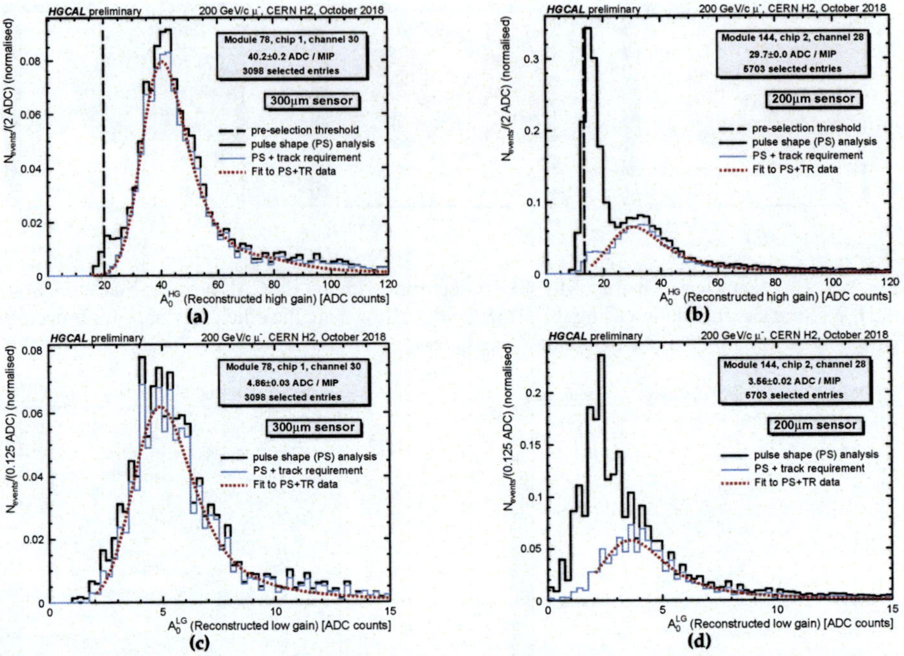

Fig. 9.6 Spectra of reconstructed MIP-induced signal amplitudes both **a**, **b** in high and **c**, **d** in low gain for two example readout channels due to incident 200 GeV/c muons. Distributions of MIP-selected cells are normalised to unity integral. The inclusive spectra where no selection is applied are scaled likewise. Module 144 consists of a sensor that has a nominal depletion thickness of 200 μm resulting in a lower response compared to the 300 μm sensors, such as in module 78

the reconstructed ADC spectra for two example channels before and after selection of MIP-induced hits. The noise contributions of these spectra are almost eliminated when requiring a hit to be part of a straight line trajectory inside the calorimeter. At the same time, only a minor number of physical hits is rejected. Equivalent energy spectra and therefore MIP calibration results *are obtained when applying the delay wire chamber-based hit selection*, cf. Fig. 9.7a. *The energy spectrum induced by MIPs is distinct even in the low gain ADC. Hence, the MIP energy calibration* is also *applied on the low gain directly, similarly to how it is anticipated for the final CMS HGCAL design*, in this thesis.

MIP Tracking with Parasitic Particle Beam

In the October 2018 beam test, due to the limited spread of the muon beam and the limited mobility of the calorimeter setup through the beam, only 31% of all cells could be calibrated individually with muons. The non-showering hadrons, in the *O1H* dataset, *could be used in addition to increase the number of calibrated cells since their energy deposition is remarkably close to the one from the muons, (...)*. *After selection of events with a MIP-like trajectory up to a certain depth, the noise rejection strategy is analogous to the analysis of the muon data. Unfortunately, given*

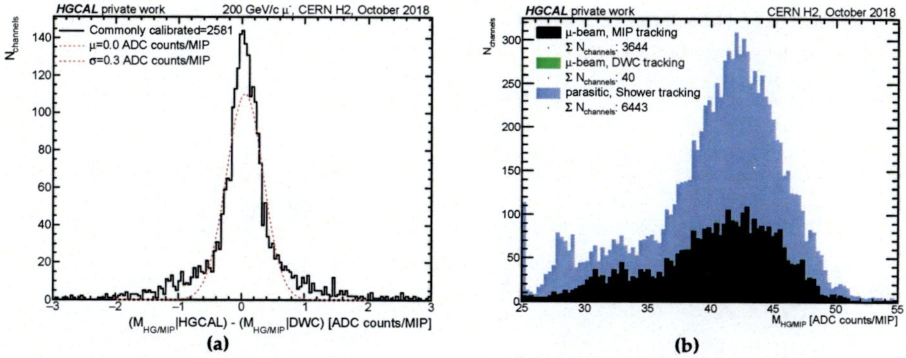

Fig. 9.7 **a** Distribution of differences in the MIP calibration constants when using the HGCAL prototype and the delay wire chambers as tracking devices. The spread amounts to 0.3 HG ADC counts/MIP and no systematic bias is obtained. **b** Summary of calibrated cells with MIPs in the October 2018 beam test campaign. The major fraction of diodes were calibrated exploiting the parasitic beam time together with the tracking of MIP signatures in the HGCAL prototype. Entries around 25 high gain ADC counts/MIP correspond to ~100 cases where the automated fitting procedure for the more than ten thousand channels in total has failed

the limited transverse size of the pion beam, cf. Fig. E.1f–h in the Appendix E, *less than ten cells per layer could be calibrated using non-showering pions. In addition, most of them had already been calibrated with muons.*

Instead, the *O3* dataset is used for the purpose of maximising the yield of calibrated channels. *Again, the core analysis and calibration strategy relies on the identification and selection of MIP-like trajectories* in the data. However, the trigger acceptance in the recording of this data was enlarged and the composition of the parasitic beam is not known. Therefore, the phase space of possible MIP trajectories inside the prototype calorimeter is much larger than in the *O1M* dataset. MIP-like trajectories could commence and disappear at any layer. Besides, they could have sizeable deviations from perpendicular orientations. The HGCAL-based "MIP Tracking" procedure would require significantly more computation time in its initial grid construction and would tend to fail identifying trajectories that start at random depths in the calorimeter. For this reason, a second, also preliminary, tracking algorithm was developed that combines calorimeter hits in order to form straight lines with arbitrary start and end point inside the calorimeter. It is designed to identify MIP-like trajectories within shower topologies, and thus it is named "Shower Tracking".

The procedure begins with the same hit selection like for the previously discussed "MIP Tracking". However, hits with sizeable signal sums in its direct neighbours as well as entire events with sizeable reconstructed signal sums in the full calorimeter are discarded a priori in order to reject signatures with too much activity in the calorimeter for which the designed Shower Tracking algorithm would otherwise be too computing intensive. Afterwards, the algorithm's core logic is divided into four steps. The sequence of these steps is iterated for all calorimeter layers starting from

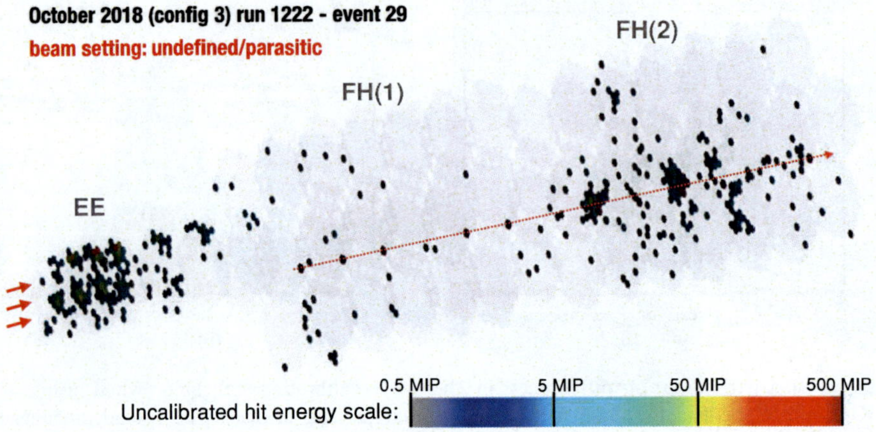

Fig. 9.8 Display of a recorded shower during the parasitic beam time in October 2018. The undefined beam enters the detector from the left-hand side. Hits associated to a MIP-like trajectory identified by the "Shower Tracking" algorithm are labelled by the dashed red line

the first layer and ending with the 17th one (19 functional calorimeter layers in the *O3* dataset). Again, the algorithm has been implemented as a CORRYVRECKAN module [14] as part of the work for this thesis.

1. The seeding of track candidates is done at the given starting layer together with the subsequent two downstream layers. For this, all possible combinations of hits in these layers are constructed. Afterwards, only those that are consistent with straight lines are kept.
2. The track candidates are iteratively extended to the downstream calorimeter layers. At most one hit for each layer is added to a given track if the straight line characteristic is maintained afterwards. Track candidates are duplicated if multiple hits in a layer are valid for the extension of the track.
3. Candidates that point to regions in the respective layer that are already traversed by a another track candidate are removed.
4. The extension of track candidates to downstream layers is stopped as soon as there are four layer gaps or when the last (i.e. 19th) layer is reached. A candidate is accepted as a valid track if it has hits from at least eight distinct layers (12 layers in FH compartment). The associated hits are removed from further association to other tracks.

Figure 9.8 depicts the event display of a shower recorded with parasitic beam for which a MIP-like trajectory in the FH layers was reconstructed. Closer investigation of Fig. 9.5a reveals that the distribution of contributing layers to such trajectories is biased towards the first FH layers. This may be interpreted as an artefact of an inefficient layer (number 15) in this setup which systematically introduces a gap to the trajectory construction. The reduced contribution by the first seven layers in the EE compartment, cf. Fig. 6.16b, can be explained by its smaller lateral coverage with

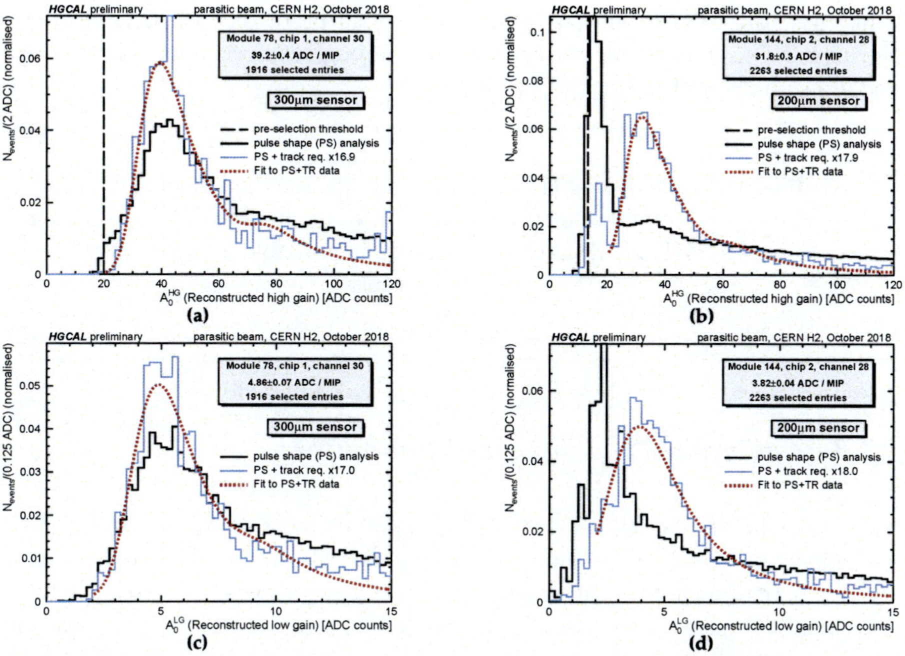

Fig. 9.9 **a**, **b** Spectrum of reconstructed MIP-induced signal amplitudes both in high- and **c**, **d** in low gain for the two readout channels of Fig. 9.6. Hits contributing to a MIP-like track in the data are selected. The inclusive spectra are normalised to unity integral. The selected spectra are scaled accordingly whereby an additional scaling, indicated in the legend, is applied for better visualisation

respect to the FH. Figure 9.5b shows that the angular distribution of the reconstructed trajectories is much broader than for the muon beam which is intuitive for MIP-like trajectories within shower topologies.

All hits with distances less than 1.5 cm to those trajectories are selected for the MIP calibration independent of their reconstructed high gain amplitude. *Figure 9.9 presents the reconstructed energy spectra with and without the track selection for the two previously exemplified channels. The spectra are purified after requiring a hit to be part of an identified track in the calorimeter.* However, the tracking algorithm appears to be inefficient as evident from the overall reduction of entries in the selected spectra compared to the inclusive ones. Detailed simulation of this test configuration with parasitic beam is beyond the scope of this work but could potentially help optimise the underlying "Shower Tracking" algorithm.

*More importantly, MIP calibration constants for another 54% of the cells could be derived from the additional dataset accumulated during the parasitic beam time, cf. Fig. 9.7*b. Only 40 out of more than 10000 cells could not be calibrated with the HGCAL-based tracking but were calibrated using the DWC-tracks for cell selection. Finally, it is emphasised that it is not claimed that the described HGCAL-based tracking algorithms are in any way optimal or superior to other currently evolving solutions. Still, this section provides experimental proof that purification of MIP signals in HGCAL can effectively be achieved through exploitation of its granularity and usage of its particle tracking capabilities.

> ✓ **PV-MIP.1** Exploiting HGCAL's granularity and the low enough noise level, the prototype calorimeter is usable for the tracking of MIPs therein.
>
> ✓ **PV-MIP.2** MIP-like signatures are identifiable in rather busy environments.

9.1.3 Calibration Stability and Uniformity

Except for the *M1* dataset, the bias voltage applied to the modules was not altered systematically during a measurement programme. Consequently, it is not surprising that variations in the high gain to MIP calibration constants ($M_{\text{HG/MIP}}$) derived individually for runs within the same experimental setup agree within the statistical precision. Nevertheless, discrepancies partially exceeding 10% between the different HGCAL prototype configurations have been observed. While for some modules, e.g. for module 76 in Fig. 9.10a, changes in the MIP calibration constants are consistent with statistical precision, substantial systematic differences have been identified for others, e.g. for module 77 in Fig. 9.10b. Data on the effective bias voltages and leakage currents for each module during the tests would have been beneficial to the understanding of the observed discrepancies. Yet, they are not available. For future beam tests, it is recommended to iterate the calibration with MIPs when modifications to the prototype calorimeter setup have been made.

For the objectives of the analysis with the given beam test data in Sect. 10, it is assumed that such variations in the readout channels' responses to MIPs should have a lower impact on the reconstruction quality than the effect of leaving a substantial set of channels uncalibrated. Therefore, the calibration results from the *O3* dataset extend the calibration database for reconstruction of the *O1** datasets whenever results from the *O1M* dataset are missing. Figure F.1 in Appendix F shows the location of the cells for which the MIP calibration constants could be derived with either method. Since no dedicated runs with muons were taken in the *O2* programme, the MIP calibration results from the other two October 2018 configurations are used for the reconstruction of its data.

The summary of the derived high gain MIP calibration constants for the *JUN* dataset is given in Fig. 9.11a. For channels on the same module, the variation in the response to MIPs is dominated by the systematic differences between the readout chips. The variation of calibration constants for channels on the same chip is, in fact, 3% and in particular relatively small compared to the chip-chip variations. Thus, the usage of chip-averaged calibration constants instead of module-averaged ones

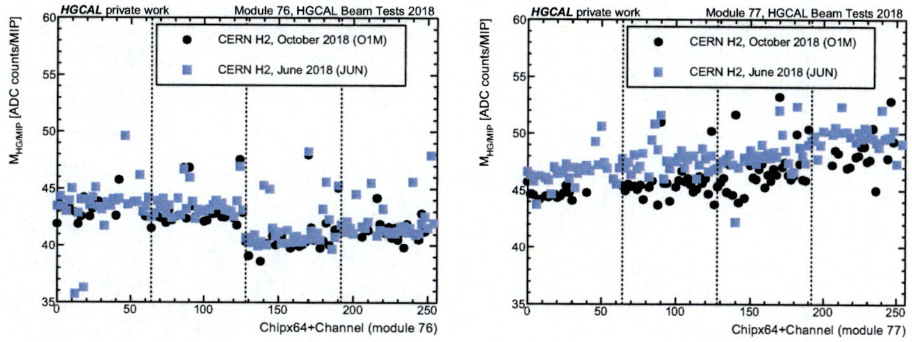

(a) **(b)**

Fig. 9.10 Comparison of the calibrated electric response in high gain between the beam test periods in June and in October 2018 (*JUN* and *O1M* datasets) **a** for module 76 and **b** module 77. Statistical error bars are smaller than the marker size. Similar discrepancies like they are shown for module 77 are present for the comparison to the *O3* dataset

for channels that could not be calibrated individually is preferred and is used in the following. Moreover, closer analysis based on the values depicted in Fig. 9.11a reveals that some chips and entire modules exhibit a more than 10% lower response to MIPs than the average from all chips. Those instances can be matched to the regions of low MIP efficiency, cf. chip 1 on module 84 in Fig. 8.13e or the full module 51 in Fig. 8.13b. The specific findings derived from this dataset are supported by analogous findings for the *O1M* dataset that is presented in Ref. [1], again for the *JUN* dataset. Therein, the summary of low gain MIP calibration constants is missing which is supplied in Fig. 9.11b. Contrary to the finding for the high gain, the spread in low gain calibration constants within a given chip is rather comparable to the variation between them. It must be noted that this fact might not necessarily be a physical effect but rather be due to the low gain's limited precision to reconstruct MIP-induced signals which leads to less sensitivity to such differences in the response.

Further analysis reveals the systematical lower response to MIPs for 200 μm than for 300 μm sensors, shown in Fig. 9.12a. *Again* [15], *the dependence on the cell's geometry is experimentally confirmed as shown in Fig.* 9.12b. The latter result is illustrated for diodes that are attached to the same chip in order to exclude the aforementioned sizeable chip-to-chip variations. The increase of the reconstructed MIP signal for smaller diode geometries and with it smaller detector capacitances can be explained by increased charge transfer efficiencies to the amplifiers [16].

Similarly, the noise is expected to decrease with smaller detector capacitances. In account of both effects, the signal-to-noise ratio is reported in Fig. 9.13a, b for different diode geometries. As can be seen, the ratio for the calibration pads is visibly enhanced when compared to the one for full hexagonal diodes. *In general, modules equipped with 300 μm sensors have S/N≈7–8 in high gain, (...) , and S/N≈4–5 in low gain, (...) . By comparison modules equipped with 200 μm sensors have S/N≈5 in high gain, (...) and S/N≈3 in low gain, (...) .*

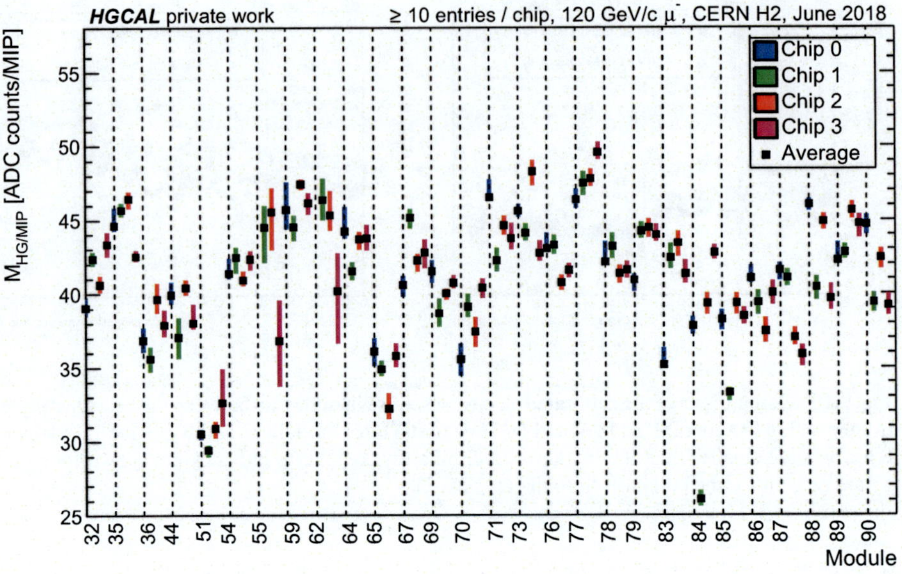

(a)

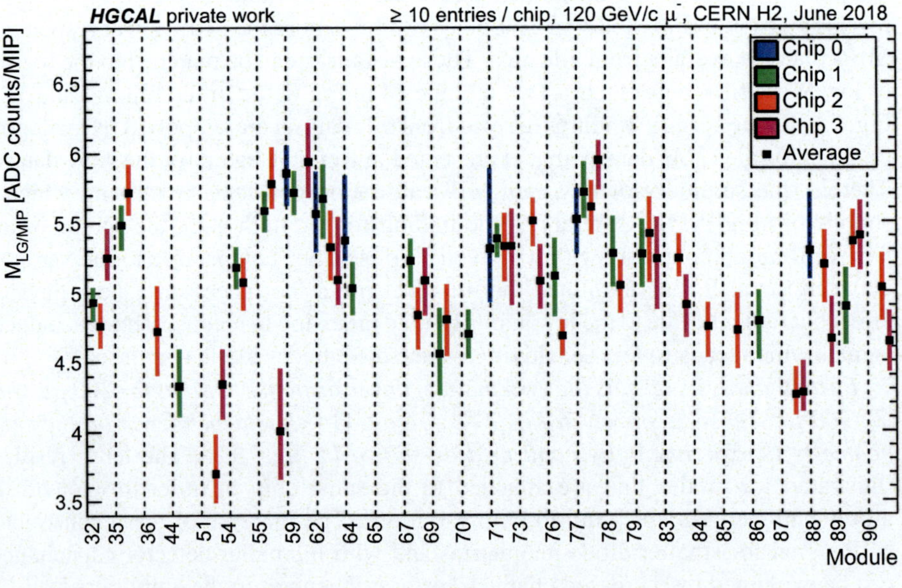

(b)

Fig. 9.11 Per-module distribution of calibrated **a** high gain and **b** low gain MIP calibration constants for the June 2018 beam test of the silicon electromagnetic calorimeter prototype (*JUN* dataset). There, all tested modules consisted of 300 μm thick silicon sensors. The bars correspond the 16% - 84% quantile range estimates of at least ten calibrated channels read out by the same chip. Only channels connected to full hexagonal sensor diodes are considered. The result for the *O1M* dataset is presented in Ref. [1]

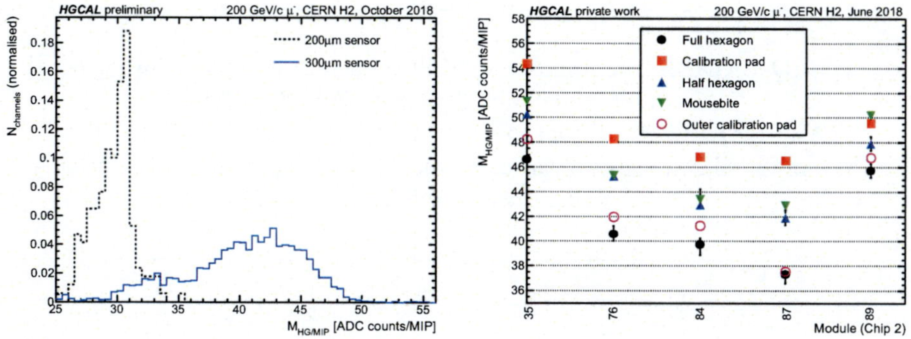

(a) (b)

Fig. 9.12 **a** Distribution of calibrated high gain MIP calibration constants including only full hexagonal diodes with different sensor thicknesses measured with 200 GeV/c muons. **b** Comparison of the response for different diode geometries

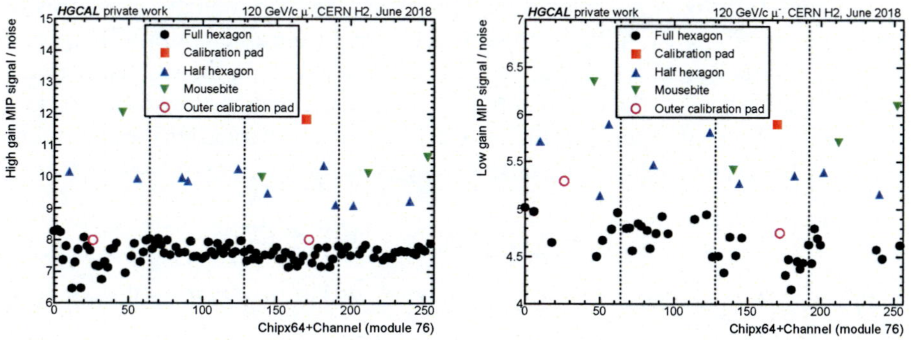

(a) (b)

Fig. 9.13 Signal due to minimum ionising particles compared to the noise level shown for module 76 **a** in high gain and **b** low gain and for different diode geometries. Noise is estimated after subtraction of common mode noise, analogous to Fig. 8.11, except for the calibration pad for which no common mode noise estimate is computed. The result for the *O1M* dataset is presented in Ref. [1]

In conclusion, calorimeter data taken with muon and undefined parasitic beams have successfully been analysed for the energy calibration with MIPs for roughly 85% of all silicon cells in the most recent beam test in October 2018. The remaining 15% use average calibration constants obtained from cells on the same readout chip. *The resulting variation of these calibration constants as well as the signal-over-noise ratio are consistent with the targeted HGCAL design values* [17].

✓ **MQ-CALIB.1** The response to MIPs does not vary more than 5% for all cells read out by the same SKIROC2-CMS.

✓ **MQ-CALIB.2** The signal-to-noise ratio (*SN*) scales according to the area of the silicon pads. In particular, the calibration pads have significantly higher SN than full hexagonal pads.

9.2 Intercalibration of the Gains

As a reminder, the SKIROC2-CMS ASIC features two gain stages (high and low) as well as one time-over-threshold (*TOT*) measurement that are used in the inference of hit energies in the HGCAL calorimeter prototypes. For high enough energy densities, the signal amplitude in a given gain saturates meaning the assumption of proportionality to the true signal size is not valid. The implications on the reconstruction are formulated in Eq. 7.8. Except for $M_{\mathrm{HG/MIP}}$, whose calibration was discussed in the previous section, all parameters therein can be derived by exploiting the approximate linearity between the gains within certain signal ranges, cf. Eqs. 9.3 and 9.4.

$$A_0^{HG} = A_0^{LG} \cdot m_{\mathrm{HG/LG}} \ , \ \ A_0^{HG} \leq TP_{\mathrm{HG}} \tag{9.3}$$

$$A_0^{LG} = (TOT - TOT_{\mathrm{offset}}) \cdot m_{\mathrm{LG/TOT}} \ , \ \ A_0^{LG} \leq TP_{\mathrm{LG}} \text{ and } TOT \text{ sufficiently large} \tag{9.4}$$

$m_{\mathrm{HG/LG}}$ and $m_{\mathrm{LG/TOT}}$ are slope constants, TP_{HG} and TP_{LG} are so-called turning points and TOT_{offset} is an offset in the low gain - TOT intercalibration. Turning points refer to the signal size threshold at which the given gain is identified to saturate and the next possible is to be used. Note that the precise knowledge of deposited energy, or rather charge, in the sensors is not necessary for the applications of Eqs. 9.3 and 9.4. Therefore, all occurring constants may be deduced in situ from beam test data.

Since this calibration was not performed as part of the work for this thesis, the procedure is only explained briefly for completeness. Sections 9.2.1 and 9.2.2 outline the methods and report on the resulting calibration constants. They are not intended to provide an extensive documentation of the procedure. More details, especially on the stability and uniformity of the gain intercalibration will be provided in the publication dedicated to the prototype modules' construction and calibration aspects [1]. Concluding the calibration necessary for the reconstruction of hit energies, an original validation of the combined gain and MIP calibration is finally presented in Sect. 9.2.3. This validation is uniquely carried out and presented in this thesis.

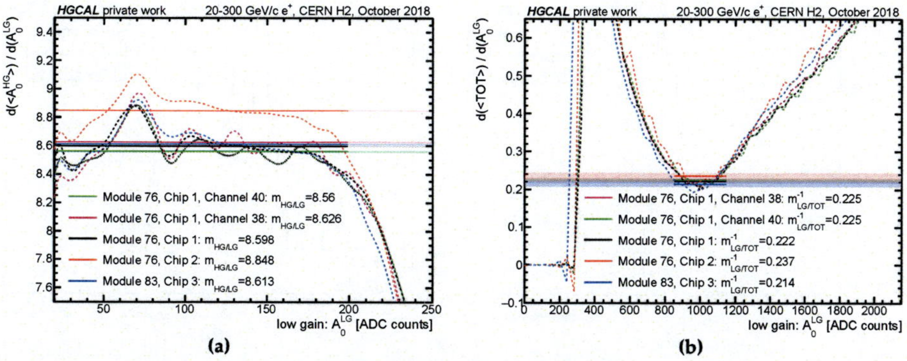

Fig. 9.14 **a** Illustration of the spline method applied for the intercalibration of the high and low gain, **b** the low gain and TOT, respectively. These explanatory graphics are made with the *O1E* dataset, whereas all available data from a given setup were used in the actual calibration. The indicated bands symbolise the 68% uncertainty which is purely statistical and does not include systematic effects

9.2.1 High Versus Low Gain

Equation 9.3 states the expected linear relationship between the reconstructed signal amplitudes in high and in low gain. Inference of the slope constants $m_{\mathrm{HG/LG}}$ for each channel is performed using a spline fit-based method. In this connection, averages of reconstructed high gain amplitudes are computed in small ranges of reconstructed amplitudes in low gain, see the profiles in Fig. 7.4a as an example. Then, splines serve as estimates for the first derivative ("d") of these profiles. Representative graphs of deduced derivatives with respect to the signal in low gain are shown in Fig. 9.14a. Their level on the left hand side, i.e. for small signal sizes before the high gain saturates around 200 low gain ADC counts, determines $m_{\mathrm{HG/LG}}$ for the calibrated entity. Turning points are inferred as the point when the derivative's derivative estimate is not compatible with zero anymore.

Intercalibration of the high and low gain does not require high energy densities in a given diode. Therefore, approximately 85% of all readout channels could eventually be calibrated with a relatively narrow beam of showering particles in the *O1E* and *O1H* datasets by this means. Figure F.2 in Appendix F shows the location of all the cells for which the high and the low gain could be intercalibrated. One finds in Fig. 9.15a that the mean slope amounts to $m_{\mathrm{HG/LG}} \approx 8.5$ whereas the RMS across all calibrated channels amounts to 0.2. The main variation of these constants are dominated by chip-to-chip differences while the slopes for channels on the same ROC are rather similar, cf. Fig. 9.14a. In optimisation studies, it was observed that the slope values were affected by almost 10% during the evolution of the pulse parameterisation, cf. Eq. 7.7.

The distribution of the turning points is presented by Fig. 9.15b. A precise assessment of the turning points is not crucial. A possible underestimation is not considered

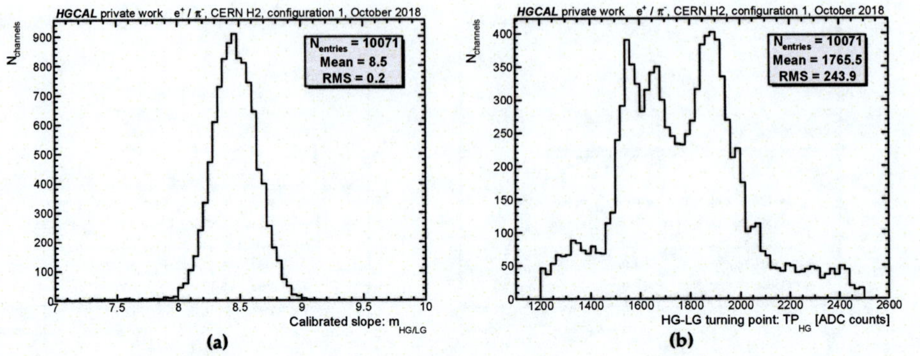

Fig. 9.15 Result of the intercalibration of the high and low gain ADC. **a** The slope ($m_{\mathrm{HG/LG}}$) and turning points (TP_{HG}) have been derived for roughly 85% of all channels with the *O1E* and *O1H* datasets

to have a major impact on the calorimetric measurements because of the low gain's good precision that is even sufficient for the detection of MIPs.

Calibration constants for randomly chosen modules were also re-derived from data taken with charge injection in the lab. The results are consistent with their in situ assessment using beam test data.

> ✓ **MQ-CALIB.3** The high and low gains in the SKIROC2-CMS exhibit the designed linearity in their signal amplitudes prior to the saturation of the high gain.

9.2.2 *Low Gain Versus Time-over-Threshold*

Equation 9.4 states the relationship between the reconstructed signal amplitude in low gain and the recorded TOT. Analogous to the high to low gain intercalibration strategy, a spline fit-based method is applied to profiles such as the ones in Fig. 7.4b. They determine the inverse of the slope $m_{\mathrm{LG/TOT}}$ and the turning point TP_{LG}. Unlike the signal sampling in the two gains, the TOT is a threshold-based measurement which is only triggered for high enough signals. For this reason, the TOT is not proportional to signal sizes for smaller charge depositions and a TOT offset $\mathrm{TOT}_{\mathrm{offset}}$ needs to be included in the relationship to the low gain.

Compared to the section before, there are two particular challenges associated to the beam test in situ calibration of the low gain to the TOT. First, the region of linearity is typically quite narrow. Figure 9.14b shows one example. Often, one

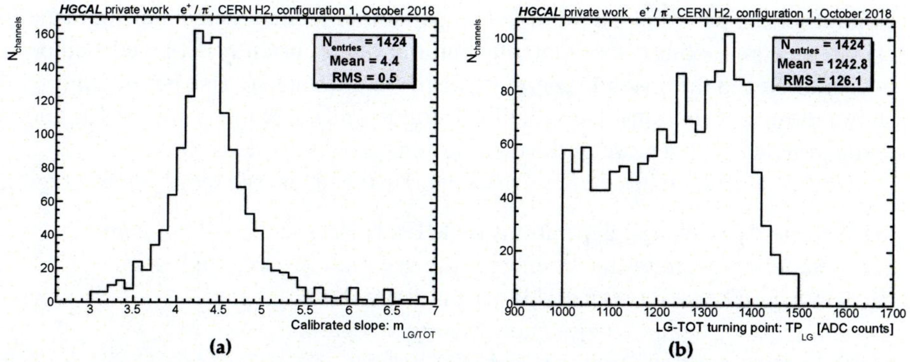

Fig. 9.16 Result of the intercalibration of the low gain ADC the time-over-threshold. **a** The slope ($m_{\mathrm{LG/TOT}}$) and **b** turning points (TP_{LG}) could be derived for only 12% of all channels with the *O1E* and *O1H* datasets due to the limited transverse beam spread with high-energetic particle showers

cannot even be sure that the minimum region in the derivatives that defines $m_{\mathrm{LG/TOT}}$ really corresponds to a real region of linearity.

In addition, high charge depositions and with it high energy densities are required to sufficiently sample the profile close to the region where the low gain saturates. Due to the limited time in the beam tests and owing to the fact that only a fraction of active diodes were exposed to dense particle showers, only 12% of all channels could be calibrated individually in the *O1E* and *O1H* datasets. Other datasets have similar deficits. Figure F.3 in Appendix F shows the location of the cells for which the low gain and the TOT could be intercalibrated. Especially the modules placed in the outer regions of the FH layers, cf. Fig. 6.16, lack the dedicated low gain to TOT intercalibration and would generally have profited from beam test data taken with wide particle beams. Figure 9.16 presents the distributions of the obtained slopes and turning points.

Again, the calibration results were validated using charge injection in the laboratory. Slopes are in good agreement to their beam test in situ counterpart while the offsets tend to differ. The latter difference might be explained by different environmental conditions. Investigations in this regard had not come to a conclusion at the time of writing this thesis. Furthermore, more sophisticated parameterisations for the low gain to TOT relationship, different from Eq. 9.4, that would account for the TOT's turn-on behaviour were being evaluated at the time of writing this thesis.

9.2.3 Reconstructed Energy Spectrum

As mentioned in the previous sections, not all cells could be calibrated in situ using the beam test data. In those cases where calibrations constants are missing, chip- or in the worst case module-averaged values are used. At the end, the full in situ energy calibration can be validated through comparison of reconstructed and simulated hit energy spectra. By this means, potential failures of the currently deployed method can be identified and the absolute energy scale in simulation may also be set. Figure 9.17 shows these distributions in double logarithmic scale separately for the EE and FH compartments. The feature of the distributions are discussed next.

For this purpose, it is beneficial to categorise the hit energy range into six regions:

I Sensitivity to energy depositions by MIPs and noise using high gain.
II Remaining range of the hit energy inferred mainly using high gain.
III Transition between high and low gain usage.
IV Inference of the hit energy mainly using low gain.
V Transition between low gain and TOT usage.
VI Inference of the hit energy mainly using TOT.

Charged pions are particularly useful for this validation. They enable the determination of the absolute energy scale in simulation by focussing on the MIP peak in region I. The most probable energy in keV deposited by muons as MIPs is inferred from GEANT4 simulation and amounts to approximately 85.5 keV for 300 μm thick silicon. This estimated conversion factor is tuned further by matching the simulated hit energy spectrum to the one obtained with the beam test data, cf. Eq. 9.5.

$$1\,\text{MIP}(300\,\mu\text{m Si}) = 85.5\,\text{keV} \rightarrow 10^{0.04} \cdot 85.5\,\text{keV} \approx 93.7\,\text{keV} \tag{9.5}$$

This global scaling of the simulated energy depositions is done for convenience and does not affect the performance validation based on relative measurements, especially the ultimate prototype calorimeter's energy resolution in Sects. 10.2.2 and 10.3.4. At this point, it is remarked that any Gaussian convolution applied to a Landau function, like for the MIP signal spectrum in Eq. 9.1, usually shifts its maximum to higher values. Hence, the best knowledge on the signal-to-noise should be injected into the simulated MIP energy spectra prior to the quantification of the MIP-to-energy scale, which was not done in this initial assessment.

After definition of the absolute MIP energy scale in simulation, a remarkable agreement between the reconstructed and simulated hit energy spectrum covering the full dynamic range is obtained. The drop in the hit rates occurring just at highest hit energies is present in the data overall validating the usage of the TOT in the measurement of high energy densities in the showers. As expected, a remaining deficit is apparent in region V corresponding to the transition between the low gain and the TOT. It is noted that the region in the low gain range where the intercalibration to the TOT is applied might be too narrow after all to render reliable calibration results.

✓ **MQ-CALIB.4** At high energy densities, the time-over-threshold is usable to reconstruct the deposited energy and to effectively cover the full range of expected shower energy densities.

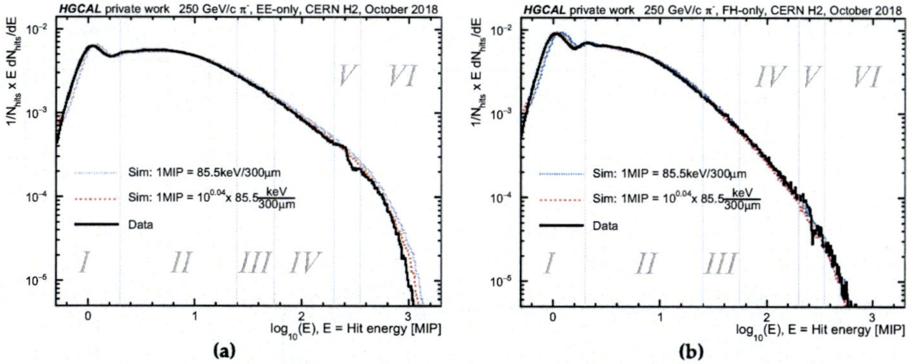

Fig. 9.17 Spectrum of reconstructed hit energy for hits **a** in the EE and **b** in the FH compartment measured with 250 GeV/c charged pions after calibration with MIPs and intercalibration of the gains. Only fully calibrated cells, i.e. those for which all gain and the MIP calibration constants could be derived, are considered. Simulated hit energies are scaled to equivalent energy depositions by MIPs. Note the double logarithmic scale. Regions I-VI label the approximate ranges of the different gain domains

9.3 Calibration and Resolution of the Time-of-Arrival

The precise timing measurement of energy depositions using HGCAL with resolutions down to $\mathcal{O}(10\,\text{ps})$ is foreseen in the final design and would represent a substantial novelty in calorimetry. Currently, the SKIROC2-CMS ASIC mounted on the beam test modules already incorporates a prototype functionality for this purpose: For high enough signal sizes, the charging of two capacitors is triggered and lasts until the next rising or falling, respectively, edges of the clock signal. Afterwards, the charge is digitised and recorded as TOA_{rise} and TOA_{fall}. The principle is sketched in the bottom half of Fig. 9.18.

Reconstruction of physical hit timestamps is explained in Sect. 7.2.5. As illustrated in Fig. 7.5a therein, it is sufficient to use one of the two definitions of the time-of-arrival (*TOA*), namely the TOA_{rise} in this thesis. Nonlinearities in the TOA response, cf. Eq. 7.12, and hit energy-dependent timewalk, cf. Eq. 7.13, pose challenges in its calibration.

Section 9.3.1 describes the calibration strategy using an external time reference detector. In the beam tests, this reference time is provided by the MCPs. It is hypothesised that the MIP timing detector upgrade for the CMS Phase 2 detector [18] could serve for this task in the final HGCAL installation within the CMS experiment. In the following, the proof of principle in the beam tests is exemplified for one readout channel and repeated for a few others. A systematic calibration of an extensive set of channels for eventual timing measurements of hadronic showers is not performed in this thesis. The presented strategy is complementary to the default method that is currently being applied by the CMS HGCAL System Test group: The default method is based on the expectation of uniformly distributed true arrival times of

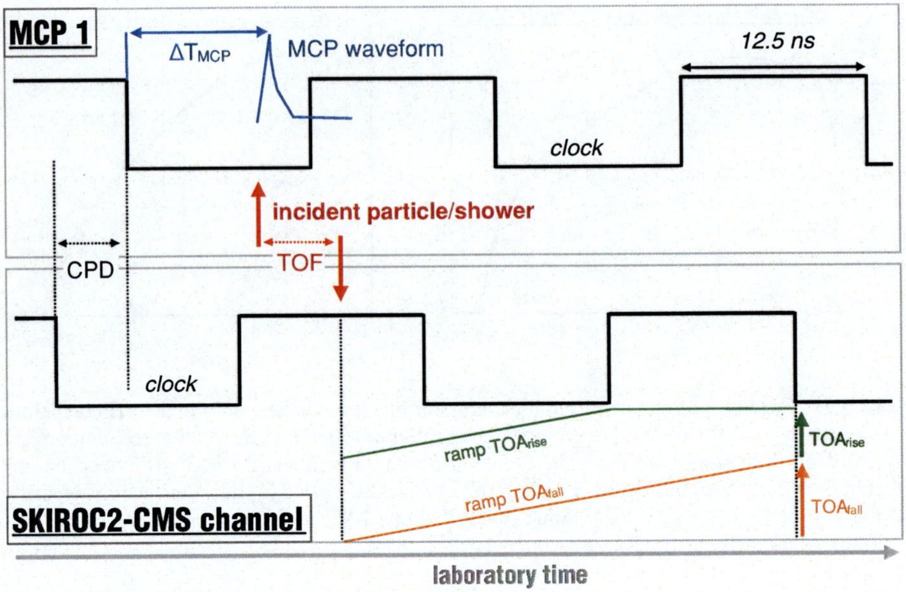

Fig. 9.18 Scheme of the different timing measurements used in the calibration and performance assessment of the TOA. *TOF* stands for "time of flight" while *CPD* is an abbreviation for "clock phase difference"

particles impinging on the calorimeter within a clock cycle, whereas the calibration with an external time reference as applied in this thesis, does not rely on such beam test-specific conditions.

Section 9.3.2 comprises a report on the obtained timing resolutions comparing the reconstructed TOA timestamp to the MCP's as well as pairwise comparisons between the HGCAL prototype readout channels. The studies qualify the TOA functionality and demonstrate that resolutions close to the SKIROC2-CMS's design value of 50 ps for high enough signal sizes are achievable.

9.3.1 Method Using an External Time Reference

The goal of the now presented timing calibration is the mapping of the TOA measurement and the reconstructed hit energy to the timestamp derived from the MCP. In this connection, it is assumed that the physical calorimeter hit timestamps (T) and the MCP timestamps (ΔT_{MCP}) differ by channel-dependent but, at least in first approximation, fixed constants. Apart from the signal path length's contribution Δl_0, cf. Eq. 7.11, those constants can be due to the time-of-flight (TOF_i) of the shower evolution from the MCP to the calorimeter cell i and the clock phase difference (CPD_i)

between the respective readout channel and the v1742 digitiser, cf. Sect. 6.4.3. Equation 9.6 describes the relationship between T and T_{MCP}.

$$T = T(TOA) = \Delta T_{MCP} + TOF_i + CPD_i \ , \quad TOF_i = \text{constant}_i, \ CPD_i = \text{constant}_i \tag{9.6}$$

With these considerations in mind, the expected relationship between the TOA, the reconstructed hit energy and the MCP timestamp relation, inserting the TOA reconstruction formula from Eq. 7.11, can be written symbolically like it is stated in Eq. 9.7.

$$\Delta T_{MCP} = f_{TOA}\left(\frac{TOA - TOA^{min}}{TOA^{max} - TOA^{min}}\right) + f_{TW}(E) + \text{constant}_i \tag{9.7}$$

Timing measurements on full particle showers would require global synchronisation of all offsets for all readout channels, implying the derivation of the constants for all channels. However, inference of the constants is not aimed for in this thesis because the bare demonstration of the SKIROC2-CMS timing capabilities does not rely on it.

Since it is the only dataset for which MCPs were available during the beam tests with a fully equipped 28-layer EE calorimeter prototype, the *O2* dataset is analysed in this study. The analysis makes use of only one MCP, namely MCP 1 which was placed more centrally in the trigger acceptance than the other one. Events are selected for which the reconstructed MCP waveform amplitudes exceed 100 ADC counts. Enforcing this requirement, the resolution of the MCP reference timestamps should be below 150 ps, cf. Fig. 7.15. Furthermore, a large range of hit energies should be covered. Therefore, the focus in this assessment is put on channels that tend to also be exposed to high energy densities, i.e. those that were located in the beam at intermediate shower depths in the EE compartment close to the maximum of electromagnetic showers.

The calibration procedure consists of four steps that, in principle, would need to be repeated for each readout channel individually.

1. Due to the clock periodicity, the measured relation between the TOA and the reference time from the MCP exhibits a discontinuity at a random (yet fixed) location. This discontinuity is first identified and corrected for. This step is shown in Fig. 9.19a. In this example, the entries in the top right corner are assigned to the previous clock cycle and thus a 25 ns offset is subtracted from ΔT_{MCP} for them.
2. The timewalk is evaluated. For this purpose, profiles of the TOA with respect to the period corrected MCP timestamp are computed for various hit energy intervals like it is illustrated in Fig. 9.19b. Note that the chosen ranges are not optimised. Especially, they do not contain the same number of entries.
3. Afterwards, the timewalk correction is quantified using the mean difference of these profiles with respect to a reference one. In this example, this reference is chosen to be the one constructed from entries with hit energies between 800–900 MIP. Then, the parameterisation for the function f_{TW} in Eq. 7.13 is fitted to

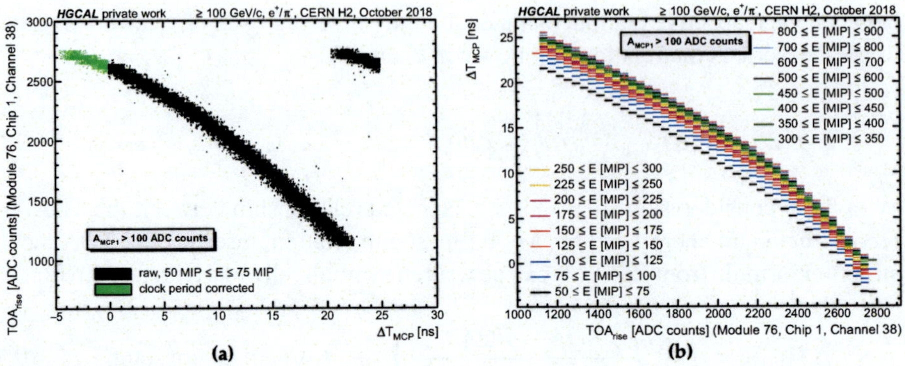

Fig. 9.19 **a** Step 1: Illustration of the clock period correction for the reference time measurement with the MCP. **b** Step 2: Inference of the timewalk by construction of the relation of the TOA to the reference time for different intervals in reconstructed hit energies. The index i for each hit is not explicitly written

the resulting differences yielding a continuous quantification of the timewalk, cf. Fig. 9.20a.

4. The timewalk is subtracted from ΔT_{MCP}. Then, the TOA profile with respect to that (by now double) corrected MCP timestamp including all hit energies is computed. In addition, the minimum and maximum TOA readings ($TOA_{\mathrm{min/max}}$) for TOA normalisation to [0,1] are retrieved. As the last step, the function f_{TOA}, according to Eq. 7.12, is fitted to the data points, cf. Fig. 9.20b. With it, the TOA calibration of the given channel is concluded.

Empiric shortcomings of this calibration method are observed in the example accompanying the procedural explanations. For instance, one finds in the particular case shown in Fig. 9.20b that $f_{\mathrm{TOA}}(0) \approx 25.8$ ns and $f_{\mathrm{TOA}}(1) \approx -1.4$ ns which deviate from the values $f_{\mathrm{TOA}}(0) = 25$ ns and $f_{\mathrm{TOA}}(1) = 0$ that should ideally be obtained. This inconsistency might be explained by variations in the TOA pedestal and in its range with time or with temperature in the considered dataset. Improvements of this calibration method would be possible through TOA normalisation separately for different time frames, e.g. for each run, and through the incorporation of constraints to the parameterisation of f_{TOA}. Nonetheless, as it is shown in the next section, despite these deficits, this calibration still yields timing resolutions close to the SKIROC2-CMS's design value of 50 ps.

✓ **MQ-TOA.1** The TOA can be calibrated using an external time reference like it could be done during HGCAL's operation in the Phase 2 CMS detector.

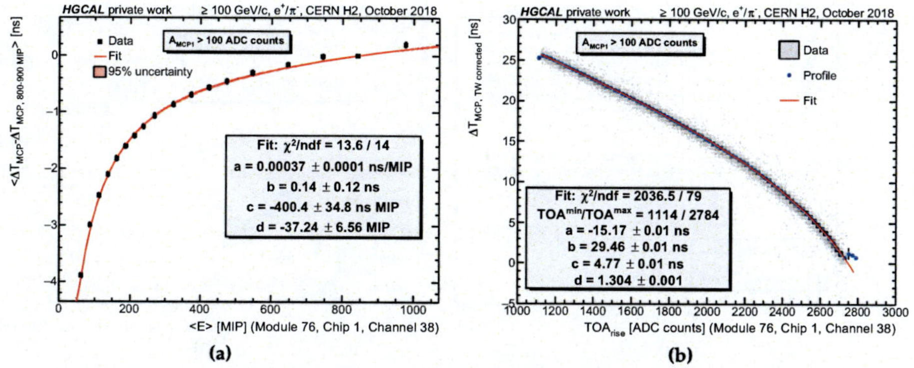

Fig. 9.20 **a** Step 3: Quantification of the timewalk correction. The fitted parameters are defined in Eq. 7.13. **b** Step 4: Relation between the TOA and the timewalk-corrected time. The fitted parameters are defined in Eq. 7.12

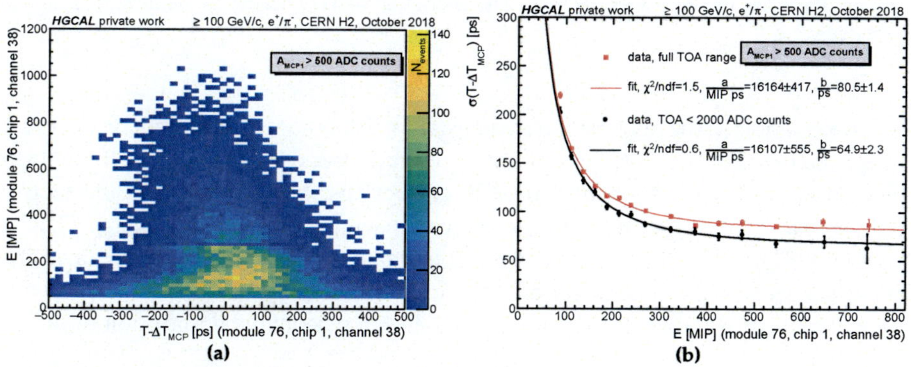

Fig. 9.21 Timing resolution for one exemplary readout channel: **a** Distribution of differences between the calibrated TOA timestamp T and the MCP reference time for different reconstructed hit energies. **b** Combined MCP and TOA resolution as a function of the hit energy for TOA values covering the the full range and for those restricted to the linear region in the $TOA - \Delta T_{\mathrm{MCP}}$ relationship. The data are well described by the parameterisation in Eq. 9.8

9.3.2 *Timing Resolution of Single Calorimeter Cells*

The initial approach in the quantification of the TOA timing performance is based on the difference between the calibrated TOA timestamp T to the time measured by the MCP. After iterating this assessment for numerous events, the spread of the resulting distribution is used as an estimate for the combined TOA and MCP timing resolution. In order to minimise the contribution by the MCP, the analysis for evaluation of the timing resolution is restricted to events in the *O2* dataset for which the reconstructed MCP waveform amplitude exceeds 500 ADC counts. This selection corresponds to keeping its timing precision below 32 ps, cf. Fig. 7.15, in this evaluation.

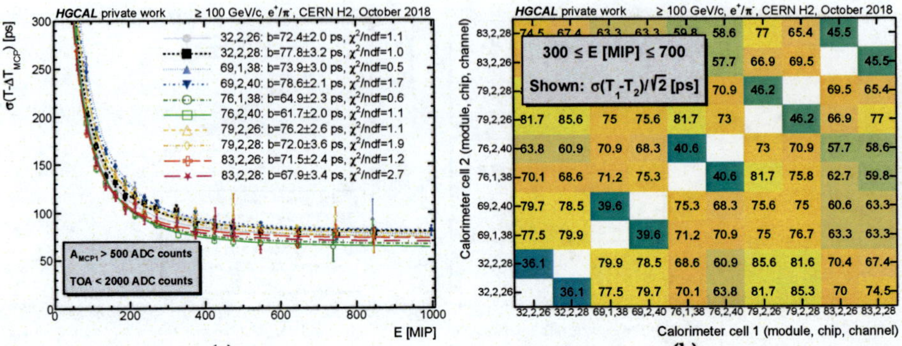

(a) **(b)**

Fig. 9.22 **a** Combined MCP and TOA resolution as a function of the reconstructed hit energy constrained to TOA values in the linear region in the $TOA_{\mathrm{rise}} - \Delta T_{\mathrm{MCP}}$ relationship for central cells (legend numbering scheme: module,chip,channel) in the EE compartment using the *O2* dataset. **b** Channel-to-channel timing resolution estimates, which are not affected by the MCP's finite resolution, for a specific hit energy interval. The restriction of TOA values to the linear region is not applied

An example of the distribution of timestamp differences for different hit energies is depicted in Fig. 9.21a. The distribution is centred around $T - \Delta T_{\mathrm{MCP}} = 0$ supporting the successful application of timewalk corrections. More relevant is the fact that the spread of the time differences decreases with increasing hit energy. The red curve in Fig. 9.21b illustrates the spread estimates ($\sigma\,(T - \Delta T_{\mathrm{MCP}})$) obtained from Gaussian fits to the timestamp difference distributions in different energy ranges. The parameterisation stated in Eq. 9.8 provides a reasonable description of the data points.

$$\sigma(T - \Delta T_{\mathrm{MCP}}) = \left(a^2 \cdot E^{-2} + b^2\right)^{0.5} \tag{9.8}$$

The constant term in this example reaches $b = 80.5$ ps and can be lowered further when constraining the TOA to the linear region in the $TOA - \Delta T_{\mathrm{MCP}}$ relationship. The linear region in the discussed example corresponds to TOA values below 2000 ADC counts, cf. Fig. 9.19a. Therein, the proportionality of one TOA bin to a physical time range is constant. Hence, it is not surprising to obtain an improved (combined) precision limit of $b = 65$ ps in that case. Assuming a quadratic contribution of the MCP time resolution of 32 ps to the combined resolution, the hereby achieved SKIROC2-CMS timing precision would amount to about 57 ps for large energy deposits which is already close to the claimed design value. Similar conclusions can be drawn for the other calibrated channels, cf. Fig. 9.22a.

Besides that, another complementary approach towards inference of the TOA timing resolution is pursued. It is based on the comparison of reconstructed hit timestamps between different readout channels. The advantage of this method is that it does not rely on the reference timing from the MCP and hence it is not affected by the MCP's finite time resolution. Again, the spread of the distribution of timestamp differences is assessed, this time pairwise between calorimeter cells. Thereafter, the corresponding value can be divided by $\sqrt{2}$ yielding an alternative estimate of the per-cell's timing resolution. Hereby, it is assumed that the timing precision of two cells is equal and that measurements between the readout channels are uncorrelated. The results for the investigated channels in this thesis are presented in Fig. 9.22b.

The, by this means estimated, resolutions for pairs of neighbouring calorimeter cells on the same modules is below 50 ps for the investigated examples. This resolution is significantly lower compared to pairs of readout channels from different modules. While the exact reason for this behaviour needs further assessment, there are at least three possible causes that could partially lead to this effect:

- First, it is noted that calorimeter cells on the same module are located at the same shower depth and therefore should record comparable signal sizes on average. With it, the timewalk correction for both channels is rather similar and does not introduce extensive variation due to slight calibration deficits.
- The assumption that TOF from the MCP to a given calorimeter cell is a constant might not be fully appropriate. Although the speed of each shower particle is in good approximation equal to the speed of light ($c \approx 30\,\mathrm{cm\,ns^{-1}}$), the TOF in Eq. 9.6 is not determined by the first shower particle in a given calorimeter cell but by the shower fragment that leads to substantial energy depositions in there to trigger the TOA measurement. Due to the longitudinal segmentation, it is imaginable that there are cases for which this threshold is just exceeded after the passing of backscattered shower particles from the downstream layers. For a layer distance of 1.5 cm and simplifying particle reactions as instantaneous, the hereby introduced delay of TOF would be 100 ps between neighbouring layers. Similar observations have been made within preliminary shower simulation studies of the October 2018 HGCAL beam test prototype. By contrast, backscattering of shower particles in the transverse direction is not significant and thus a TOF variation due to backscattering for cells on the same module is not as significant.
- Jitter in the clock distribution to the modules could cause variations in *CPD* that would affect the comparison between channels on different modules but not for cells on the same one.

In summary, as it has been discussed, there are clear ideas on how to further enhance this TOA calibration approach. Also, possible explanations on the differences between the obtained timing resolution estimates have been given. The presented TOA timing resolution with respect to the external time reference is well below 80 ps for the highest hit energies for all investigated channels including the MCP reference time resolution of ~30 ps. There are plausible indications that these values are rather upper limits and further systematic effects should be corrected for to reach the SKIROC2-CMS's design value of 50 ps. Despite of that, the timing functionality of the SKIROC2-CMS ASIC in beam conditions may be considered as experimentally confirmed. More detailed studies on other approaches for the beam test-specific situation were ongoing during writing of this thesis. As an outlook, the accumulated dataset should allow for measurements in the mid-term future on the time evolution of hadronic showers with unprecedented timing precision in calorimetry.

✓ **MQ-TOA.2** The timing resolution provided by the prototype SKIROC2-CMS ASIC is close to its design value of 50 ps for the highest signal sizes.

References

1. Acar B et al (2021) Construction and commissioning of CMS CE prototype silicon modules. JINST 16:T04002. https://doi.org/10.1088/1748-0221/16/04/T04002
2. Landau L (1944) On the energy loss of fast particles by ionization. J Phys (USSR) 8:201–205. https://cds.cern.ch/record/216256
3. Quast T (2019) Geant4 based event display & standalone simulation. https://github.com/ThorbenQuast/HGCal_TB_Geant4/. Tag: v1.0. Accessed 12 Aug 2019
4. Rivera EC (2017) Advanced silicon sensors for future collider experiments. PhD thesis, IFCA (CSIC-UC), Spain. https://cds.cern.ch/record/2291517
5. Aleksa M (2018) Calibration techniques and strategies for the present and future LHC electromagnetic calorimeters. JINST 13. https://doi.org/10.1088/1748-0221/13/02/c02032
6. Patrignani C, et al (Particle Data Group) The review of particle physics. Chin Phys C 40 100001. https://doi.org/10.1088/1674-1137/40/10/100001
7. Bichsel H (1988) Straggling in thin silicon detectors. Rev Mod Phys 60:663–699. https://doi.org/10.1103/RevModPhys.60.663
8. Brun R, et al. (2018) ROOT data analysis framework - user's guide
9. Wigmans R (2017) Calorimetry - energy measurements in particle physics, No. 107, 2nd edn. Oxford Science Publications, Oxford
10. Quast T (2018) Construction and beam-tests of silicon-tungsten prototype modules for the CMS high granularity calorimeter for HL-LHC. JINST 13:C02044. https://doi.org/10.1088/1748-0221/13/02/c02044
11. Contardo D, et al. (2015) Technical proposal for the phase-II upgrade of the compact muon solenoid. CERN-LHCC-2015-10 (CMS-TDR-15-02). https://cds.cern.ch/record/2020886
12. Quast T (2019) MIP tracking. https://gitlab.cern.ch/tquast/corryvreckan/tree/master/src/modules/HGCalMIPTracking. Tag: HGCAL-2018. Accessed 18 Nov 2019
13. The Corryvreckan Authors (2019) Corryvreckan - the maelstrom for your test beam data." https://gitlab.cern.ch/corryvreckan/corryvreckan. Accessed 18 Nov 2019
14. Quast T (2019) Shower tracking. https://gitlab.cern.ch/tquast/corryvreckan/tree/master/src/modules/HGCalShowerTracking. Tag: HGCAL-2018. Accessed 18 Nov 2019
15. Akchurin N et al (2018) First beam tests of prototype silicon modules for the CMS high granularity endcap calorimeter. JINST 13:P10023. https://doi.org/10.1088/1748-0221/13/10/p10023
16. Spieler H (2019) Advanced concepts in the readout and signal processing. Advanced Summer School in Radiation Detection Measurements, Munich, 2011. https://www-physics.lbl.gov/~spieler/TUM_2011/Advanced_Concepts_in_Readout_and_Signal_Processing.pdf. Accessed 28 Nov 2019
17. CMS Collaboration (2017) The phase-2 upgrade of the CMS endcap calorimeter. CERN-LHCC-2017-023 (CMS-TDR-019). https://cds.cern.ch/record/2293646
18. CMS Collaboration (2017) Technical proposal for a MIP timing detector in the CMS experiment phase 2 upgrade. CERN-LHCC-2017-027 (LHCC-P-009). https://cds.cern.ch/record/2296612

Chapter 10
Performance Validation of the Silicon-Based Calorimeter Prototype

After the calibration of the energy scale in the previous chapter, the reconstructed calorimeter hits correspond to localised measurements of the energy density of the recorded particle showers. Calorimeter hits of the same event can be visualised using event displays like in Fig. 10.1. They are the basis for the performance validation of the silicon-based HGCAL prototype to reconstruct the properties of the incident particles which is studied in this chapter.

The studies in this chapter address four distinct topics:

1. HGCAL prototype's calorimetric performance in the beam tests.
2. Comparison of the measured performance to the expectation from the GEANT4 simulation.
3. Performance enhancement through application of sophisticated reconstruction methods exploiting the calorimeter's granularity.
4. Impact of using averaged MIP calibration constants instead of the per-cell MIP calibrations, or when the high gain information is ignored. These approaches mimic, to some extent, the functionality of HGCAL's ultimate readout chip design.

In general, an essential requirement for addressing these topics with beam test data is the availability of well-defined experimental conditions. The electron data from the HGCAL beam test in June 2018 at the H2 beam line do not fulfill this requirement because of the reduced quality of the test beam at that campaign. Just after the data taking, the majority of the recorded showers in the investigated EE prototype were found to show the signature of preshowering test electrons, attributed to unwanted upstream material in the beam line. The event display in Fig. 10.2 is a representative example. While this finding strongly motivated the extension of the online data quality monitoring system, cf. bottom graphics in Fig. 6.5, the electron data recorded in the June 2018 HGCAL beam test are not used in this thesis. Instead, the prototype performance is evaluated exclusively using the *O1E* and *O1H* datasets.

T. Quast, *Beam Test Calorimeter Prototypes for the CMS Calorimeter Endcap Upgrade*, Springer Theses, https://doi.org/10.1007/978-3-030-90202-5_10

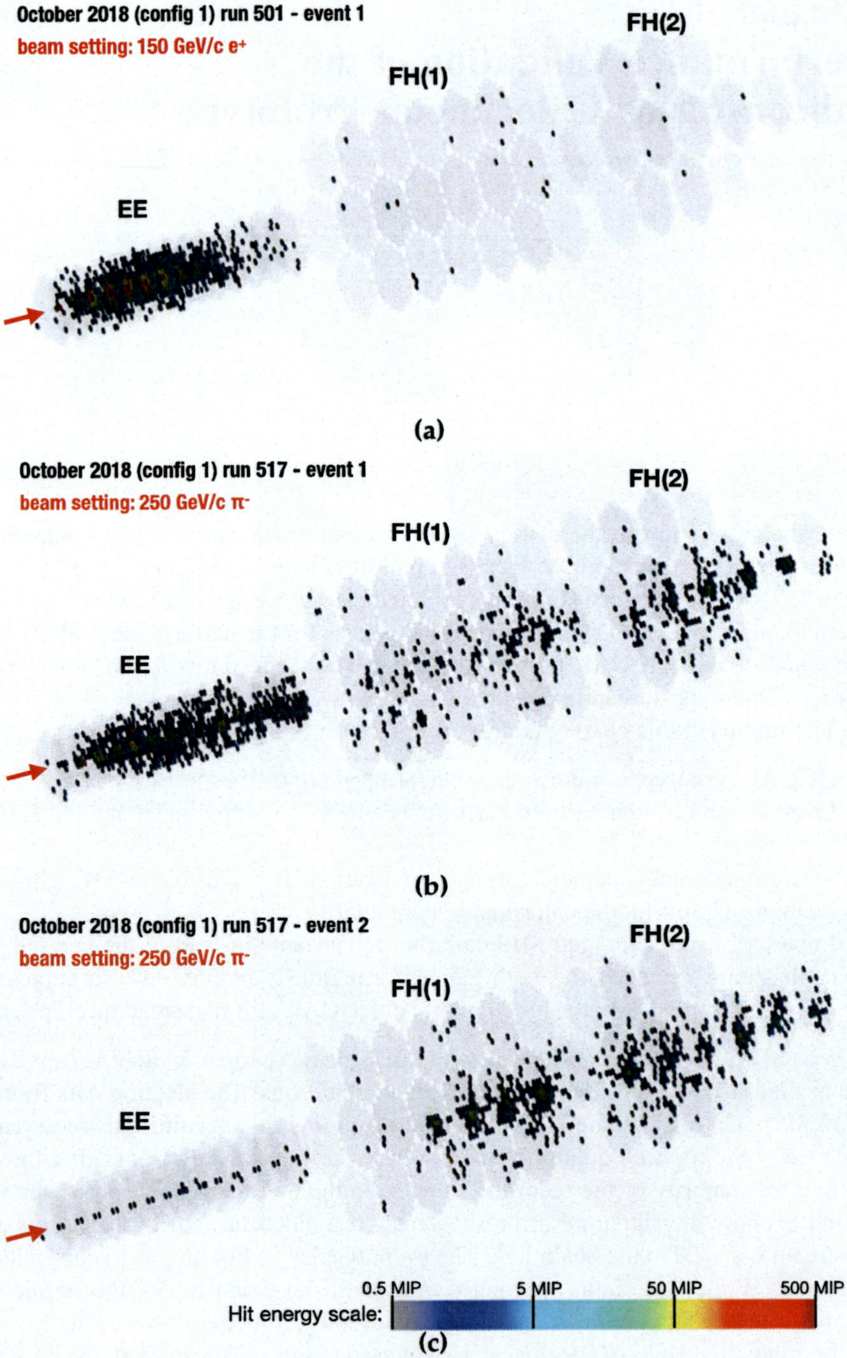

Fig. 10.1 **a** 150 GeV/c positron-induced shower recorded in the *O1E* programme, and **b** 250 GeV/c charged pion-induced shower from the *O1H* dataset with a shower start in the EE compartment and **c** another one with a shower start in the FH compartment. The HGCAL beam test prototype event display tool [1] was developed as part of the work for this thesis

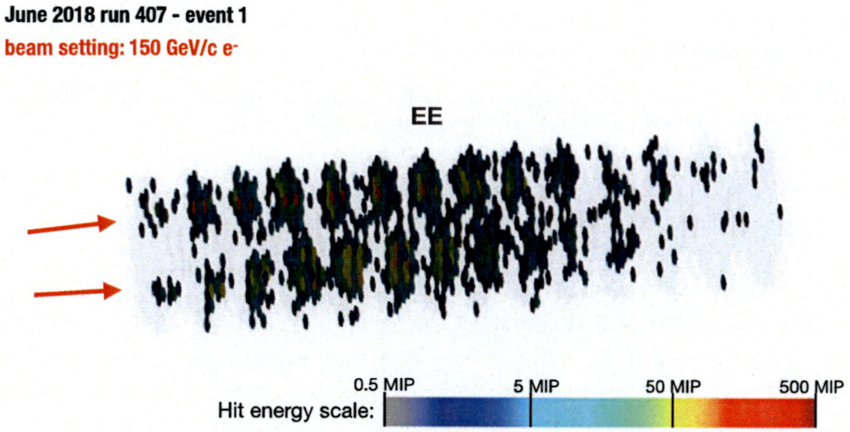

Fig. 10.2 A typical event in the June 2018 HGCAL beam test at the CERN H2 beam line. There, the HGCAL prototype consisted of an EE compartment with 28 layers. Unwanted material upstream the H2 beam line caused preshowering of the electrons resulting in multiple showering particle incidences per readout. The electron data recorded in the June 2018 beam test at the H2 beam line at CERN are not analysed in this thesis

General remarks on the simulated samples, hit- and event selections as well as the definitions of the alternative hit energy reconstructions are given in Sect. 10.1. Section 10.2 deals with the performance of the EE compartment with positron-induced showers concentrating on the agreement to the simulation. The FH compartment is included in the measurements with hadronic showers that are presented in Sect. 10.3. Comparisons between the data and the Monte Carlo are provided therein. However, they are not discussed in as much detail as for the electromagnetic showers, due to the incomplete calibration of the full FH compartment which prevents a reliable comparison to GEANT4 simulation. As part of the studies with hadronic showers, the application of advanced energy reconstruction methods, that are partially designed to exploit the calorimeter's granularity, is studied. State-of-the-art machine learning techniques are used in this connection. The same is attempted for the subsequent separation of electromagnetically and hadronically induced showers in the EE compartment. The latter is described in Sect. 10.4 which at the end explains the experienced limitations of those machine learning-based methods when they are applied to realistic calorimeter data.

10.1 General Remarks

Simulated Samples

The storage location of the simulated samples that are used in the various sections in this chapter are defined in Table 10.1. The simulation model of the HGCAL beam test prototype is implemented within the CMSSW (version 10_1_2) framework. Simulation of the calorimeter data is performed using GEANT4 version 10.4 therein.

Table 10.1 Location of the simulated samples used in this thesis for comparison to the experimentally inferred HGCAL prototype's performance

Dataset	Sections
/eos/cms/store/group/upgrade/HGCAL/simulation/2018/Oct/V2/8To17Oct_withMCP/mc/CRAB_PrivateMC/crab_Ele_*	10.2.1, 10.2.2, 10.4
/eos/cms/store/group/upgrade/HGCAL/simulation/2018/Oct/V2/noSmearingSamples_beamNearHGCALWithAngle/8To17Oct_withMCP/mc/CRAB_PrivateMC/crab_Ele_*	10.2.3
/eos/cms/store/group/upgrade/HGCAL/simulation/2018/Oct/V2/8To17Oct_withMCP/mc/CRAB_PrivateMC/crab_Pi_*	9.2.3, 10.3, 10.4

The FTFP_BERT_EMM physics list [2] is chosen as reference because it has provided the best agreement between the 2016 HGCAL beam test data and the corresponding simulation [3]. In the simulation model, the silicon diodes are sensitive volumes. The associated hit energy is obtained from summing all energy depositions from all simulation steps in its volume. The sum is then scaled by a sensor thickness-dependent factor to MIP equivalents, cf. Eq. 9.5. Any effects due to charge propagation in the silicon, signal digitisation or cross-talk are neglected.

In the simulation setup, the GEANT4 particle gun is placed at the beginning of the PPE172 area, i.e. more than 30 m upstream with respect to the actual calorimeter. By this means, beam particle interactions with the upstream material in that area are included in the simulation. The studies of potential beam distortions due to the presence of the NA61 experiment in front of the HGCAL setup have not concluded at the time of writing this thesis and are not considered in the samples in Table 10.1. Furthermore, the misalignment of the sensitive layers of the HGCAL prototype is not included in the simulations. By contrast, the actual transverse beam spread, the beam impact position onto the calorimeter as well as its angle with respect to the EE compartment derived from the beam test data are incorporated.

Delay wire chambers in the model are not defined as sensitive and hence their data is not simulated by GEANT4. Instead, the beam gun position is used to mimic their measurements and a Gaussian smearing according to their estimated intrinsic resolution of $\sim$500 µm, cf. Sect. 7.4, is applied.

Event and Hit Selection

Data events with impact measurements from all three delay wire chambers are exclusively selected. Any additional event selections are specified in the corresponding sections. Hits are selected such that noise contributions are further suppressed: Hits with hit energies E_i below 0.5 MIP are always rejected, as well as those where the reconstructed signal amplitude in high gain significantly exceeds the low gain-based assessment as long as $E_i < 2.0$ MIP. The latter selection primarily is applied to reject unphysical energy depositions from hits of one particular readout chip in the very first layer in the EE compartment that was affected by an extraordinarily high level of noise. This particular selection affects the studies related to the longitudinal shower development.

Alternative Hit Energy Reconstructions

The relevance of the level of detail in the MIP calibration, and the necessity of the high gain for the final calorimetric performance indicators are assessed. For this purpose, the hit energies are re-reconstructed using averaged MIP calibration constants or neglecting the high gain information utilising Eq. 7.10. Most of the analyses are iterated with these variations and the results are often reported together throughout this chapter. The naming convention is as follows:

1. ***Data*** Nominal hit energy reconstruction using all gains and per-cell MIP calibration constants $M_{\mathrm{HG/MIP}}$. Uncalibrated channels are assigned chip- or module-averaged values. The same is done for the constants from the gain intercalibration.
2. ***MIP scale up/down*** Impact of the nominal result when all $M_{\mathrm{HG/MIP}}$ are simultaneously shifted by their statistical uncertainty. This implies full correlation in their derivation. It should be understood as a worst-case scenario of the uncertainty in the calibrated energy scale.
3. ***MIP chip average*** All readout channels served by the same SKIROC2-CMS are assigned one common, i.e. chip-averaged, value for $M_{\mathrm{HG/MIP}}$.
4. ***MIP module average*** All readout channels on the same module are assigned one common, i.e. module-averaged, value for $M_{\mathrm{HG/MIP}}$.
5. ***Low gain MIP*** Hit energy reconstruction as defined by Eq. 7.10. By this means, the high gain information is ignored. When available, the respective low gain-based MIP calibration constants $M_{\mathrm{LG/MIP}}$ are taken for each readout channel individually. Otherwise, chip or module averages are taken.

10.2 Studies with Electromagnetic Showers

In this section, typical calorimeter observables for electromagnetic showers are computed from reconstructed calorimeter hits exclusively from the 28-layer HGCAL EE prototype whose total depth amounted to 27.7 radiation lengths (X_0), cf. Fig. 6.16a. Identical analyses are applied on simulated positron showers for comparison to the experimental results. A reasonable prototype performance and a basic agreement to simulation for the following quantities, accessible in the beam test experiments, is considered essential for the verification of the HGCAL design choices:

- Longitudinal shower evolution and shower shapes, discussed in Sect. 10.2.1.
- Energy linearity and resolution, discussed in Sect. 10.2.2.
- Position and angular resolution, discussed in Sect. 10.2.3.

The observable definitions are given in the appropriate sections. Apart from the position reconstruction with HGCAL, those definitions are not motivated extensively due to the relative simplicity of electromagnetic showers.

In this thesis, the following systematic effects were not yet incorporated into the simulated samples in full detail that may impact the agreement to the beam test data:

- It is known that the choice of the underlying physics list in GEANT4 may cause substantial shifts in the simulated energy scale.
- The upstream material in the entire H2 beam line leads to additional energy losses of the incident beam particles before they impinge onto the calorimeter prototype.
- Different transverse beam profile shapes may result in discrepant levels of transverse leakage in the simulated and the tested prototype calorimeter.
- Misalignment of the layers in the EE compartment can amount to a fraction of a millimetre in the experimental setup which may affect transverse shower shapes.

Precise study of all these effects is a time consuming task which is currently distributed among multiple collaborators to the HGCAL beam test data analysis efforts. The corresponding studies were not concluded at the time of writing this thesis. As consequence, quantification of the data-simulation agreement is not provided here. The corresponding discussions are limited to qualitative statements in view of the overall functionality of this prototype calorimeter.

The *O1E* dataset, used subsequently, contains data from positron-induced showers with momenta ranging from 20–300 GeV/c. For earlier runs with high-momentum positrons therein, all SKIROC2-CMS ASICs were operated with an identical configuration which appeared to be inappropriate for a few of them. For some events, the reconstructed, visible energy appears to be lowered systematically. The location of the affected readout chips inside the calorimeter can be roughly assessed by the beam profile of events with lower energy sums as it is depicted in Fig. 10.3a. The exact implication of that wrongful SKIROC2-CMS configuration to the calorimeter hit reconstruction has not been established.

In the following, those affected events are rejected in a twofold manner. First, the runs taken with the common SKIROC2-CMS configuration are discarded from the analysis. In addition, events are selected based on their impact onto the calorimeter derived from the DWCs in order to circumvent the affected region, cf. Fig. 10.3a. Figure 10.3b illustrates the effect on the energy spectrum imposing the latter requirement to the "bad" runs with common SKIROC2-CMS configuration. Despite being a redundant selection criterion after rejecting all affected bad runs, defining a common beam impact window in both data and simulation is helpful in general. It renders the comparison less sensitive to differences in the transverse leakage due to shifted beam impact positions in the experiment and in the simulation. Therefore, the DWC-based selection is applied consistently for all the studies in this section.

For some positron beam momenta, the contamination of the positron beam with early-showering pions amounts to $\mathcal{O}(1\%)$. This fraction is considerable for some analyses as it may cause distortions in some of the investigated observables. In particular, such events are rejected from the analysis of the shower shapes and the energy resolution using the discrimination technique that is described in Sect. 10.4.

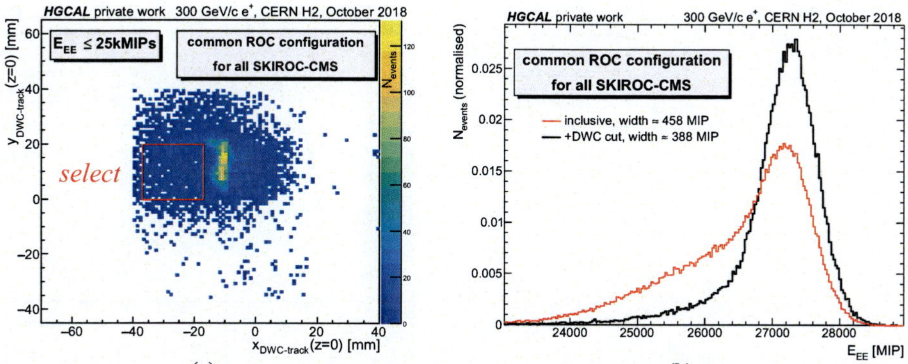

(a) (b)

Fig. 10.3 **a** Beam profile for the first runs with 300 GeV/c positrons in the *O1E* dataset taken with a common configuration of all SKIROC2-CMS ASICs as measured by the delay wire chambers. Events are selected for which the calorimeter's signal sum E_{EE}, cf. Eq. 10.1, is observed to be particularly low. **b** Normalised distribution of E_{EE} for the same runs before and after reduction of the beam acceptance offline. Note that runs with the common configuration of the readout chips (ROCs) are not used in this analysis

Further and independent studies of electromagnetic showers using this prototype calorimeter are currently carried out by the CMS collaboration. The principal findings will be summarised in a dedicated publication [4].

10.2.1 Shower Profiles

For characterisation of the longitudinal shower evolution, the total energy sum in the entire EE compartment E_{EE}, cf. Eq. 10.1, as well as the summed energy in each layer E_{layer}, cf. Eq. 10.2, are calculated first.

$$E_{\text{EE}} := \sum_{i=1}^{N_{\text{hits}} \in \text{EE}} E_i \tag{10.1}$$

$$E_{\text{layer}} := \sum_{i'=1}^{N_{\text{hits}} \in \text{ layer}} E_{i'} \tag{10.2}$$

Afterwards, the mean summed energy in a given calorimeter layer can be visualised as a function of the layer depth as it is done for two representative positron samples in Fig. 10.4a, b. Altogether, the obtained profiles from the beam test data are qualitatively reproduced by the FTFP_BERT_EMM physics list of GEANT4. In particular, the measured shower maximum and the decrease in the sampled shower energy with increasing depth inside the calorimeter match the expectation from simulation. A

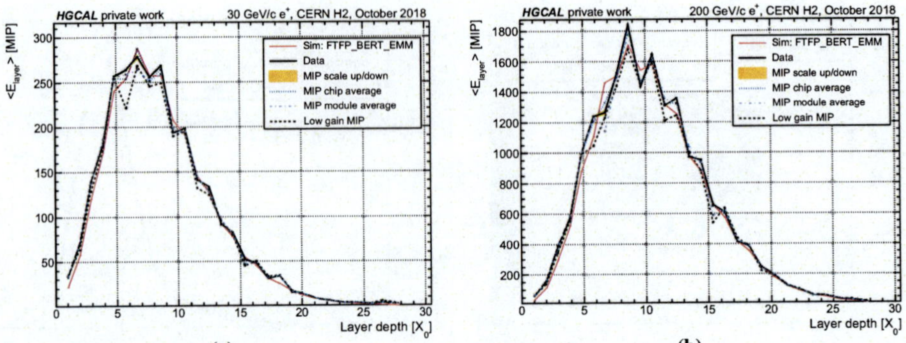

(a) (b)

Fig. 10.4 **Longitudinal shower profiles for electromagnetic showers induced by incident positrons** with a momentum of **a** 30 GeV/c and **b** 200 GeV/c

prominent exception is the agreement for layer 7 around 6.6 X_0 whose module has an energy response that is known to be lower with respect to other modules due to a malfunctioning SKIROC2-CMS. Apart from that, the most remarkable deficit is depicted by the oscillations between subsequent layers on the same EE cassette around the maximum which is more pronounced in data than in GEANT4. Ongoing investigations are targeted at the possibility of backscattering shower constituents between the calorimeter layers whose modelling might not be optimised in this particular physics list.

The longitudinal shower evolution and its dependence on the energy is assessed further. For this purpose, the longitudinal centre of gravity COG_z, cf. Eq. 10.3, commonly referred to as *shower depth*, is derived for each event.

$$COG_z := \sum_{\text{layer}=1}^{28} \frac{E_{\text{layer}} \cdot z_{\text{layer}}[X_0]}{E_{\text{EE}}} \tag{10.3}$$

Figure 10.5a shows the distribution of this observable for 120 GeV/c positrons. As can be seen, the COG_z is overestimated in simulation which can be attributed to missing beam line material therein [3]. As a result, the reported average shower depths in Fig. 10.5b tend to be slightly larger in simulation with respect to the beam test measurement. Nonetheless, an overall logarithmic scaling with respect to the incident positron momentum is obtained, analogous to the scaling in Eq. 3.5. The scaling behaviour is comparable between the beam test data and the simulation. Thus, it can be interpreted that the longitudinal evolution of electromagnetic cascades is well described in the calorimeter simulation of this HGCAL prototype.

An approach towards global characterisation of the transverse and longitudinal shower extent is pursued next. The aim is to quantify the energy spread from calorimeter hits in a shower along each spatial dimension inside the calorimeter with respect to the energy-weighted centre of gravity. In this connection, Eq. 10.4 states the formula of a tensor $(I)_{ab}$ that is evaluated for each shower. The indexes a, b are placeholders for the spatial dimensions x, y, z.

$$\bar{x}^a := \sum_{i=1}^{N_{\text{hits}} \in \text{EE}} \frac{E_i \cdot x_i^a}{E_{\text{EE}}}, \text{ defining: } x_i^1 = x_i,\ x_i^2 = y_i,\ x_i^3 = z_i$$
$$(I)_{ab} := \sum_{i=1}^{N_{\text{hits}} \in \text{EE}} \frac{E_i \cdot \left(x_i^a - \bar{x}^a\right) \cdot \left(x_i^b - \bar{x}^b\right)}{E_{\text{EE}}} \quad (10.4)$$

The computation of $(I)_{ab}$ is analogous to a principal component analysis method [5] deploying the hits' relative contributions to the summed energy as the weight. Then, the minimal and maximal eigenvalues (EV) of $(I)_{ab}$ are defined as proxies for the transverse or longitudinal shower spread, respectively.

$$\text{"longitudinal shower spread"} := \sqrt{\max EV((I)_{ab})}$$
$$\text{"transverse shower spread"} := \sqrt{\min EV((I)_{ab})} \quad (10.5)$$

The distributions of these quantities are shown in Fig. 10.6 for 120 GeV/c positrons. In this example, the longitudinal spread amounts to several centimetre corresponding to the typical extent of the shower cascade, whereas the transverse spread is confined to less than 1.8 cm consistent with the Molière radius R_M of the passive absorber materials, cf. Table 3.1. One can observe that the mean transverse spreads of simulated and real showers inside this prototype calorimeter are slightly different. However, it must be remarked that this might not necessarily be a shortcoming of the used physics list. As it will be shown in Sect. 10.2.3, the calorimeter layers are misaligned with respect to each other by $\mathcal{O}(0.5\text{ mm})$. Neither is misalignment incorporated in the simulation, nor are the hit coordinates corrected in the data for this study. In addition, noise may also cause a systematic shift of the transverse shower spread to higher values. This hypothesis is supported when artificially increasing the signal-to-noise ratio by incorporating the low gain-based energy reconstruction using Eq. 7.10. In this case, the distributions in the data are indeed shifted to slightly higher values. The impact of such a deficit in the shower shape description on sophisticated machine learning-based inferences, namely particle identification, is revisited in Sect. 10.4 of this chapter.

✓ **PV-POS.1** The longitudinal development of electromagnetic showers is in agreement between the simulation and the beam test measurements. In particular, the longitudinal shower depth scales logarithmically with the incident positron energy.

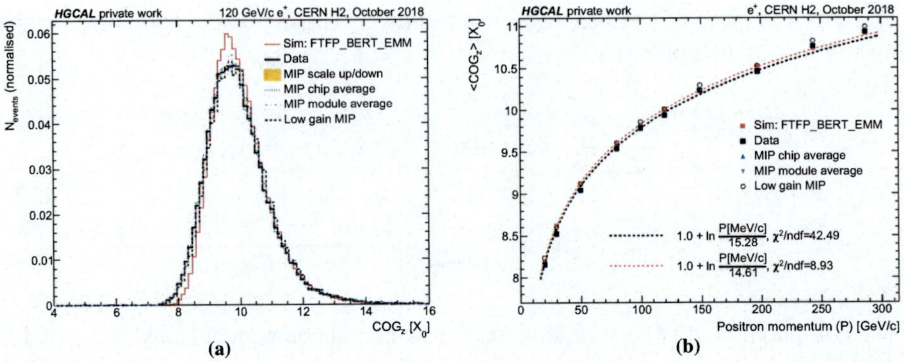

Fig. 10.5 **a** Distribution of the longitudinal centre of gravity computed for 120 GeV/c positron-induced showers and **b** the logarithmic dependence of its mean on the incident particle momentum

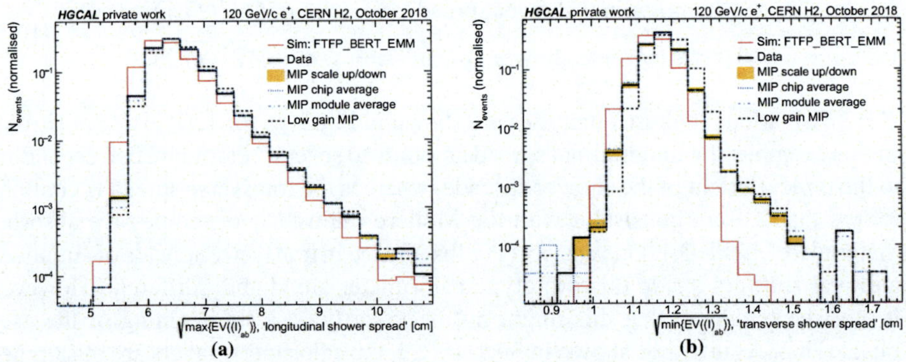

Fig. 10.6 **a** Longitudinal and **b** transverse shower spread for 120 GeV/c positron-induced showers derived from a principal component analysis-inspired method, cf. Eq. 10.4. Note the logarithmic scale of the vertical axis

10.2.2 Energy Resolution

In other studies of electromagnetic showers with this sampling calorimeter prototype, the so-called *dE/dx* method is often used for inference of the beam energy [4]. There, the idea is to interpolate the sampled shower energy in two adjacent layers to the passive absorber material in between them. Subsequently, the energy deposited in the absorber is estimated by scaling that interpolated energy according to the absorber's thickness. By this means, the invisible energy deposited in the passive material can be added to the total visible energy. During the analysis of the beam test data in 2016 [3], the energy reconstruction benefitted from this algorithm due to the non-uniform sampling configuration of the employed HGCAL prototype back then.

By contrast, the EE compartment considered in this thesis comprised an almost uniform sampling. As consequence, the dE/dx weights are approximately identical, which explains why the dE/dx method would not necessarily yield improvements in the energy resolution. Furthermore, the scheme can anyways be misleading since it is constructed based on the simplified picture that the shower consists of a collection of MIPs, which is not the case. In particular, it is generally not guaranteed that dE/dx, with weights derived from first principles, provides a direct mapping of the sampled energy to the full shower energy. For simplicity, the usage of the bare energy sums in the energy reconstruction of electromagnetic showers is preferred in this thesis.

Provided that the arrangement of sampling layers is uniform, which is the case for the investigated prototype calorimeter, and the respective calorimeter is large enough to fully contain the showers, the visible energy E_{EE}, cf. Eq. 10.1, is anticipated to be proportional to the beam energy. As it is illustrated in Figs. 10.4 and 10.6b for 120 GeV/c positrons, representatively also for higher momenta, the latter condition is met. Hence, Eq. 10.6 may be used for the derivation of the incident beam energy.

$$E_{\mathrm{beam}} = m^{-1} \cdot E_{\mathrm{EE}} \tag{10.6}$$

This relation is commonly referred to as *energy linearity* and a calorimeter is considered to be *linear* if this relationship is applicable. m denotes a calibration constant that determines the overall energy scale. E_{beam} stands for the incident beam particles' energy. Since the test positrons' were highly relativistic, their momentum is practically equal to their energy (when natural units are used). Equation 10.7 explicitly states the relative difference between the momentum and the energy for 20 GeV/c positrons.

$$1 - \frac{c \cdot P_{20\,\mathrm{GeV/c\ e^+}}}{E_{20\,\mathrm{GeV/c\ e^+}}} = 1 - \beta_{20\,\mathrm{GeV/c\ e^+}} = 3 \cdot 10^{-10} \tag{10.7}$$

Figure 10.7 shows the distributions of reconstructed, visible energies for all positron energies that were recorded in the *O1E* programme. The central regions of these spectra can be characterised by Gaussians. Contributions to lower values in the spectra are consistent with the presence of additional passive material in the H2 beam line which is supported by preliminary simulations of the beam line. In order to reduce the bias by those low energy entries on the assessment of the calorimetric performance, Gaussians are fitted iteratively to the distributions in an asymmetric range of -1.5σ to $+2.5\sigma$ around their mean. Apart from the statistical fitting uncertainty on the retrieved width σ_e and on the mean μ_e, a relative systematic error of 2% on σ_e and of 0.1% on μ_e, respectively, are assumed in this procedure to reflect the uncertainty in the choice of the fitting range.

The evaluation of the energy linearity is presented by Fig. 10.8a. It illustrates the estimated mean as a function of the positron energy. At this point, precise knowledge on the true particle energy arriving at the calorimeter prototype is important. If one were to use the nominal beam energies as reference, the assigned reference energy (E_{beam}) would be overestimated and saturation behaviour would artificially be introduced in the linearity assessment in Fig. 10.8a. Here, preliminary estimates of the actual beam energies under consideration of synchrotron losses are incorporated.

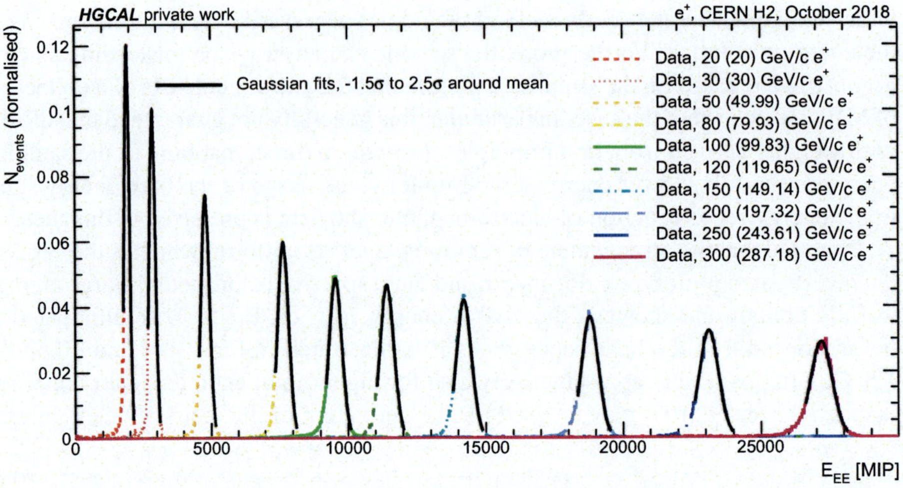

Fig. 10.7 Distributions of the reconstructed, visible energy inside the EE compartment of the HGCAL prototype tested in October 2018 for all positron momenta available in the *O1E* dataset. All distributions are normalised to unity integral. The distributions are well described by Gaussian curves (black solid lines) when fitting the central regions. The numbers in brackets state the preliminary estimates of the actual positron momenta in the H2 beam line that are used in the following

Altogether, the applicability of Eq. 10.6, and with it energy linearity, is confirmed as can be seen from the χ^2/ndf computed from the straight line fit to the data points in Fig. 10.8a. An extra offset parameter b is allowed in the fit which turns out to be consistent with zero, both in the beam test measurement and in the GEANT4 simulation. The proportionality constants in the beam test is approximately 4% higher than expected from the FTFP_BERT_EMM physics list. Discrepancies in the total energy scale have already been observed in the 2016 beam tests [3] and are known to be different for distinct physics lists. Noteworthily, the slope is in better agreement when using the low gain and TOT only for the reconstruction of hit energies, cf. Eq. 7.10. Further investigations on this matter remain to be pursued in the future.

Fortunately, most of the previously considered observables characterising the shower shapes are by definition not affected by the absolute energy scale. Another example of such an unaffected observable is the *energy resolution*. It is defined by Eq. 10.8 as the ratio of the visible energy ($E := E_{\text{EE}}$) distribution's width to its mean.

$$\text{Energy resolution} = \frac{\Delta E}{E} := \frac{\sigma_e}{\mu_e} \tag{10.8}$$

Its dependence on the incident positron energy is shown in Fig. 10.8b. For the highest positron energies, resolutions below 1.5% are obtained. Especially for high-energetic positrons, the agreement between the beam test data and the simulation is remarkable. The increasing discrepancy for low-energetic positrons may be attributed to insufficient simulation of the upstream interactions of the incident particles. The

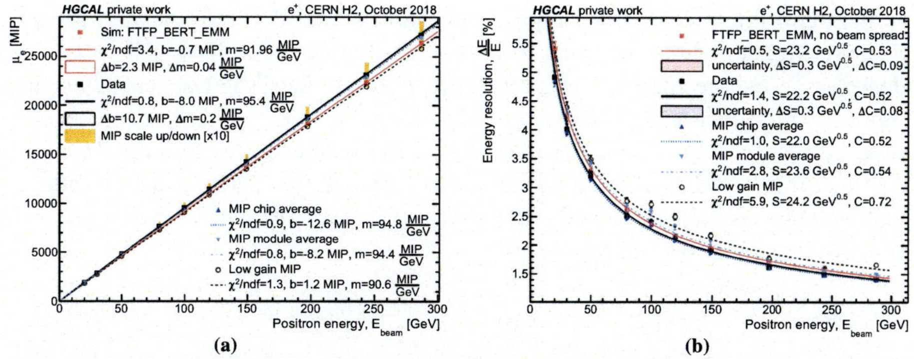

Fig. 10.8 **a** Linearity and **b** resolution assessment of the 28-layer HGCAL-EE prototype calorimeter's response to electromagnetic showers induced by positrons. The applied beam spread is subtracted in quadrature from the results on the resolution derived from simulation. The Δ's symbolise uncertainties on the final parameter values

results follow the parameterisation in Eq. 10.9. C represents the constant term and $S \cdot E_{\text{beam}}^{-0.5}$ the stochastic term. A term proportional to E_{beam}^{-2}, which would describe the electronic noise, is not included to avoid possible correlation to the stochastic term. Another argument for neglecting any noise term is that noise is not part of the simulation. Its inclusion only in the beam test data but not in simulation would complicate the comparability of the results.

$$\left(\frac{\Delta E}{E}\right)^2 = C^2 + \frac{S^2}{E_{\text{beam}}} \tag{10.9}$$

Ultimately, fitting the parameters to the beam test data yields $C = (0.52 \pm 0.08)\%$ while the stochastic term amounts to $S = (22.2 \pm 0.3)\%\ \text{GeV}^{0.5}$. Both of them are in good agreement with the expectation from simulation. The results are almost identical when chip-averaged MIP calibration constants in the hit energy reconstruction are used, suggesting that the corresponding extensive calibration of each readout channel individually might not even be necessary. By contrast, the calorimetric performance degrades when the sizeable variations in the MIP calibration constants between the chips are neglected, namely through the usage of module-averaged constants. The resolution is also worsened when the low gain is used directly corresponding to an increase of the signal-to-noise ratio.

In summary, the response of the EE prototype calorimeter compartment to electromagnetic showers is, in fact, linear and its energy resolution obeys the typical scaling law of a sampling calorimeter. The resolution at the highest energy is close to 1.5% and slightly smaller than expected from simulation. It should be mentioned that this performance could not be achieved without the successful energy calibration of this calorimeter, as it was demonstrated in Sects. 9.1 and 9.2.

For a concrete yet simplified application of the measured calorimeter performance, the hereby obtained resolution (ΔE) may be assumed to be independent of the incidence angle and the angular measurement precision of the CMS tracker shall be negligible. In this scenario, the mass m_Z of a Z boson with $m_Z = 91.1\,\text{GeV}/\text{c}^2$, decaying to an (isolated) electron and a (isolated) positron with energies $E_1 = 250\,\text{GeV}$ and $E_2 = 275\,\text{GeV}$ (relative angle between the particles $\theta = 20°$)[1] detected independently in the calorimeter endcap, should be reconstructable with a relative precision of $\Delta m_Z/m_Z \approx 1.0\%$, cf. Eq. 10.10.

$$\left(\frac{\Delta m_Z}{m_Z}\right)^2 = \left(\frac{\Delta E_1}{2 \cdot E_1}\right)^2 + \left(\frac{\Delta E_2}{2 \cdot E_2}\right)^2 = \frac{C^2}{2} + \frac{S^2}{4} \cdot \left(\frac{1}{E_1} + \frac{1}{E_2}\right) \tag{10.10}$$

✓ **PV-POS.2** The mean visible energy measured by the prototype calorimeter is proportional to the incident positron energy.
✓ **PV-POS.3** The energy resolution is in good agreement with the simulation, especially at high energies. The constant term is below 1%.

10.2.3 Position and Angular Resolution

Remark: Italic passages in this section are substantively quoted from the author's original contribution to the 2016 HGCAL beam test publication [3].

Accurate measurements of the incidence position and direction of showers are key inputs to the Particle Flow performance of the upgraded CMS detector. The prototype calorimeter's capability to localise positron-induced particle showers has been investigated using the *O1E* dataset in this thesis. *Information on each cell's coordinate and deposited energy are used to reconstruct the shower main impact position on a calorimeter layer. Subsequently, residuals to external reference measurements are computed. The width of the residual distribution can be considered to be the position resolution of a given sensor* if the reference precision is negligible. *Similar studies have already been performed for other highly-granular calorimeter prototypes, e.g. in Refs.* [6, 7].

This section is divided into two conceptional parts. First, the position resolution study is performed individually for single layers, hereby qualitatively reproducing the findings with the initial 2016 HGCAL beam test prototype [3]. Having a fully equipped EE compartment in 2018, the applied method is then extended to reconstruct the shower axis comprising the measurement from numerous calorimeter layers

[1] Decay of a particle with mass m into two relativistic particles with energies E_1 and E_2 flying at an angle of θ to each other: $m = 2 \cdot (E_1 \cdot E_2)^{0.5} \cdot \sin(\theta/2)$.

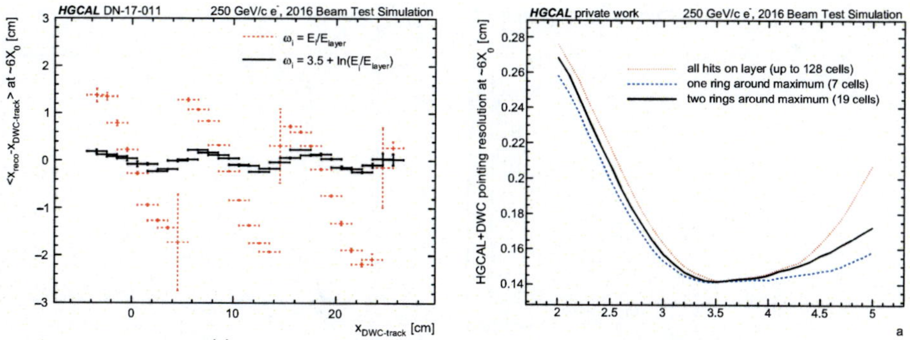

(a) (b)

Fig. 10.9 Optimisation of the position reconstruction on the layer close to the shower maximum in the 2016 HGCAL beam test prototype [3] shown for simulated 250 GeV/c electrons: **a** Mean residual as a function of the impact positions ($x_{\text{DWC-track}}$) for a linear energy weighting (black) and a logarithmic weighting (red). The bias towards preferred coordinates, such as the cell centres or boundaries, is minimised when the logarithmic weighting is used. **b** Optimisation of parameter a in Eq. 10.12 for the logarithmic cell summing scheme using different sets of cells. For $a \approx 3.5$, the residual width is minimal for all considered cell sets. The 19-cells scheme is chosen because it yielded slightly better performance in the analysis of the 2016 beam test data [3]

simultaneously. In this context, the prototype calorimeter's orientation with respect to the test beam as well as interalignment corrections of its layers are inferred.

Part 1: Position Reconstruction for each Layer

Equation 10.11 defines the reconstructed shower impact position on a given layer.

$$\vec{x}_{\text{reco}} := \frac{\sum_{i \in M} \omega(E_i) \cdot \vec{x}_i}{\sum_{i \in M} \omega(E_i)} \tag{10.11}$$

where:

- *The set M comprises either all cells on the investigated layer or cells within one (two) rings around the cell with maximum deposited energy.*
- $\omega_{i(E_i)}$ *represents an energy weighting function.*

Two degrees of freedom have been varied in order to both optimise the resolution and to minimise the reconstruction bias towards preferred coordinates on the sensor (...). It was found that a logarithmic weighting as in Eq. 10.12 using two rings around the cells with maximum deposited energy (= 19 cells) maximises the precision for 250 GeV/c electrons in the first layer of the 2016 HGCAL prototype, cf. Fig. 10.9a, b.

$$\omega(E_i) := \max\left(0,\ a + \ln\left(\frac{E_i}{E_{\text{layer}}}\right)\right),\quad a := 3.5 \tag{10.12}$$

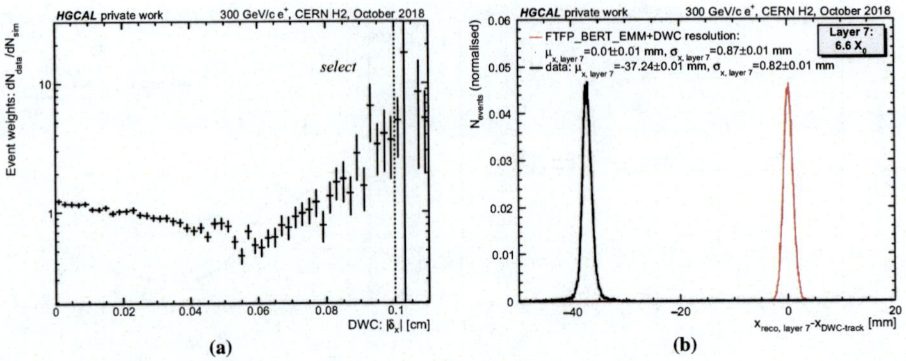

Fig. 10.10 **a** Event weights for simulated 300 GeV/c positron samples to incorporate non-Gaussian effects of the DWC's pointing resolution into the simulation. The analysis is restricted to events with $|\delta_{x(y)}| < 0.1$ cm. **b** Residual distribution of the horizontal shower impact position measured in layer 7 and the DWC extrapolation for incident 300 GeV/c positrons. DWC event weights are applied in the simulation

As the sensor geometry of the 2016 prototype was identical to the one deployed in 2018, the optimisation of the cell summing scheme is fully transferable to the 2018 HGCAL prototype.

The delay wire chambers (*DWCs*) in the beam line are used for reference of the particle incidence. The simulation of its data, discussed in Sect. 10.1, is applicable only if the multiple scattering of the simulated particle travelling from the beam gun position to the calorimeter is negligible compared to the expected position resolution of the prototype. As this is not necessarily the case for lower momentum positrons travelling through the more than 30 m long simulated beam line, the particle gun was placed 50 cm in front of the calorimeter in the simulation used for this section. The deflection due to multiple scattering of 20 GeV/c positrons in 50 cm air amounts to less than 0.01 mm whereas it would be more than 3 mm for 30 m of air. Non-Gaussian effects in the DWC's pointing resolution are incorporated into the simulation through event weights. Event weights are constructed such that the distribution of the quadratic sum, cf. Eq. 10.13, of the biased DWC residuals (Δx_P^b analogous to Eq. 7.14) in Eq. 10.13 are identical between the beam test data and the simulation. They are inferred and applied separately for the x- and y-coordinates. Figure 10.10a shows the event weights for the simulation with 300 GeV/c positrons.

$$\delta_x^2 = \sum_{P}^{\text{ext,D,E}} \left(\Delta x_P^b\right)^2, \quad \text{analogous for y} \tag{10.13}$$

An example distribution of the position residuals for one layer close to the shower maximum of 300 GeV/c positrons is presented in Fig. 10.10b. Analogous distributions are computed and evaluated for all other layers and positron energies. The core of the position residual distributions are characterised by iterative Gaussian fits in

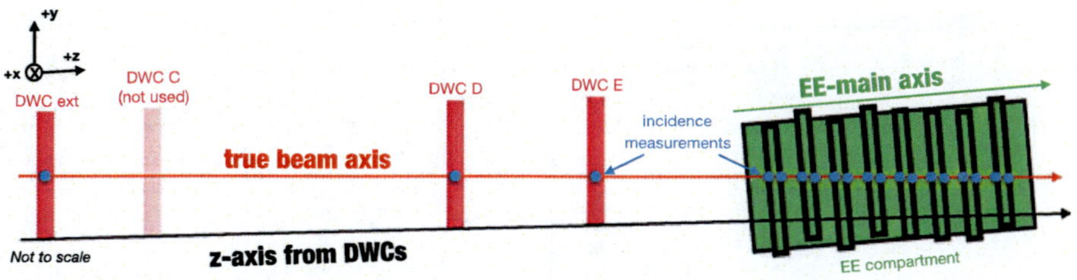

Fig. 10.11 Schematic illustration of the a priori different orientations of the involved z-axes

the symmetric range of -2.0σ to $+2.0\sigma$ around the mean. Systematic errors on the obtained parameters are the same as for the quantification of the energy distributions in Sect. 10.2.2. The Gaussian standard deviation σ serves as proxy for the combined DWC+HGCAL position resolution whereas μ can be used for inference of the calorimeter-DWC alignment. For a sketch of the realistic, i.e. misaligned setup, see Fig. 10.11.

The fitted μ parameter of each layer is visualised as a function of the corresponding depth in Fig. 10.12a. Therein, the exhibited trend is consistent with what is inferred from the analysis of the muon data, cf. Fig. 8.12, and reveals a misorientation of the DWCs' and the EE compartment's coordinate systems. A straight line fit to the mean residuals is used for quantification of the relative angle $\langle\alpha_{x(y)}\rangle$. The average angle of beam particle trajectories as measured with the DWCs in their respective reference frame amounts to 0.4 mrad in x and -0.5 mrad in y. Both values are negligible compared to the angle between the DWC's z-axis and the EE's that is quantified by that slope in Fig. 10.12a. As consequence, the slope $\langle\alpha_{x(y)}\rangle$ can be understood as the angle of the calorimeter's z-axis with respect to the true beam axis which has a value of $\langle\alpha_x\rangle \approx 13$ mrad. A comparable value is retrieved when the same setup is used for the tracking of MIPs in the muon beam using the calorimeter itself, cf. the blue distribution in Fig. 9.5b. Due to the cassettes' flexible placement into the hanging file structure, each layer is misaligned differently with respect to the beam axis. Layerwise misalignment with respect to the beam axis is only relevant when the information of each layer is combined. Such a correction is applied only in the second part of this section.

Figure 10.12b illustrates the pointing resolution of the layer, whose residual distribution has been shown previously in Fig. 10.10b, for all positron energies. Like it was reported elsewhere before, e.g. in Ref. [6], the resolution may be parameterised according to Eq. 10.14. Inclusion of an additional noise-like term in order to describe conceivable energy-dependent effects in the DWC extrapolation does not improve the agreement of the model to the data. The energy-dependent stochastic term $a \cdot E_{\text{beam}}^{-0.5}$ can be motivated by the sampling fluctuations in the contributing diodes that, in first approximation, scale with $E_{\text{beam}}^{-0.5}$ as well.

$$\sigma^2_{x(y),\ \text{HGCAL-layer}} = a^2 \cdot E_{\text{beam}}^{-1} + b^2 \tag{10.14}$$

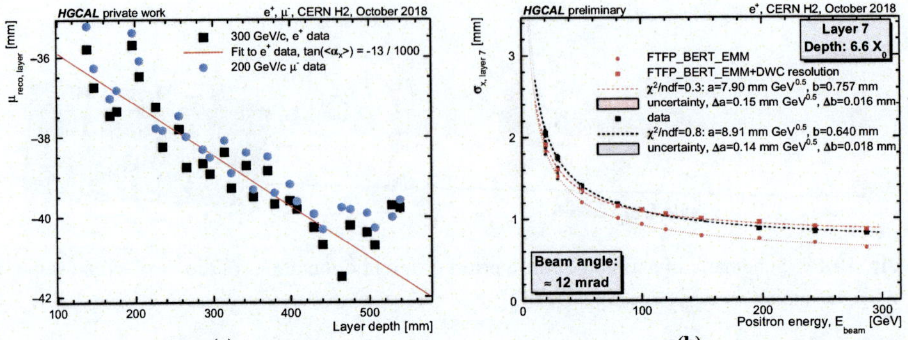

(a) **(b)**

Fig. 10.12 **a** Mean position residuals as a function of the layer depths from which the calorimeter's orientation with respect to the DWC coordinate system can be assessed. A fit to the mean residuals reveals the orientation of the prototype's z-axis to the true beam axis which was incorporated into the simulation. Note that the same may be done with the position residuals using muons, cf. Fig. 8.12. **b** Combined HGCAL and DWC pointing resolution at a depth of 6.6 X_0 as a function of the incident positron energy. After inclusion of the finite DWC precision into the simulation, a reasonable agreement between the data and the simulation is observed. Hence, the results for the simulation without smearing of the reference measurement may be interpreted as the calorimeter-only position resolution

Although the silicon diodes on the sensor are $\sim$1 cm wide, the shower incidence can be reconstructed with a more than 10$\times$ better precision at high energies. Even for the lowest energy point of 20 GeV, the precision is still better than the binary resolution of $\sim 1.1\,\text{cm}/\sqrt{12}$. After inclusion of the finite DWC tracking precision into the simulation, the agreement between the data and the simulation is remarkable for the high energies. By contrast, the resolution for the lower energy points is approximately 10% underestimated in the simulation. It might hint at a deficit in the calorimeter simulation with GEANT4 and/or at a too optimistic estimate of the actual DWC tracking precision due to multiple scattering in the data. As discussed in Sect. 7.4, multiple scattering is not relevant for the higher beam energies. The quadratic difference between the beam test data and the simulation without application of the DWC smearing at the highest accessed positron energy amounts to approximately $\sqrt{(0.83^2 - 0.66^2)}\,\text{mm} \approx 0.50\,\text{mm}$. This is only $\sim$ 0.07 mm more than the idealised DWC tracking resolution in Figure 7.12b indicating a good modelling of the DWC measurements in the simulation at high energies.

Since the beam test data can be well reproduced by the simulation, the simulation-based results without the DWC resolution applied, i.e. with values below 0.7 mm for positron energies above 200 GeV, may be interpreted as the prototype's true pointing resolution at the depth of 6.6 X_0, close to the electromagnetic shower maximum.

Part 2: Reconstruction of the Shower Axis

In the following, the reconstructed points on all the layers are combined to compute the axis of a shower. The shower axis is computed from a straight line fit to

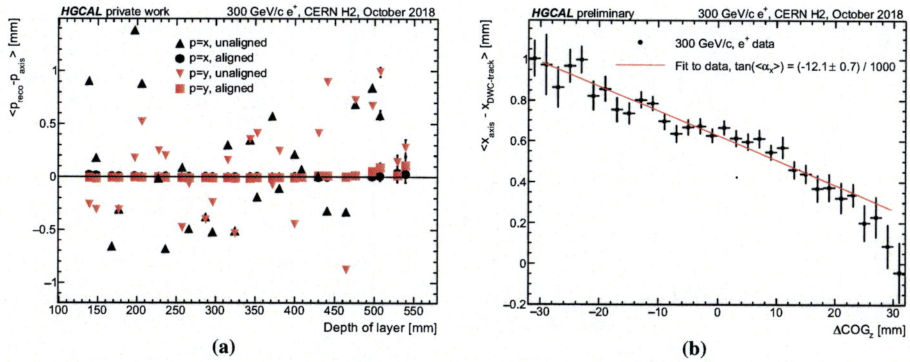

Fig. 10.13 **a** Illustration of the interalignment of the calorimeter prototype layers using position residuals with respect to the shower axis derived with the unaligned system (red markers) and with the system after alignment (blue markers). **b** Shower axis position residuals with respect to the delay wire chambers evaluated at different shower depths. ΔCOG_z represents the longitudinal centre of gravity in millimetres with respect to its average in the given dataset. It yields an alternative measurement of the calorimeter's orientation with respect to the beam

all reconstructed impact points including all the layers in which 1% to 95% of the total visible shower energy is deposited. In order to improve its accuracy, the shower axis is used beforehand to infer alignment corrections of the individual layers with respect to each other (*interalignment*). Hereby, the interalignment allows for simultaneous shifts and tilts of the two layers on the same cassette. By this means, the mechanical constraint that two layers are installed onto the same cassette is respected. Figure 10.13a demonstrates its successful application by overlaying the mean position residuals at each layer with respect to the shower axis $x(y)_{\text{axis}}$ derived from the unaligned together with the ones derived from the aligned system. Since the calorimeter setup was not altered within the *O1E* programme, this procedure is performed only for the 300 GeV/c positron dataset and its corrections are applied to all others.

The shower axis may also be used for inference of the beam angle in yet another way compared to the first part of this section. For this purpose, the axis' location at the position of the longitudinal centre of gravity is evaluated and compared to the DWC extrapolation as shown in Fig. 10.13b. The result remains unchanged: The beam angle, and with it the rotation of the calorimeter coordinate system with respect to the reference frame, amounts to $\sim$12 mrad which is consistent with the previous assessment.

After correcting the shower axis residual distributions by that rotation, cf. Fig. 10.13b, their widths are determined as before and are interpreted as the shower axis pointing resolution of the HGCAL prototype. Due to the combination of measurements from multiple layers, the hereby obtained resolution is better than the one for single layers. The evolution of the shower axis pointing resolution as a function of the positron energy is shown in Fig. 10.14a. Again, Eq. 10.14 describes the data points reasonably well and no improvement is achieved when a noise-like term is

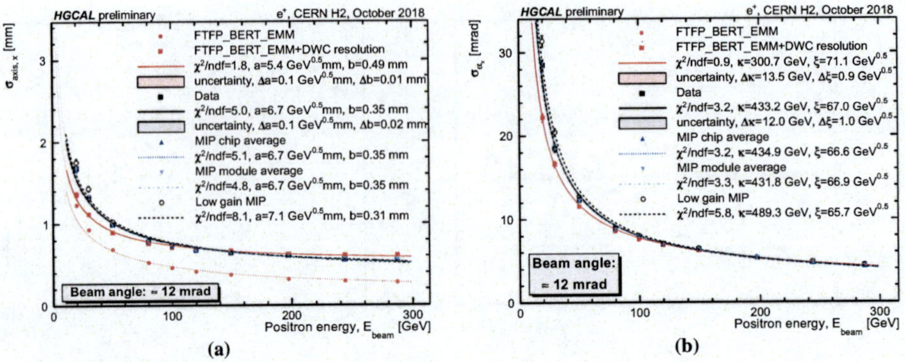

Fig. 10.14 **a** Pointing and **b** angular resolution of reconstructed positron-induced shower axes inside the HGCAL prototype for different energies. The resolution of the reference measurement has not been subtracted from the data. Due to the large distance between the DWCs, their intrinsic measurement precision has negligible effect onto the reference angle measurement

added. Similar to the case for the layer-wise assessment, the data is in reasonable agreement with the simulation at high energies when the finite DWC precision is included. However, it should be noted that an extrapolation to energies beyond the shown scale cannot be reliable since the fitted constant term is below the resolution of the DWC reference. The quadratic distance between the data points at the highest energy to the simulation without the DWC resolution is consistent with the DWC tracking precision. A discrepancy between the data and simulation becomes evident with decreasing energy, which is expected given that the underlying pointing resolution at each layer also shows a similar behaviour, cf. Fig. 10.12b.

Finally, the shower axis angle ($\alpha_{x(y)}$) is compared to the trajectory from the DWCs. In this context, the implications of the intrinsic DWC resolution on the trajectory angle is practically irrelevant due to the much larger spacing between the DWCs in the setup. This is illustrated by the two overlapping simulation results on the angular resolution for different positron energies in Fig. 10.14b. The angular resolution at the highest energy point is below 5 mrad which is reproduced with the beam test data. For lower shower energies, the disagreement between the data and simulation increases like it was observed for the pointing resolutions. It is found empirically that the parameterisation in Eq. 10.15 with noise- and stochastic-like constants κ and ξ provides a suitable description of the angular resolution's dependence on the electromagnetic shower energy.

$$\left(\frac{\sigma_{\alpha,\, x(y)}}{\text{mrad}}\right)^2 = \frac{\kappa^2}{E_{\text{beam}}^2} + \frac{\xi^2}{E_{\text{beam}}} \tag{10.15}$$

One should be aware that the hereby reported values refer only to (almost) perpendicularly impinging particles onto the calorimeter since the divergence of the beam angle, or equivalently the range of tested incidence angles, in the beam test is below 1 mrad. In order to assess the full range of incidence angles, future HGCAL prototypes would have to be rotated in the test beam like it was already done by other R&D experiments, e.g. by the CALICE collaboration [8].

For both the shower axis position (Fig. 10.14a) and the angular measurement (Fig. 10.14b) using the HGCAL calorimeter prototype, the performance does not degrade when chip- or module-averaged MIP calibration constants are used in the hit energy reconstruction. This fact can be explained by the small transverse shower spread such that the contributing channels tend to be read out by the same chip where the MIP calibration constants anyways exhibit minor variation. Moreover, the performance degrades only slightly for lower positron energies when the high gain is excluded from the hit energy reconstruction. This observation is consistent with the statement that hits with significant amount of reconstructed energy, inferred from either the TOT or the low gain, dominate the position measurement.

Summary

In conclusion, it has been established that the calorimeter's granularity together with a suitable energy weighting scheme can in fact be exploited to achieve shower position resolutions of less than 1 mm. The angular resolution to reconstruct the shower axis of perpendicularly impinging positrons is less than 5 mrad above 250 GeV. Both precisions scale with the incident particle energy. The DWCs used in the beam tests are essential for this assessment. After integrating their finite resolution into the simulation, the data points are well described for the higher energy points. The discrepancy at lower beam energies hints at deficits in the calorimeter simulation and/or is related to the neglect of multiple scattering in the reference tracking model. It requires further assessment with accurately modelled DWC data in the simulation in the future.

Revisiting the application example at the end of the Sect. 10.2.2, the decay of the Higgs boson into two photons is considered now. By contrast to charged particles, photons are not detected in the CMS tracker. Instead, their flight direction can be reconstructed in the calorimeter. In addition to the prior simplifications, one may assume the measured angular resolution ($\Delta\theta := \sigma_{\alpha_x}$) to be independent of the incidence angle. Furthermore, the calorimeter performances for photons are assumed to be identical to the ones for electrons/positrons. It has been verified with the beam test data that $\Delta\theta$ and ΔE are uncorrelated. Then, the relative precision in the mass measurement of a Higgs boson with mass $m_H = 125.0\,\mathrm{GeV}/c^2$, decaying into two (isolated) photons with energies $E_1 = 350\,\mathrm{GeV}$ and $E_2 = 370\,\mathrm{GeV}$ (relative angle between the particles $\theta = 20°$) which are detected in HGCAL, should be reconstructable with a relative precision of $\Delta m_H/m_H \approx 1.7\%$, cf. Eq. 10.16.

$$\left(\frac{\Delta m_H}{m_H}\right)^2 = \left(\frac{\Delta E_1}{2 \cdot E_1}\right)^2 + \left(\frac{\Delta E_2}{2 \cdot E_2}\right)^2 + \left(\frac{\cos\frac{\theta}{2}}{2 \cdot \sin\frac{\theta}{2}}\right)^2 \cdot \left(\Delta\theta(E_1)^2 + \Delta\theta(E_2)^2\right) \tag{10.16}$$

✓ **PV-POS.4** The pointing resolution to reconstruct the shower axis amounts to a fraction of a millimetre.

✓ **PV-POS.5** The resolution to reconstruct the shower axis angle is below the level of 10 mrad at high energies.

10.3 Studies with Hadronic Showers

In the previous section, electromagnetic showers were analysed exclusively using the HGCAL prototype's EE compartment. Now, the validation is extended to hadronic showers, induced by negatively charged pions. For this, the FH compartment is indispensable. The primary goal of the studies in this section is the demonstration that the HGCAL calorimeter prototype is functional to reconstruct typical signatures of hadronic showers using its granularity.

As discussed in Sect. 3.2, hadronic showers are conceptually different from electromagnetic showers. The CALICE collaboration has performed many beam test studies on hadronic showers with granular calorimeter prototypes over the last years [9]. A study of track segments inside hadronic showers [10] or the characterisation of their spatial development [11] are just two recent examples. The level of sophistication and attention to systematic uncertainties therein is remarkable. To repeat equivalent studies with the HGCAL prototype is beyond the scope of this thesis.

The previous section has established the functionality of the prototype modules that were placed in the EE compartment for calorimetric measurements. There is no obvious reason why the same should not hold for the modules in the FH compartment. In fact, signals from minimum ionising particles are properly reconstructable in the modules in the FH, cf. Sect. 9.1. Nevertheless, there are a few known and decisive issues that prevent a fully conclusive comparison of hadronic showers to GEANT4 simulation at this time:

- While the performed studies use the configuration 1, cf. Fig. 6.16, the MIP calibration constants for the outer modules in the FH were mostly derived from configuration 3 that was exposed to the parasitic beam (*O3* dataset) in the October 2018 beam test. It is not guaranteed that this calibration is fully transferable to configuration 1 (i.e. to the *O1** datasets), cf. Sect. 9.1.3.
- Most of the outer modules in the FH compartment lack dedicated intercalibration of the low gain to TOT, cf. Sect. 9.2.2 and especially Fig. F.3 in Appendix F.
- Besides charged pions, the test beam may, in principle, contain anti-protons and kaons whose showering signature in the calorimeter is known to differ slightly [12]. Since hadron identification with the provided Cherenkov counters in the H2 beam line was not possible at this energy realm, the exact particle composition of the test

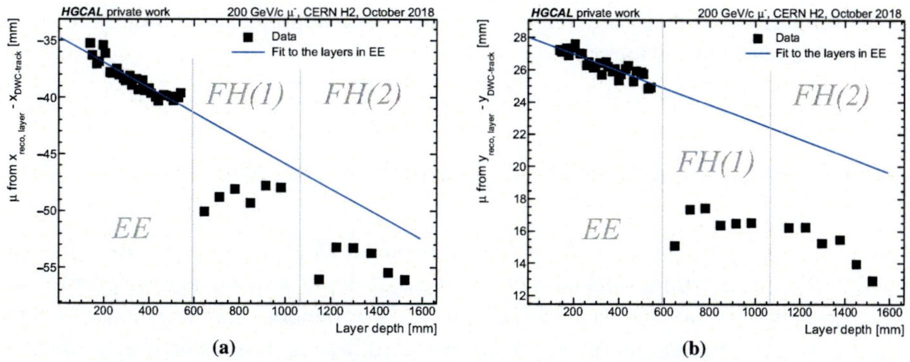

Fig. 10.15 Mean position residuals at the various layer depths computed with 200 GeV/c muons for the full calorimeter prototype in **a** x- and **b** in y-direction. It illustrates the misalignment of the FH compartment's mechanics with respect to the EE

beam could not be measured. Its assessment through simulation of the H2 beam line has not concluded at the time of writing this thesis.

- The two mechanical hanging file structures of the FH compartment were shifted with respect to the EE by more than half the extent of a full hexagonal diode ($\sim$0.5 cm), cf. Fig. 10.15. This shift is not incorporated into the used simulation model that was available at the time of writing this thesis. Thus, this deficit may lead to systematic differences in the transverse leakage and shower shapes between the data and the simulation.
- Even if all above mentioned issues were treated properly, the choice of the most suitable physics list in GEANT4 is not trivial, see e.g. Refs. [13, 14].

In summary, the current simulation is not expected to be in perfect agreement to the experimental data with pion beam, and the studies aim only at the validation of two aspects:

1. The typical signatures of hadronic showers in the HGCAL prototype calorimeter should be qualitatively observed in the data.
2. It remains to be confirmed that the usage of machine learning techniques and with it the exploitation of the granularity can enhance the calorimetric performance.

In contrast to the electromagnetic shower data, potential effects due to the transverse spread of the charged pion beam are assumed to have a negligible impact on the calorimeter performances. Therefore, no dedicated event selection on the transverse acceptance is performed here.

The prototype's capability of reconstructing the first nuclear interaction initiating the particle shower will be used throughout the studies and is elaborated in Sect. 10.3.1. Shower shapes of hadrons showering exclusively in the FH compartment are presented in Sect. 10.3.2. Section 10.3.3 gives insight into the prototype's default energy reconstruction performance of hadronic showers and explains its limitations. The studies on hadronic showers conclude in Sect. 10.3.4 with the motivation and comparison of various machine learning-based approaches in order to enhance the calorimetric performance.

10.3.1 Reconstruction of the Primary Interaction

In theory, the probability $1/N \cdot dN/d\lambda$ of an incident hadron to start showering within the infinitesimal thickness element $d\lambda$ at depth λ in the calorimeter follows an exponential decay with scale λ_0, cf. Eq. 3.8. For incident charged pions, λ_0 corresponds to the pion interaction length λ_π. In the following, it is tested if this behaviour can be detected in the HGCAL prototype calorimeter. Hereby, the number of events dN_l with the primary interaction happening between layers $l-1$ and l is inferred first. In the following, the corresponding layer l is called the *primary interaction layer*. Apart from its dependence on the layer's depth λ_l in the calorimeter, this number in general also relates to the thickness $\Delta\lambda_l = \lambda_l - \lambda_{l-1}$ of the preceding passive material, according to Eq. 10.17.

$$dN_l \propto \exp(-\frac{\lambda_l'}{\lambda_\pi}) \cdot \left(\exp(\frac{\Delta\lambda_l}{2 \cdot \lambda_\pi}) - \exp(-\frac{\Delta\lambda_l}{2 \cdot \lambda_\pi})\right), \quad \lambda_l' := \frac{\lambda_l + \lambda_{l-1}}{2}, \quad \lambda_{l=0} := 0 \tag{10.17}$$

$\Delta\lambda_l$ is not a constant in this calorimeter prototype, but varies. Therefore, the difference term in the brackets is not constant, either, but depends on the depth. In this case, $\Delta\lambda_l$ is assumed to be at least sufficiently small compared to λ_π. In this approximation, the ratio $dN_l/\Delta\lambda_l$ should again follow the exponential decay simplifying the evaluation, cf. Eq. 10.18.

$$\exp(\pm\frac{\Delta\lambda_l}{2 \cdot \lambda_\pi}) \approx 1 \pm \frac{\Delta\lambda_l}{2 \cdot \lambda_\pi} \rightarrow \frac{dN_l}{\Delta\lambda_l}(\lambda_l') \propto \exp\left(-\frac{\lambda_l'}{\lambda_\pi}\right) \tag{10.18}$$

How is the primary interaction layer reconstructed? There are various, application-specific algorithms for the identification of the first calorimeter layer downstream the primary interaction, e.g. for the CALICE AHCAL see Ref. [15]. The specific procedure used in the following is described below:

1. For each layer, the sum of hit energies E_l from hits in the direct vicinity (1.4 cm radius) to the extrapolation from the delay wire chamber track is computed. The restriction of using hits only around the incident particle trajectory renders the algorithm less sensitive to fake hits.
2. If $E_1 > 20$ MIP, the first layer is identified as the primary interaction layer (*PI*).
3. Otherwise, if $E_2 > 20$ MIP and $E_2 > 2 \cdot E_1$, the second layer is identified as the PI.
4. Otherwise, if $E_l > 20$ MIP and $E_l > 2 \cdot E_{l-1}$ and $E_l > 2 \cdot E_{l-2}$ for any of the remaining layers, the lth layer is identified as the PI.
5. Otherwise, no layer is identified as the PI.

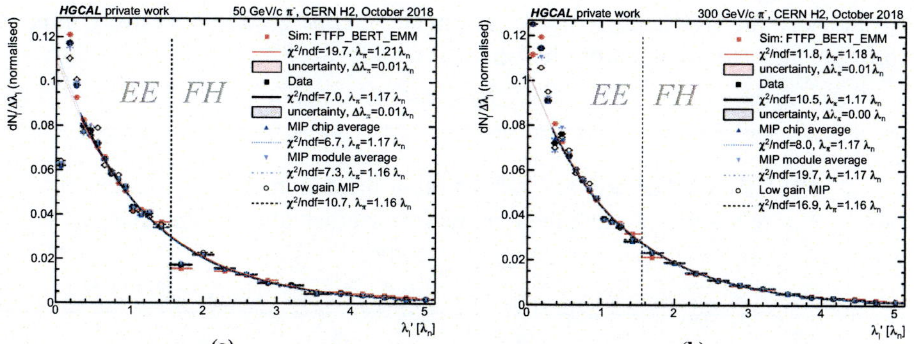

(a) (b)

Fig. 10.16 Normalised distribution of the depth of identified primary interactions, for **a** 50 GeV/c and **b** 300 GeV/c charged pion-induced showers. The depth is given in units of the nuclear interaction length λ_n

The fraction of events with the 50 GeV/c pion beam for which no shower start is identified amounts to roughly 13.0%. For 300 GeV/c, the fraction amounts to 2.5%. These fractions exceed the expectation for non-showering pions in this calorimeter ($\approx e^{-5/1.2} \approx 1.6\%$) which can partially be explained by muon contamination in the beam, e.g. from in-flight pion decay. Since the expected fraction of muons was not known at the time of writing this thesis, this issue is not discussed further.

$\Delta\lambda_l$ is obtained from the simulation model in units of the nuclear interaction length λ_n. The conversion into pion interaction length equivalents is up to ~10% different for the involved main absorber materials (λ_π/λ_n for iron:lead:copper $= 1.22 : 1.13 : 1.21$, cf. Table 3.1). To account for the resulting inhomogeneity in the EE compartment, two calorimeter layers mounted on the same cassette are defined as one sensitive unit in this assessment.

Figure 10.16a presents the measurement of $dN_l/\Delta\lambda_n$ for 50 GeV/c pions, and Fig. 10.16b for 300 GeV/c, respectively. Altogether, the experimental data with showering pions is well described by the exponential decay in Eq. 10.18. The only discrepancies to the model, especially for the lower beam momenta, are observed in the first layers and in the transition between the EE and FH compartments around $\sim 1.55\ \lambda_n$. They hint at the possibility of further improvement of the deployed primary interaction finder algorithm. The incorporation of beam energy- and layer-dependent identification thresholds or the usage of sophisticated machine learning-based methods are conceivable in this regard. After rejection of these outliers, the retrieved scale is $\lambda_\pi = 1.17 \pm 0.01$ (stat.) λ_n in this data. It is consistent with the simulation and, within the approximations, is consistent with the ratio of the pion-to-nuclear interaction lengths of the involved materials. Besides, the distributions exhibit only little impact by the application of alternative MIP calibration constants or when hit energies are inferred without the high gain.

In short, the HGCAL prototype calorimeter can be used to effectively infer the primary interaction of hadron showers therein.

✓ **PV-PION.1** The primary interaction of incident hadrons is reconstructed reliably exploiting the longitudinal segmentation of the HGCAL prototype.

10.3.2 Shower Profiles

The computation of longitudinal shower profiles is identical to the procedure described in Sect. 10.2.1. The representative result for 300 GeV/c pions showering at the beginning of the EE compartment is illustrated in Fig. 10.17a. Like for electromagnetic cascades, the shower maximum is reached within that compartment. As expected due to the long range of nuclear cascades, substantial energy is still deposited in the FH compartment. Figure 10.17b presents the analogous measurement for pions interacting in the first layer of the FH compartment showing that the mean energy in the last layers does not converge to zero. It demonstrates that the total longitudinal depth of this prototype calorimeter is insufficient to fully contain hadronic cascades. It is evident in both examples that the response of the last three layers is lower in the data with respect to the simulation. In the considered configuration, these layers housed only one sensitive module making them particularly sensitive to the aforementioned inconsistencies in the alignment between the experimental setup and the simulation model. Thus, this is not necessarily a deficit of the associated energy calibration.

For completeness, the maximum and minimum extent in each spatial dimension measured exclusively in the FH are reported in Fig. 10.18a, b, respectively. Hereby, the entries for I_{ab}, cf. Eq. 10.4, are computed from the hits in the FH only and the total energy therein E_{FH}, as defined by Eq. 10.19, is used.

$$E_{\mathrm{FH}} := \sum_{i=1}^{N_{\mathrm{hits}} \in \mathrm{FH}} E_i \tag{10.19}$$

Similar to the electromagnetic showers in the EE, those observables show basic agreement to the simulation. The average maximum spread is slightly higher than measured in the beam test. Assuming the maximum spread relates to the longitudinal spread along the beam axis, it may be attributed to the different mean summed energy in the last layers. Neglected noise and misalignment in the simulation might render

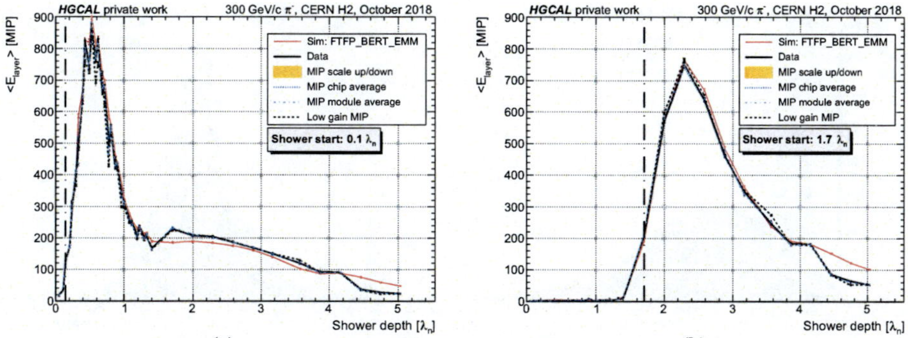

Fig. 10.17 **Longitudinal shower profiles for 300 GeV / c charged pions where the shower is initiated (a) in the first layer of the EE (b) and in the first layer of the FH compartment**

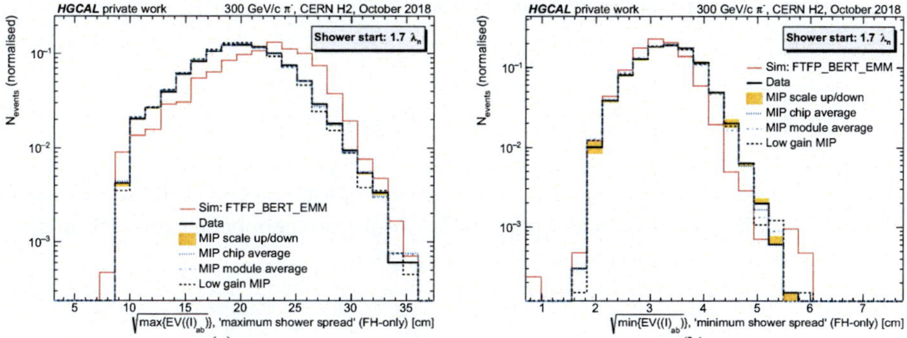

Fig. 10.18 **a** Longitudinal and **b** transverse shower spread for 300 GeV / c charged pions where the shower is initiated at the beginning of the FH compartment. The values are derived from a principal component analysis-inspired method, cf. Eq. 10.4. Mind the logarithmic scale

the minimum spread, interpretable as the transverse shower spread, underestimated in the simulation. Neither of these issues is investigated further in this thesis.

10.3.3 Energy Reconstruction

After the demonstration that basic characteristics of hadronic showers are reconstructed properly in the calorimeter prototype, the focus is shifted towards its main purpose, i.e. the energy measurement. The considered pions in the *O1H* dataset are highly relativistic and the difference between their momentum and their energy is negligible, cf. Eq. 10.20. Hence, the beam momentum is set equal to the particle energy using suitable units.

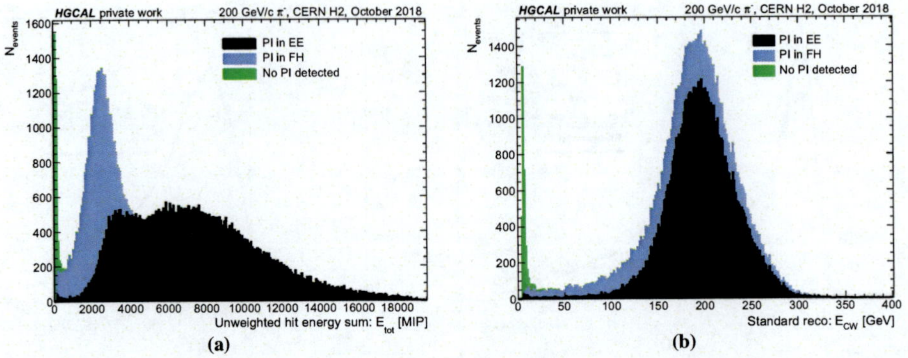

Fig. 10.19 Reconstructed shower energy spectrum for incident 200 GeV/c charged pions **a** without and **b** with incorporation of different compartment weights in the energy sum. Side remark: Events where no PI is detected correspond to small energy sums which is consistent with the signature from non-showering pions or muon contamination in the beam

$$1 - \frac{c \cdot P_{20\,\mathrm{GeV/c}\ \pi^-}}{E_{20\,\mathrm{GeV/c}\ \pi^-}} = 1 - \beta_{20\,\mathrm{GeV/c}\ \pi^-} = 2 \cdot 10^{-5} \qquad (10.20)$$

Usage of the total energy sum E_{tot} from both compartments, cf. Eq. 10.21, as proxy for the particle energy is inadequate for hadronic showers because of the different sampling structure of the EE and FH compartments.

$$E_{\mathrm{tot}} := \sum_{i=1}^{N_{\mathrm{hits}} \in \mathrm{EE,FH}} E_i \qquad (10.21)$$

One instructive example of the spectrum of this unweighted hit energy sum (E_{tot}) is depicted in Fig. 10.19a. It illustrates the necessity for a compartment-specific treatment of energy sums from each compartment when they are summed to a quantity that is related to the true incidence energy.

In this connection, a minimal shower energy reconstruction scheme is used in this section. It shall serve as the default procedure and is subsequently referred to as *standard reco*. It should comprise a weighting of energy sums from each compartment as input to the total energy sum E_{CW} in the calorimeter. In this context, one weight would be sufficient to correct for the different sampling. However, two weights w are introduced instead to also provide the conversion into the final energy scale, cf. Eq. 10.22.

$$E_{\mathrm{CW}} := w_{\mathrm{EE}} \cdot E_{\mathrm{EE}} + w_{\mathrm{FH}} \cdot E_{\mathrm{FH}} \qquad (10.22)$$

The weights w_{EE} and w_{FH} are retrieved from a regression on independently simulated shower samples of 5–350 GeV/c pions with a shower start in the calorimeter. The true energy of the incident particle is known in the simulation. For each data point,

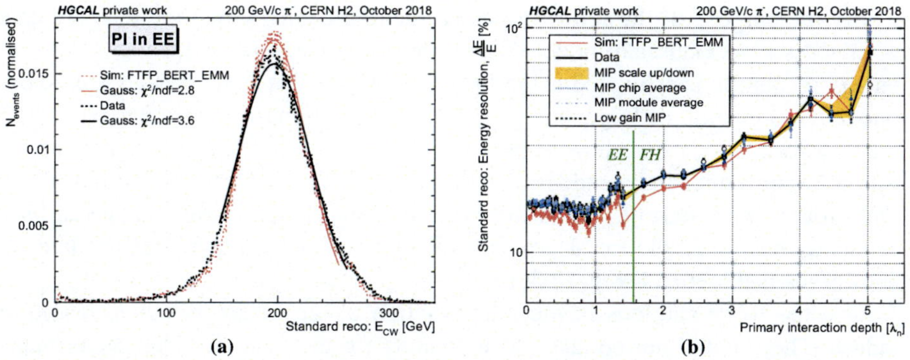

Fig. 10.20 **a** Distribution of compartment-weighted energy sums for 200 GeV/c charged pions whose showers are initiated in the EE. The distribution's core is not accurately described by a Gaussian. **b** Energy resolution estimates for charged pion showers that are initiated at different depths. The increase for showers that are initiated later is due to increasing longitudinal leakage

the agreement between the reconstructed shower energy to the truth is quantified using the χ_E-measure as it is defined by Eq. 10.23.

$$\chi_E = \chi_E(w) := \frac{(E_{\text{CW}}(w) - E_{\text{true}})^2}{E_{\text{true}}} \tag{10.23}$$

Circumventing a potential bias from outliers, the weight optimisation, formulated in Eq. 10.24, targets at minimising the 95%-quantile of the χ_E-distribution instead of its average.

$$w = \text{argmin}_{\chi_E}\{\int_0^{\chi_E} \frac{dN}{d\chi'_E} d\chi'_E = 0.95\} \tag{10.24}$$

The result from this regression is $w_{\text{EE}} = 0.015\,\text{GeV/MIP}$ and $w_{\text{FH}} = 0.079\,\text{GeV/MIP}$. Figure 10.19b presents the previously discussed energy distribution after incorporation of these weights which now exhibits only one core as intended. The tail towards small values of E_{CW} hints at the presence of shower leakage meaning the calorimeter is not large enough to fully contain and measure all shower constituents. The entries at low values for E_{CW} are due to non-showering pions in the calorimeter and the presence of muons in the beam of charged pions.

Figure 10.20a focusses on the normalised distribution of reconstructed shower energy for pions that primarily interact in the EE compartment. Applying iterative Gaussian fits is found to be unsuitable for the quantitative characterisation of its core which is visible from the large χ^2/ndf from the fit. In case of a non-compensating calorimeter as this one, non-Gaussian energy distributions are generally expected, cf. Sect. 3.2.2. When using this minimal energy reconstruction scheme, any differences between this calorimeter's responses to electromagnetic and hadronic shower constituents are not corrected for. Alternative approaches aiming at compensation for each event are studied in the next section.

For the purpose of the following assessment, the energy distributions are characterised with a Gaussian model despite their non-Gaussian behaviour. Then, the energy resolution is estimated using the convention in Eq. 10.8. It is approximately one order of magnitude larger compared to the resolution for electromagnetic showers, cf. Sect. 10.2.2. For showers that are initiated in the EE, it is found that the estimates are rather independent of where exactly the primary interaction occurs. Only for showers starting within the FH, increasing longitudinal leakage and subsequently energy resolution degradation is observed which increases the later the shower starts therein, cf. Fig. 10.20b.

One has to be cautious in the interpretation of the hereby obtained resolution values. They should not be taken as indicators for the final HGCAL performance. The calorimetric performance of the beam test prototype is significantly impaired by transverse and longitudinal shower leakage. To some extent, longitudinal leakage can be minimised in this assessment by also including the data from the CALICE AHCAL prototype which acted as a backing calorimeter for the HGCAL prototype. The event numbers between the two prototype calorimeters have been synchronised for the *O1H* programme as work for this thesis using the algorithm in Sect. 6.2.4.

Despite the aforementioned deficits, the simulation describes the calorimetric performance in the beam test data rather reasonably. This basic agreement is a prerequisite for the studies in the next section.

10.3.4 Advanced Energy Reconstruction Methods

By construction, the energy reconstruction procedure in the previous section has not benefitted neither from the HGCAL prototype's longitudinal segmentation nor from its transverse granularity. In this section, it is investigated if these features can help in improving the energy measurement for hadronic showers. In the following, three additional algorithms are first introduced and then compared to the standard reco.

All algorithms contain internal parameters subject to optimisation. The optimisation is analogous to Eq. 10.24 and in particular makes use of the same simulated pion dataset as for the applied tuning of w_{EE} and w_{FH} in the standard reco. The algorithms are implemented as graph computations using the TensorFlow [16] software package. TensorFlow provides tools, such as the *ADAM* method [17] used in this work, for the simultaneous optimisation of $\mathcal{O}\left(10^6\right)$ free parameters in the deployed algorithms.

1. Layer Weighting

The first tested algorithm utilises the longitudinal segmentation of the HGCAL prototype and is basically an extension of the standard reco. Instead of applying only one weight per compartment, each layer energy E_{layer} is multiplied by a distinct weight w_{layer} in the sum E_{lw}, cf. Eq. 10.25.

$$E_{\mathrm{lw}} := \sum_{\mathrm{layer}=1}^{40} w_{\mathrm{layer}} \cdot E_{\mathrm{layer}} \tag{10.25}$$

This approach was already studied for the HGCAL design in the "Trigger simulation and performance" section of the technical design report [18]. Figure 10.21a visualises the obtained weights for the beam test prototype in comparison to the previously discussed compartment weights, cf. Sect. 10.3.3. Apart from the first and last layers, the values roughly follow the expectation from the energy loss in the passive material between them. The high value in the very last layer may be interpreted as an effect of longitudinal leakage.

2. Software Compensation

The second method is intended to also profit from the calorimeter's transverse granularity in order to achieve compensation of the generally different responses to electromagnetic and hadronic shower constituents. For each energy deposition, a characteristic is computed that relates to the dominating constituent type that generated it. Determined by this attributed characteristic, the hit energy is accordingly weighted and summed to the total signal. The CALICE collaboration has originally proposed this scheme and refers to it as "local software compensation" because it considers each energy deposition locally. In their applications, it is reported that the calorimetric performance for hadronic showers is significantly improved (hadronic energy resolution by $\mathcal{O}(20\%)$) when this scheme is applied [19, 20]. The algorithm to compute the software compensated energy sum E_{SW} that is used in this thesis is defined by Eq. 10.26. It may be considered as a simplified version of the CALICE scheme.

$$E_{\mathrm{SW}} := w_{\mathrm{EE}} \cdot \sum_{i=1}^{N_{\mathrm{hits}} \in EE} E_i \cdot w_{\mathrm{SW}}\left(\frac{E_i}{E_{\mathrm{tot}}}\right) + w_{\mathrm{FH}} \cdot \sum_{i=1}^{N_{\mathrm{hits}} \in FH} E_i \cdot w_{\mathrm{SW}}\left(\frac{E_i}{E_{\mathrm{tot}}}\right) \tag{10.26}$$

The starting point is the standard reco. It is extended by the introduction of a function w_{SW} which takes the hit's relative energy contribution to the total signal sum (E_i/E_{tot}) as input. The physics intuition is that calorimeter hits are equivalent to the measurement of local shower energy densities. For a non-compensating calorimeter, electromagnetic and hadronic shower constituents generate different energy densities on average and should be separable by this means. Other approaches, such as the one described in Ref. [21], define the absolute hit energy as input to the weight function w_{SW} but simultaneously allow also for an energy dependence of it. For the proof of principle here, the division of the hit energy by the total shower signal should render this algorithm less sensitive to the total shower energy.

w_{SW} is approximated as a step function with eleven steps. The binning was chosen such that all bins are filled in an average shower, though its choice lacks systematical assessment. Figure 10.21b presents the result of the optimisation. Following the physical intuition for a calorimeter with $h/e < 1$, lower energy densities that relate

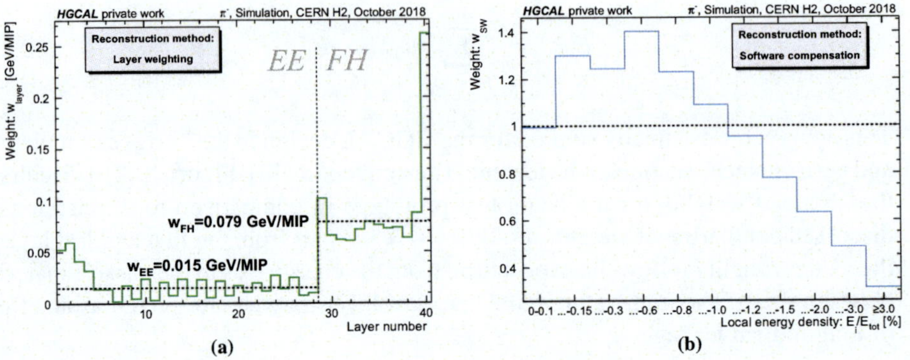

Fig. 10.21 **a** Layer weights for the two compartments in comparison to the layer weights obtained in the standard reco (green). **b** Software compensation weights that are determined by a hit's relative contribution to the summed energy of the shower. The values are obtained from the optimisation of Eq. 10.24

to hadronic shower constituents are multiplied with a relatively higher weight in the sum for E_{SW}. Hence, a compensating effect is demonstrated.

3. 2D-Convolutional Deep Neural Network

Driven by the broad availability of deep learning [22] methods, an "explosion of applications in particle and event identification and reconstruction in the 2010s" [23] using machine learning has happened. Also for the final HGCAL design, sophisticated deep learning-based algorithms, in particular variations of convolutional and graph neural networks [24], are currently being studied in regards of their possible application.

The idea in the context of the prototype's beam test data analysis is as follows: Reconstructed hits are arranged as a 3D shower image $\{(x,\ y,\ z,\ E)_i\}$. For this purpose, the hexagonal, layer-wise coordinate system is mapped to a Cartesian one pursuing the strategy as illustrated in Fig. G.1a in the Appendix G. Afterwards, the shower image is input to a deep neural network for inference of the incident particle's energy E_{DNN}, cf. Eq. 10.27.

$$E_{\mathrm{DNN}} := f_w(\{(x,\ y,\ z,\ E)_i\}) \qquad (10.27)$$

The chosen network is denoted as f_w, the main building blocks are two-dimensional convolutions and its exact architecture is documented in Table G.1. It comprises $\mathcal{O}\left(7 \cdot 10^5\right)$ free parameters in total which are optimised like before. Through this so-called *network training*, f_w may learn to incorporate abstract representations of the incident particle energy with respect to the input shower image. In theory, a neural network with sufficient capacities is able to approximate any arbitrary functional relationship with arbitrary precision, cf. Fig. G.1b. The hereby learnt representations by f_w may not always be intuitive. However, they tend to fully exploit the available information which in turn often translates into an improved measurement precision.

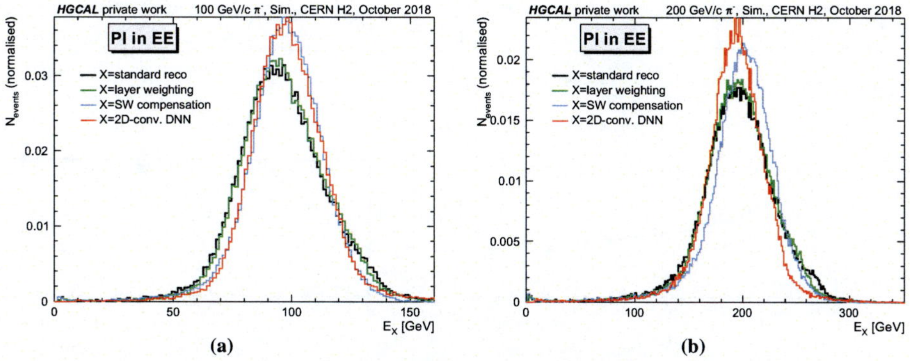

Fig. 10.22 Distribution of the computed shower energy with the different energy reconstruction methods discussed in this section **a** for 100 GeV/c charged pions and **b** for 200 GeV/c charged pions. The evaluation is based on simulated charged pions with a shower start anywhere inside the EE compartment

Successful applications of this concept in calorimetry have already been reported elsewhere, e.g. for collider physics [25] or for astroparticle physics [26].

Method Summary

In summary, besides the standard reco there are three alternative methods tested. The first one uses the longitudinal segmentation, the second one uses the transverse granularity in addition and is motivated by physics intuition, whereas the third processes the full shower image with the downside of being potentially untraceable. The optimisation of all involved parameters is identical to Sect. 10.3.3: It is based on the objective formulated by Eqs. 10.23 and 10.24. Moreover, it makes use of simulated samples of 5–350 GeV/c pions with a shower start in the calorimeter.

Performance Comparison and Discussion

While a uniform distribution of shower energies was used for the training, the following evaluation is based on independent and energy-binned, simulated samples. Unless otherwise stated, the evaluation is limited to pion showers that are initiated anywhere inside the EE compartment. Exemplary distributions of reconstructed shower energies are given in Fig. 10.22. Gaussian fits to those distributions are applied for the quantification of the calorimetric performance. The corresponding parameters are denoted as μ_h and σ_h to clearly distinguish them from their electromagnetic counterpart in Sect. 10.2.2. Using again the definition from Eq. 10.8, Fig. 10.23a presents the energy resolutions for all four methods. The mean deviation from the nominal particle energy, i.e. the assessment of the energy linearity, is shown in Fig. 10.23b.

Introducing weights for each layer does not improve the energy resolution with respect to the standard reco, whereas the energy linearity shows slight improvement. By contrast, significant enhancement of the energy resolution is achieved both through software compensation and the convolutional neural network. The precision at the highest energy point is improved by more than 15% compared to the standard

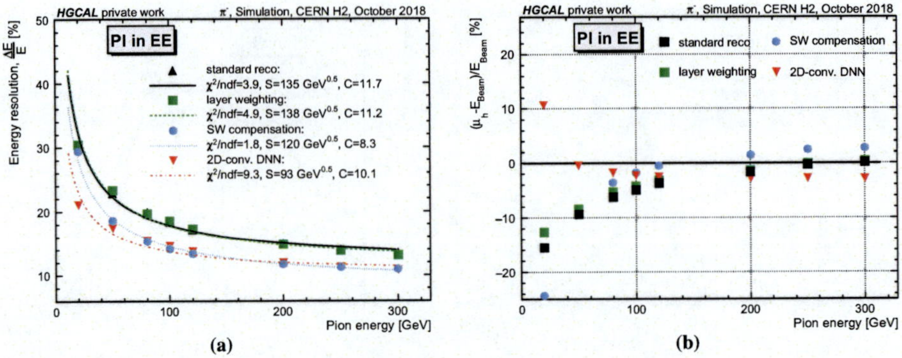

Fig. 10.23 **a** Comparison of the energy resolution and **b** the linearity for the four energy reconstruction methods discussed in this section. The evaluation is based on simulated charged pions with a shower start anywhere inside the EE compartment. The definition of the fit parameters in the legend is analogous to Sect. 10.2.2 and the values are given in %

reco. This improvement comes at the cost of a worsened linearity at low energies. The systematic underestimate of the reconstructed particle energy for low beam energies using software compensation is likely related to the neglect of any shower energy dependence of w_{SW}. Due to the limited extent of a low energetic hadron shower in the calorimeter, the number of hits is systematically lowered and every hit tends to carry a higher fraction of the total sum. Therefore, the energy depositions tend to be assigned to the higher energy density bins and are down-weighted in the sum. By contrast, the overestimate for the DNN may qualitatively be explained by positive bias weights in the last network layers.

Nevertheless, it is remarkable that the physically inspired software compensation method competes well with the performance of the, in many applications more powerful, deep neural network approach. This statement holds true even after testing two alternative neural network architectures (based on layer-wise 2D convolutions or 3D-convolutional architectures, not presented here) in this regard. This finding suggests that the relevant calorimeter information is interpretable with physics intuition alone. Consequently, the subsequent assessments are focused on the results from the software compensation method.

The derived weights from the regression on the simulated samples are applied to the beam test data. Then, the cores of the distributions of reconstructed energy, in fact, become more Gaussian-like than for the standard reco when using the software compensation, cf. Figs. 10.24a and 10.20a. Again, Fig. 10.24b shows the degradation of the resolution for later starting cascades in the FH compartment.

Despite the listed, known deficits at the beginning of Sect. 10.3, the agreement of the simulation to the experimental data is reasonable. The slope of the energy linearity, and with it the energy scale, in Fig. 10.25a for the data and the simulation is in better agreement than what was obtained for the electromagnetic showers. There is a sizeable offset b of $-1.1\,\mathrm{GeV}$ ($-5.9\,\mathrm{GeV}$) in the measured (simulated) response towards small pion energies. The offsets derived from beam test data and

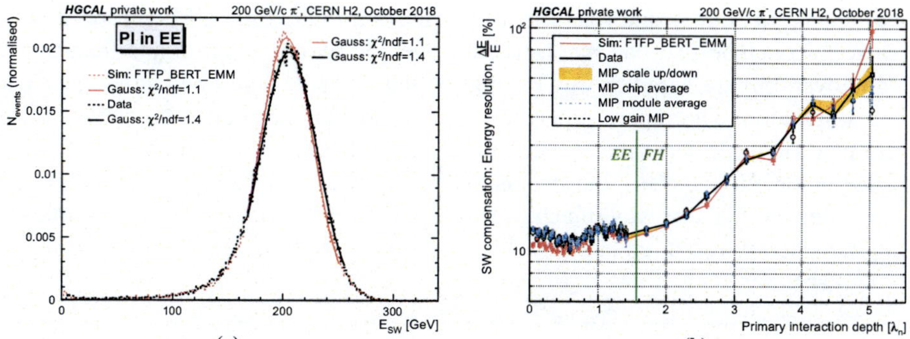

(a) (b)

Fig. 10.24 **a** Distribution of reconstructed particle energy using software (*SW*) compensation for 200 GeV/c charged pions. This method compensates for the different response to electromagnetic and hadronic showers. Thus, the core of the distribution becomes more Gaussian with respect to the standard reco. **b** Energy resolution for charged pion showers initiated at different depths. The increase for later showers is due to the limited longitudinal coverage of the HGCAL beam test prototype

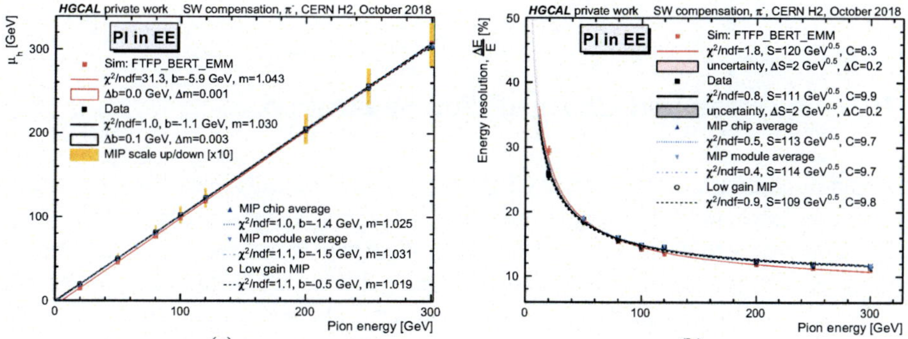

(a) (b)

Fig. 10.25 **a** Energy linearity and **b** resolution of the HGCAL prototype calorimeter for charged pions with a shower start anywhere inside the EE compartment using a variation of the CALICE inspired, local software compensation. Data from the CALICE AHCAL prototype are not included in these results. The underlying evaluation of the reconstructed shower energy distributions is identical to the procedure in Sect. 10.2.2

the simulation disagree. Among many possible explanations, this discrepancy may imply that the energy density profile for lower particle energies is not simulated realistically. On the other hand, the resolution is well described by the simulation, cf. Fig. 10.25b. The measured constant term C amounts to roughly 10% for showers that are initiated inside the EE compartment. Furthermore, the measured energy resolution is not affected by alternative MIP calibrations or when the hit energy reconstruction without high gain is applied.

It is emphasised that the presented energy reconstruction approaches in this part should be understood as prototype procedures. There are clear ideas concerning the current shortcomings and for further optimisation in the future. Therefore, it is too early to conclude that one reconstruction method is more appropriate than others. Moreover, the measured energy resolution and linearity is in good agreement with the simulation. The reported performance parameters are not to be quoted as the ultimate HGCAL performances due to the limited size of the prototype in the *O1H* programme. Nonetheless, the results of this study at this stage are consistent with the statement that one way of achieving response compensation and with it improving the energy resolution can be achieved through exploitation of the calorimeter's granularity.

✓ **PV-PION.2** The energy resolution for hadron-induced showers is improved when compensating for the different responses to electromagnetic and hadronic constituents.

✓ **PV-PION.3** The energy resolution and linearity of hadron-induced showers exhibit basic agreement to the simulation.

10.4 Electron and Charged Pion Separation

An additional task of the current CMS electromagnetic calorimeter is particle identification, especially the separation of electrons and charged pions [27]. A similar functionality is foreseen for the HGCAL [18]. Early showering pions can resemble the typical signature from electrons and their distinction in the calorimeter often requires sophisticated analysis of the shower data. This section assesses if separation of electrons and pions in the test beam can be achieved with the HGCAL prototype calorimeter.

Manual Classifier Engineering

A large fraction of charged pions interact and create particle showers downstream the first layers of the EE compartment, whereas electromagnetic showers are more likely to emerge at the beginning. Hence, the majority of charged pions should be identifiable by the presence of a MIP-like trajectory in the first layers. Specifically, the PI identification described in Sect. 10.3.1 could be applied to reject more than 85% of the pions if events with a PI downstream the first three layers were excluded. But how can those pions that shower earlier than that be identified? As elaborated in Sect. 3.2, hadronic showers tend to be more sparse, more spread, and in particular tend not to have a compact shower core compared to the electromagnetic ones. In general, the calorimeter's granularity should be helpful in detecting such differences. The manual engineering of meaningful summary statistics given the granular calorimeter data is not always straightforward and it shall not be attempted here.

Usage of Deep Learning Classifiers

Instead, state-of-the-art image recognition techniques based on deep learning for the particle identification are investigated in the following. Classification algorithms using deep learning are already deployed for the reconstruction of particle jets in the CMS detector [28] and attempts in the context of calorimetry have also been published already, e.g. in Ref. [29]. The subsequently described approach, that is used in this work, is not a conceptual novelty. Complementing the existing studies that are largely based on idealised calorimeter simulation, the second part of this section will give insight into the applicability of such deep learning methods to realistic calorimeter data. Furthermore, the relevance of the calorimeter's granularity for the task at hand will be addressed. Therefore, the data from the FH compartment are excluded from the classification analysis because the sum of energy depositions therein would otherwise be the most dominant handle for the decision. Significant activity in the EE compartment based on the number of hits in the EE ($N_{\text{hits, EE}} \geq 60$) is required to reject non-interacting pions therein from the analysis.

The classification can be formulated as it is done in Eq. 10.28.

$$D_{e\text{ vs. }\pi,\, d} := DNN_d(\{(x,\ y,\ z,\ E)_i\ ,\ i \in EE\})\ ,\quad D_{e\text{ vs. }\pi,\, d} \in [0, 1]\ ,\quad d = \{1, 2, 3\} \tag{10.28}$$

By definition, values of $D_{e\text{ vs. }\pi,\, d}$ close to 0 shall relate to electron-induced showers whereas values close to 1 shall imply the incidence of a charged pion. The processing of the calorimeter hits in the EE compartment to shower images $\{(x,\ y,\ z,\ E)_i\ ,\ i \in \text{EE}\}$ is analogous to Sect. 10.3.4. Like f_w for the energy regression, DNN_d denotes a deep neural network with a several hundred thousand free parameters which is referred to as *classifier*.

Three different classifier architectures, differentiated by the index d, are tested. DNN_1 applies a layer-wise evaluation and combines the information in a feedforward structure, DNN_2 contains 2D convolutions and the building blocks of DNN_3 are 3D convolutions. The detailed description of the network architectures is provided by Tables G.2–G.4 in the Appendix G.

Network training is performed by batch-wise minimisation of the so-called cross-entropy L_d in Eq. 10.29 using the ADAM method [17] implemented in TensorFlow [16].

$$L_d = -\sum_{\text{events}}^{\text{batch}} y_{\text{true}} \cdot \ln\left(D_{e\text{ vs. }\pi,\, d}\right) + (1 - y_{\text{true}}) \cdot \ln\left(1 - D_{e\text{ vs. }\pi,\, d}\right)$$
$$y_{\text{true}} := 0,\ \text{if incident particle is true electron (positron), else: } y_{\text{true}} := 1 \tag{10.29}$$

Simulated electrons and charged pions with uniformly distributed incidence momentum between 5–350 GeV/c serve as the training sample. y_{true} is taken from the truth. Event weights are introduced to provide the same yield of both particle types in the network training. The training of each network d is stopped when L_d reaches a constant plateau.

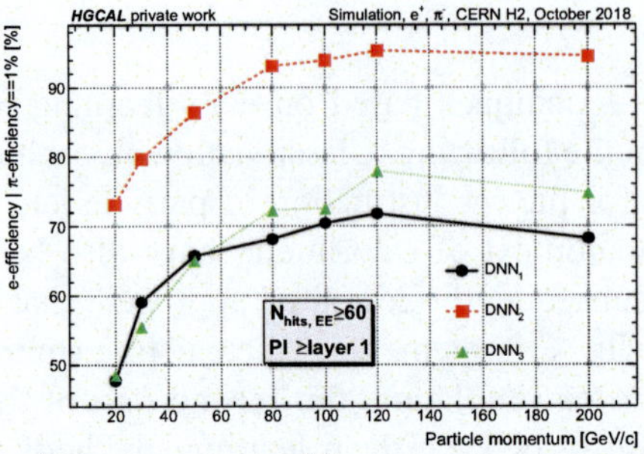

(a)

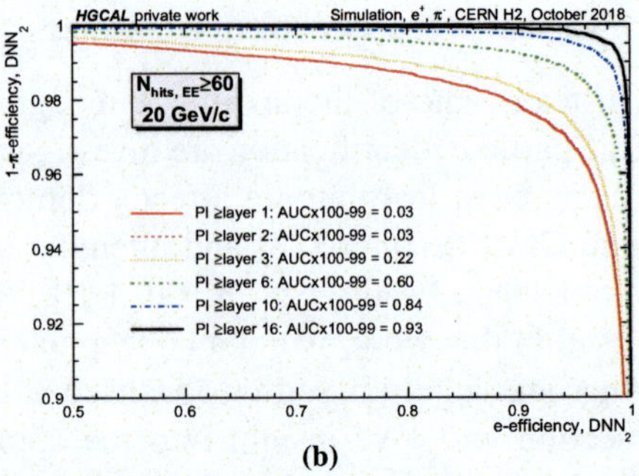

(b)

Fig. 10.26 **a** Electron signal efficiency at the medium working point for the three studied DNN-based classifiers as a function of the incident particle momentum. **b** Receiver operating characteristic (*ROC*) curves for the preferred classifier DNN_2 for different selections of the earliest PI. *AUC* stands for area under the curve

Performance Comparison of the Classifiers

For the subsequent evaluation, independent and energy-binned, simulated samples are used. Events with associated values of $D_{e \text{ vs. } \pi,\ d}$ below a certain threshold are classified as being electron-like, whereas all others are identified as pion-like. The medium working point is defined as the threshold for which the misidentification efficiency amounts to 1%. Defining pions as background, the electron signal efficiency at the medium working point for the three trained network architectures is presented in Fig. 10.26a. The efficiency shows a dependence on the beam momentum which is not discussed further here. The sizeable performance difference between the three investigated architectures is most notable. DNN_2 with 2D convolutional building blocks in which the kernels cover all layers at once (using the analogy to typical image recognition, the calorimeters layers can be interpreted as colours), appears to be the most powerful one in this attempt. Certainly, further studies are necessary to generalise the finding from this example application to a generic statement on preferred or deprecated network architectures in the context of electron-pion identification. For the purpose of brevity, the following report is limited to DNN_2 unless it is stated otherwise.

Interpretation of the Classifier Inference

Figure 10.26b provides visual support for the initially outlined physics intuition that separation can be achieved through identification of the PI. It shows the receiver operating characteristic (*ROC*) curves for different selections of the earliest PI layer. The ROC curve integral, and with it the separation power of this classification, monotonously increase the later the shower is initiated.

Nevertheless, there are indications that the classifiers are also sensitive to more subtle differences between both types of showers. As an example, the two event displays in Fig. 10.27 visualise the reconstructed data of two events in a beam of 200 GeV/c positrons for which the first layer is reconstructed as the PI. The shower in Fig. 10.27a is characterised by all three classifiers as electron-like which is in

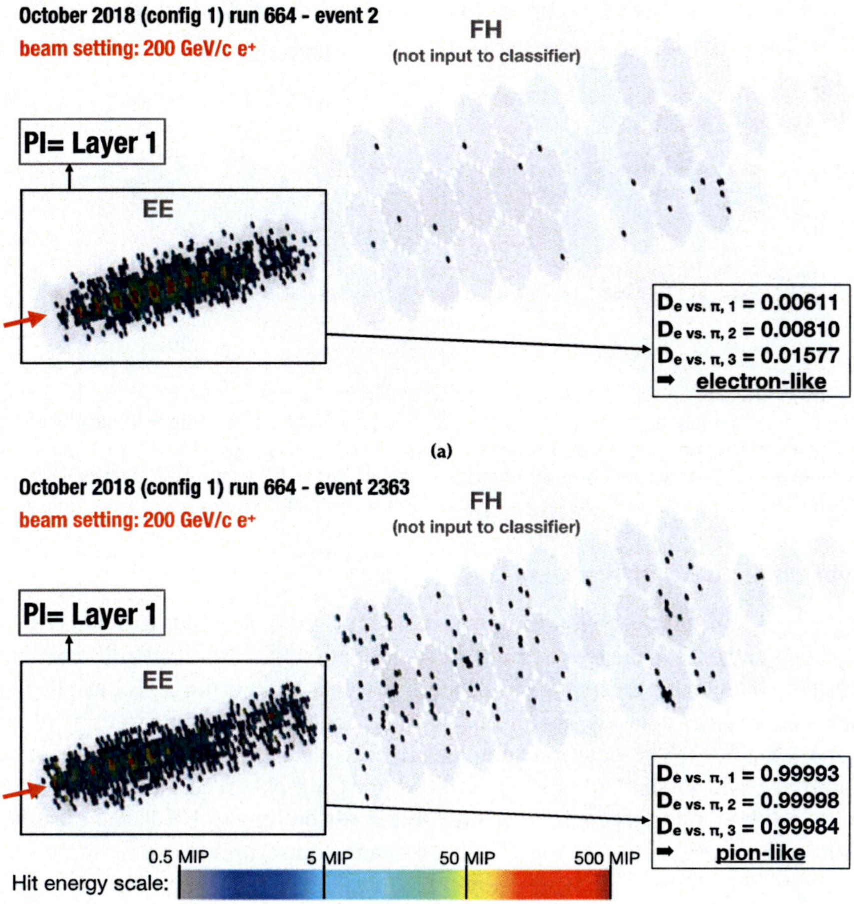

(a)

(b)

Fig. 10.27 Example of **a** an electron-like and **b** a pion-like shower recorded with test beam of positrons. The reconstructed shower start corresponds to the first layer. Only the hits from the EE compartment are input to the DNN classifiers

agreement with the absence of hits in the FH compartment. By contrast, the shower displayed in Fig. 10.27b causes numerous hits in the FH which hints at the presence of an incident pion. Considering only the EE, the different classifiers are able to identify the pion-like signature in this event. Due to the granularity in the measurement, differences in the shower evolution between both examples can also be spotted visually. Both example showers exhibit different levels of sparsity, spread and centrality of their cores. Without manually quantifying these differences beforehand, the classifiers appear to have learnt an abstract representation of this behaviour.

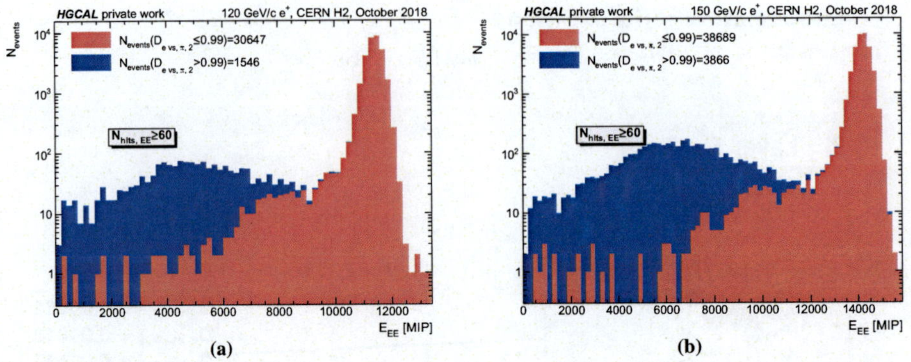

Fig. 10.28 Reconstructed energy sum in the EE compartment for **a** 120 GeV/c and **b** 150 GeV/c particles from a run with positrons. Electron-identified contributions are shown in red and and pion-identified contributions are blue. Together with the 100 GeV/c sample, these two instances show the relatively highest amount of pion-like signatures inside the EE in the *O1E* dataset

Application to Beam Test Data

For this reason, this deep learning-based electron-pion classifier is deployed in the following rather than the reconstructed PI information for particle identification in the beam test data in this thesis. DNN_2 has been evaluated accordingly for all showers in the *O1E* and *O1H* datasets. By this means, it is found that a sizeable fraction of events in the 100-150 GeV/c positron samples are classified as pions. A notable contamination of positron signatures in the pion beam is not observed. Pion contamination in the positron samples causes considerable contributions to EE-based observables such as the energy sum, cf. Fig. 10.28a, b. Thus, a discriminator maximum of 0.99 corresponding to the loose working point, equivalent to 10% pion efficiency, for simulated 80 GeV/c electrons was tacitly applied for positron purification in the studies of Sect. 10.2.

Despite their separation power, the applicability of these classifiers for inference of the beam composition is limited. This limitation becomes evident when attempting to infer the fraction of electrons and pions using the measured distribution of $D_{e\text{ vs. }\pi,\,d}$. The fitting of simulated templates, exemplified in Fig. 10.29a, is not reliable because the phase space of intermediate classifier values is underrepresented in the simulation. The addition of simulated templates for anti-protons or kaons in the beam did not yield any improvement.

A potential cause for this discrepancy in the intermediate classifier values is the deficit in the simulation of the electromagnetic shower shapes. For assessment of this hypothesis, the underrepresented frequency of transverse shower spread values above 1.3 cm in the simulation, cf. Fig. 10.6b, is revisited. After positron purification based on the hit occupancy in the FH compartment ($N_{\text{hits, FH}} \leq 80$), the discriminator distribution for the positron data is shown as the black curve in Fig. 10.29b. The frequency of positron-like signatures, as identified by DNN_2, in this sample is two orders of magnitude higher than of pion-like signatures. This difference is reduced

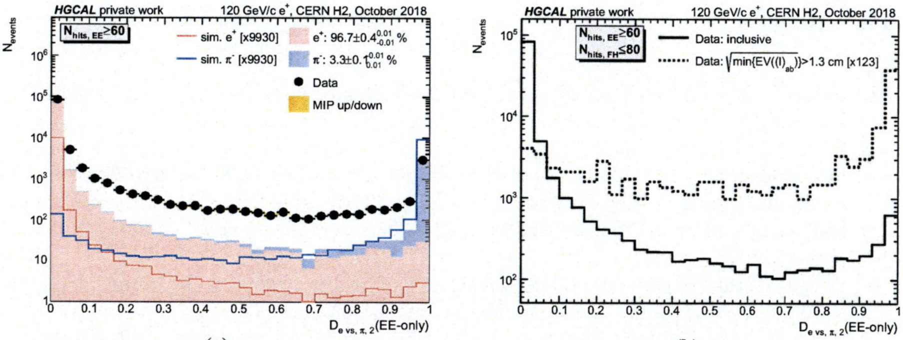

(a) (b)

Fig. 10.29 **a** Distribution of electron-pion classifier values for 120 GeV/c positrons. Simulated templates are stacked and scaled to match the frequency in the data. The indicated errors correspond to the statistical uncertainty from the template fit and to the change in the fitted fractions when shifting the MIP calibration scale up and down by its uncertainty. **b** Discriminator distributions for the 120 GeV/c positron dataset. The sample is purified with positrons selecting events with minimal activity in the FH compartment. The solid curve corresponds to the inclusive dataset while the dotted one corresponds to an additional selection of showers with a transverse shower spread characteristic measured in the EE that is underrepresented in the simulation. Note that the y-axes in the graphics are logarithmic

(and even reverted) when the analysis is restricted to events with transverse shower spread beyond 1.3 cm. Then, showers with that particular shower spread are not properly classified as electron-like, hereby revealing a weakness of this deep learning-based classification approach.

Summary

In summary, it is demonstrated that state-of-the-art deep learning methods are in principle applicable for the distinction of electromagnetically and hadronically induced cascades in the HGCAL prototype calorimeter. This section has exemplarily shown that its granularity may add relevant input to this task. In the context of the ongoing beam test analysis, the tested methods indicate a small contamination of pions in the *O1E* dataset. A first study revealed that the deficit in the simulation of showers with intermediate classifier values is correlated to insufficient modelling of electromagnetic shower shapes. It emphasises the importance of accurate calorimeter simulation, when sophisticated methods are designed with Monte Carlo generated samples and then are intended for inference on realistic data.

✓ **PV-EPI.1** The classifier is able to separate electromagnetic and charged hadron-induced signatures also when it is blinded to the measurements in the hadronic calorimeter compartment.

∘ **PV-EPI.2** The classifier is not fully applicable to realistic calorimeter data.

References

1. Quast T (2019) Geant4 based event display & standalone simulation. https://github.com/ThorbenQuast/HGCal_TB_Geant4/. Tag: v1.0. Accessed 12 Aug 2019
2. Banerjee S (2017) Validation of physics models of Geant4 using data from CMS experiment. J Phys: Conf Ser 898:042005. http://dx.doi.org/10.1088/1742-6596/898/4/042005
3. Akchurin N, et al. (2018) First beam tests of prototype silicon modules for the CMS High granularity endcap calorimeter. JINST 13:P10023. http://dx.doi.org/10.1088/1748-0221/13/10/p10023
4. Anagul S, et al. Response of a CMS HGCAL silicon-pad electromagnetic calorimeter prototype to 20-300 GeV positrons. In preparation for submission to JINST
5. Bohm G, Zech G (2005) Einführung in Statistik und Messwertanalyse für Physiker. Verlag Deutsches Elektronen-Synchrotron
6. Acosta D, et al. (1991) Localizing particles showering in a spaghetti calorimeter. Nucl Instrum Meth A 305:55–70. http://dx.doi.org/10.1016/0168-9002(91)90519-V
7. Gorodnichev VB, et al. (1994) Study of position resolution and electron-hadron separation of electromagnetic calorimeter with a silicon structure. Nucl Instrum Meth A 343:470–475. http://dx.doi.org/10.1016/0168-9002(94)90226-7
8. CALICE Collaboration (2008) Design and electronics commissioning of the physics prototype of a Si-W electromagnetic calorimeter for the international linear collider. JINST 3. http://dx.doi.org/10.1088/1748-0221/3/08/P08001
9. Sefkow F, et al. (2016) Experimental tests of particle flow calorimetry. Rev Mod Phys 88:015003. http://dx.doi.org/10.1103/RevModPhys.88.015003
10. Deng Z, et al. (2017) Tracking within hadronic showers in the CALICE SDHCAL prototype using a hough transform technique. JINST 12:P05009. http://dx.doi.org/10.1088/1748-0221/12/05/P05009
11. Eigen G, et al. (2016) Hadron shower decomposition in the highly granular CALICE analogue hadron calorimeter. JINST 11:P06013. http://dx.doi.org/10.1088/1748-0221/11/06/P06013
12. Bilki B, et al. (2015) Pion and proton showers in the CALICE scintillator-steel analogue hadron calorimeter. JINST 10:P04014. http://dx.doi.org/10.1088/1748-0221/10/04/p04014
13. Bilki B, et al. (2015) Testing hadronic interaction models using a highly granular silicon-tungsten calorimeter. Nucl Instrum Meth A 794:240–254. http://dx.doi.org/10.1016/j.nima.2015.05.009
14. Adloff C, et al. (2013) Validation of GEANT4 Monte Carlo models with a highly granular scintillator-steel hadron calorimeter. JINST 8:P07005. http://dx.doi.org/10.1088/1748-0221/8/07/p07005
15. CALICE Collaboration (2011) Pion showers in the CALICE AHCAL prototype. tCAN-026. https://twiki.cern.ch/twiki/pub/CALICE/CaliceAnalysisNotes/CAN-026.pdf. Accessed 05 Nov 2019
16. Abadi M, et al. (2016) TensorFlow: large-scale machine learning on heterogeneous distributed systems. arXiv:1603.04467 [cs.DC]
17. Kingma DP, Ba J (2014) ADAM: a method for stochastic optimization. arXiv:1412.6980 [cs.LG]
18. CMS Collaboration (2017) The phase-2 upgrade of the CMS endcap calorimeter. CERN-LHCC-2017-023 (CMS-TDR-019). https://cds.cern.ch/record/2293646
19. Adloff C, et al. (2012) Hadronic energy resolution of a highly granular scintillator-steel hadron calorimeter using software compensation techniques. JINST 7:P09017. http://dx.doi.org/10.1088/1748-0221/7/09/p09017
20. Repond J, et al. (2018) Hadronic energy resolution of a combined high granularity scintillator calorimeter system. JINST 13:P12022. http://dx.doi.org/10.1088/1748-0221/13/12/p12022
21. Israeli Y (2018) Energy reconstruction in highly granular calorimeters for future electron-positron colliders. PhD thesis, Technische Universität München, Germany. http://mediatum.ub.tum.de/?id=1459326

22. LeCun Y, Bengio Y, Hinton G (2015) Deep learning. Nature 521:436–444. http://dx.doi.org/10.1038/nature14539
23. Albertsson K, et al. (2018) Machine learning in high energy physics community white paper. http://arxiv.org/abs/1807.02876 [physics.comp-ph]
24. Qasim SR, et al. (2019) Learning representations of irregular particle-detector geometry with distance-weighted graph networks. Eur Phys J C 79:608. http://dx.doi.org/10.1140/epjc/s10052-019-7113-9
25. Hooberman B, et al. (2017) Calorimetry with deep learning: particle classification, energy regression, and simulation for high-energy physics. https://dl4physicalsciences.github.io/files/nips_dlps_2017_15.pdf. Accessed 05 Nov 2019
26. Erdmann M, Glombitza J, Walz D (2018) A deep learning-based reconstruction of cosmic ray-induced air showers. Astropart Phys 97:46–53. http://dx.doi.org/10.1016/j.astropartphys.2017.10.006
27. Barney D (1998) A pedagogical introduction to the CMS electromagnetic calorimeter. CMS-CR-1998-004. https://cds.cern.ch/record/473761
28. Stoye M (2018) Deep learning in jet reconstruction at CMS. J Phys: Conf Ser 1085:042029. http://dx.doi.org/10.1088/1742-6596/1085/4/042029
29. de Oliveira L, Nachman B, Paganini M (2018) Electromagnetic showers beyond shower shapes. arXiv:1806.05667 [hep-ex]

Chapter 11
Fast Generative Modelling of Electromagnetic Calorimeter Showers

Remark: Italic passages in this chapter are substantively quoted from Ref. [1]. The original text has been written by Martin Erdmann, Jonas Glombitza (both RWTH Aachen) and the author of this thesis.

Towards the end of the last chapter, deep learning-based methods have been used to enhance the HGCAL prototype's performance in the energy reconstruction of hadronic showers and in the separation of electrons and charged pions. In the latter study, it has been observed that slight differences between the actual calorimeter data and the simulated data may lead to diverging performances that could affect the interpretation of the results obtained from those methods. Deep learning has already proven to provide suitable tools in order to correct for non-trivial differences between the data and the simulation [2]. However, the simulation of calorimeter showers sequentially with GEANT4 or other expert-engineered tools is still required in such a solution.

As it was explained in Sect. 3.5, the sequential shower simulation is computing intensive and may take a couple of seconds for one electromagnetic shower with energies at the order of 100 GeV. At HL-LHC, detailed calorimeter simulation will pose a significant challenge because "computing systems will have to cope with greatly increased data samples and data acquisition and processing rates" [3]. At the same time, a proportional increase in the funding of computing resources is not foreseen. Hence, the contribution of detailed Monte Carlo simulations to the computing budget for HL-LHC should in general be as minimal as possible.

In this chapter, it will be explained how deep learning can help in the fast and still precise simulation of electromagnetic particle showers in the HGCAL topology without expert knowledge on the evolution of particle showers therein. A principal change of paradigm hereby is the interpretation of calorimeter data as shower *images*. In this analogy, calorimeter cells are regarded as pixels. Then, the principal idea of the approach followed here is depicted in Fig. 11.1 and can be introduced as follows:

The most recent approach for simulating particle showers in calorimeters are so-called generative adversarial networks (GAN) [4–8]. *According to this concept, the temporal sequence of the shower development is first marginalised by training*

T. Quast, *Beam Test Calorimeter Prototypes for the CMS Calorimeter Endcap Upgrade*, Springer Theses, https://doi.org/10.1007/978-3-030-90202-5_11

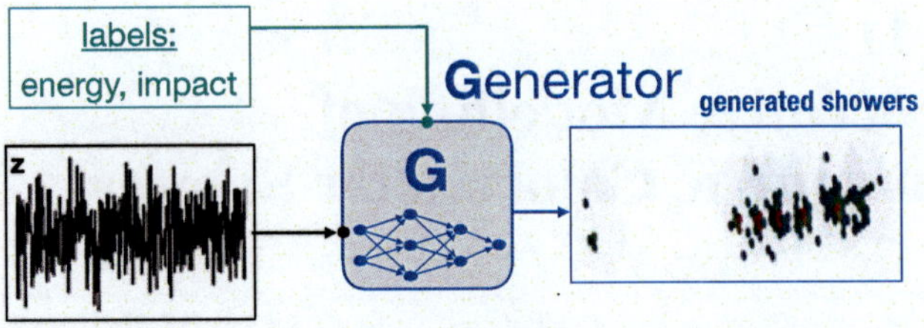

Fig. 11.1 Principle of fast calorimeter simulation using a generative neural network, aka. *generator*. The generator can be understood as a multidimensional function which maps random noise together with certain labels to calorimeter shower images. It typically contains $\mathcal{O}\left(10^6\right)$ free parameters that are determined through adversarial training with reference data

a generator and the three-dimensional spatial distribution of the energy depositions is then generated directly. At first glance, this ansatz appears similar to common detector-specific parameterisations of detailed simulations which are usually developed by experts in the field. With the GAN concept, however, a high-dimensional probability distribution for spatial energy depositions is obtained automatically either directly from measured data, or alternatively from the above-mentioned detailed simulations. (...)

Unlike the computing intensive, sequential shower simulations, *which calculate many stochastic processes of individual interactions, the entire spatial energy depositions of the shower are determined in a single evaluation of the generator network. In order to obtain the realisation of a single shower, the stochastic process is incorporated through a set of random numbers as input to the probability distribution coded in the network. These random numbers ensure that none of the generated particle showers look alike. The speed for calculating a shower realisation is several orders of magnitude faster than detailed shower simulations, since in the network only a fixed sequence of linear algebra operations is carried out together with the evaluation of the activation functions.*

In Sect. 11.1 the developed simulation framework based on the Wasserstein distance is explained. Then, the framework is applied for the simulation of calorimeter images from a simplified benchmark dataset which is defined in Sect. 11.2. It follows the specification of the training strategy in Sect. 11.3. A basic familiarity with machine learning and especially deep neural networks (see e.g. Ref. [9]) may be beneficial in the understanding of all the aspects in that section. The detailed assessment of the generator's performance in simulating physical calorimeter data is documented in Sect. 11.4. The procedure and the findings up to this point have already been published in Ref. [1]. Text passages and graphics are mostly copied from that publication. Stylistic modifications are made for consistency with the style of this thesis. The last two sections complement the originally published results. Section 11.5 summarises the attempts at further optimisation. Finally, the developed framework is used to simulate test beam data of a fully equipped prototype of EE compartment in Sect. 11.6. In this context, it is exemplified that the fast simulation with deep generative neural networks potentially leads to a better agreement to the beam test data than with GEANT4.

11.1 WGAN Framework for Fast Calorimeter Simulation

The GAN concept consists of two networks working in opposition to one another. The generator network is meant to learn the probability distribution which is encoded in realistic datasets. The second network is used to evaluate the differences between the generated datasets and the realistic datasets. The feedback of the second network to the generator network is used to improve the probability distribution encoded in the generator. Conversely, the second network is trained to distinguish between ever smaller remaining differences between the generated datasets and the realistic datasets. In this dual training process, the probability distribution of energy depositions is sampled from realistic data and transferred to the generator network.

Even though traditional GANs show impressive results, the training process is unstable and hard to monitor. Furthermore, GANs often suffer from mode collapsing when the generator is only able to generate data in a subspace of the real distribution. The (...) Wasserstein GAN [10] *(WGAN) and its improvement* [11] *allows for a stabilised training procedure by delivering adequate gradients to the generator, providing a meaningful loss metric not being susceptible to mode collapsing.*

In this section, the principle of Wasserstein GANs is introduced. Then, the method of label conditioning in order to incorporate initial conditions into the generation process is presented. Figure 11.2 gives an overview of the full training procedure.

11.1.1 Wasserstein Generative Adversarial Networks

In Wasserstein GANs, the Wasserstein-1 metric is used as a similarity measure between the generated samples $\tilde{x} = G(z)$ and the real samples x. This distance is also known as earth mover's distance, because in a figurative sense it defines the cost for moving a distribution onto a target distribution using optimal transport. In the adversarial framework the Wasserstein loss is constructed using the Kantorovich-Rubinstein duality:

$$L = sup_{f \in \mathrm{Lip}_1} \left(\mathbb{E}[f(x)] - \mathbb{E}[f(\tilde{x})]\right). \tag{11.1}$$

Here, "$sup_{f \in \mathrm{Lip}_1}$" states that the supremum is over all the 1-Lipschitz functions f after application on the real samples x and generated samples $\tilde{x}$. During the adversarial training, the 1-Lipschitz functions f which fulfill Eq. 11.1 are approximated by the discriminator network C. It is called critic because it is trained to allow for an estimate of the Wasserstein distance instead of being able to discriminate between real and generated samples. To allow for the approximation of the 1-Lipschitz functions using a neural network, the Lipschitz constraint is enforced by the gradient penalty [11] *which extends the objective function to:*

$$L = \mathbb{E}[C(x)] - \mathbb{E}[C(G(z))] - \lambda\, \mathbb{E}[(||\nabla_{\hat{u}} C(\hat{u})||_2 - 1)^2]\,. \tag{11.2}$$

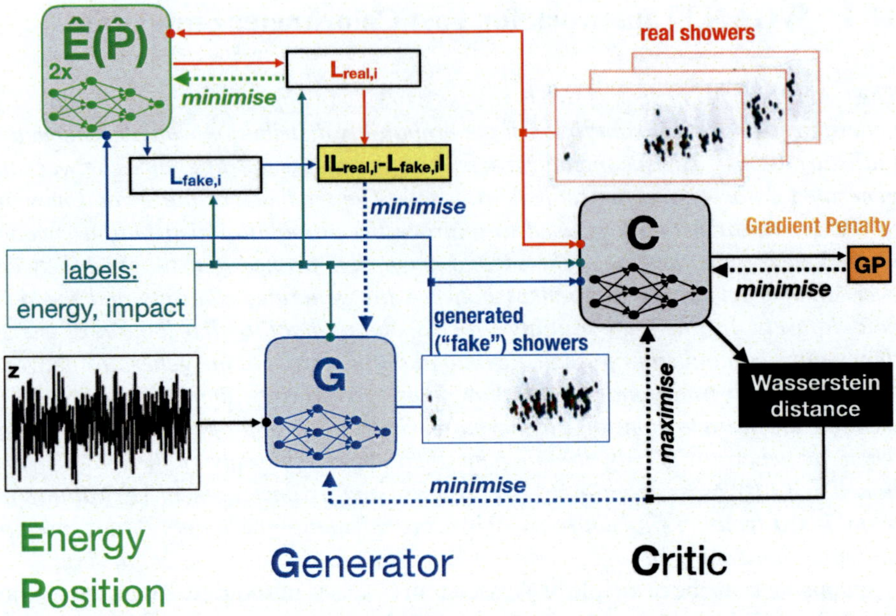

Fig. 11.2 Overview of the adversarial WGAN framework for the training of the generator. The critic network C estimates the Wasserstein (= earth mover's) distance between real and generated, i.e. fake, shower images. Two auxiliary networks $a_1 = \hat{E}$ and $a_2 = \hat{P}$ are used in order to assist the incorporation of the initial shower conditions (energy and impact position) into the generation process

Here, λ is a hyperparameter for scaling the gradient penalty. The mixture term

$$\hat{u} = \varepsilon x + (1 - \varepsilon)\tilde{x} \tag{11.3}$$

states that the Lipschitz constraint is enforced by sampling on straight lines between pairs of generated samples $\tilde{x}$ and real samples x. The random sampling is performed by sampling ε from a uniform distribution $\mathcal{U}(0, 1)$. To ensure accurate gradients for the generator, the critic is usually trained for several iterations before one generator update is applied. Thus, in Wasserstein GANs the generator attempts to minimise Eq. 11.2 between the generated and the real samples, while the Wasserstein distance is approximated using the critic network by maximising Eq. 11.2. This differs to the traditional GAN setup where, under the assumption of an optimal discriminator, the generator attempts to minimise the Jensen-Shannon divergence.

11.1.2 Label Conditioning

For calorimeter simulations, generated samples must reflect certain label characteristics according to physics laws. The labels define the initial state of the simulation such as the incident particle's kinematics and the degree of possible background activity (pile-up) in the calorimeter. However, this label dependency is not ensured for samples generated by a generator which is trained using the WGAN approach. To be able to generate samples which can be associated with explicit labels, the concept of label conditioning introduced by Auxiliary Classifier GANs (AC-GANs) [12] *is a widely used concept for generative approaches in physics simulations* [2, 6, 7]. *To advance the WGAN concept to label conditioning the concept of Ref.* [2] *is adapted. In this specific configuration, the initial state is determined by the electron energy and its impact position. Accordingly, the generator dependency is modified to* $G = G(z, E, P)$, *and in this framework, besides noise* z, *the generator is given the physics labels of the electron energy* E *and the impact position coordinates* $P = (P_x, P_y)$ *as input. Furthermore, the label information is also provided to the critic.*

To constrain the generator and to evaluate how well the label characteristics are reflected in the generated samples, two constrainer networks a_i *are used. These constrainer networks are trained under supervision to reconstruct the impact position and the electron energy respectively using the real (labelled) samples. The mean squared error*

$$L_{\text{real},i} = [y_i - a_i(x)]^2 \tag{11.4}$$

is used as an objective function for the constrainer networks. Here, y_i *is one label associated with the real sample* x *and* $a_i(x)$ *denotes the respective reconstruction by the constrainer network. The constrainer networks are trained under supervision during the critic training and are fixed during the generator training. To enforce label conditioning of the generator, the generator loss is extended by*

$$L_{\text{aux}} = \sum_i^n \kappa_i |L_{\text{real},i} - L_{\text{fake},i}|, \tag{11.5}$$

where κ_i *is a hyperparameter to scale the respective auxiliary loss. The loss*

$$L_{\text{fake},i} = [\alpha_i - a_i(\tilde{x})]^2 = [\alpha_i - a_i(G(z, E, P))]^2, \tag{11.6}$$

states how well the input labels $\alpha_1 = E,\ \alpha_2 = P$ *for the generation can be reconstructed from the generated shower by the constrainer networks. In summary, the generator is trained to minimise the Wasserstein distance, cf. Eq.* 11.2, *and the auxiliary loss, cf. Eq.* 11.5, *provided by the constrainer networks. The absolute difference between both loss terms in Eq.* 11.5 *ensures that the label reconstruction of the generated and real samples remains on the same scale.*

11.2 Benchmark Dataset

The WGAN framework in this thesis was developed prior to the principal HGCAL beam test campaigns in 2018. For its development, the benchmark dataset was constructed from a calorimeter setup inspired by a previous HGCAL prototype (that is not discussed in this thesis). The assumed calorimeter is equal to the prototype of the EE compartment that was tested at CERN in September 2017 in the PPE172 area [13]. It was made from seven modules that were arranged as seven sensitive layers covering 2.8–16.2 X_0 in depth. A simplified sketch of the EE prototype is shown in Fig. 3.4.

The deployed 6″ sensor geometry was the same as for the 2018 prototypes. Like it was done already in the context of the other deep learning applications in this thesis, cf. Sects. 10.3.4 and 10.4, the hexagonal coordinates on each sensor are converted according to Fig. G.1a for the computation of shower images. Cells that are connected to neither a full nor a half hexagonal diode are not processed to pixels. By this means, the calorimeter hit locations can be indicated in a $12 \times 15 \times 7$ Cartesian-like frame. Two representative shower images after this transformation are shown in Fig. 11.3.

Although real data from this HGCAL prototype are available, a high level of common mode noise due to grounding issues impaired its calorimetric performance during the beam tests in 2017. Therefore, the usage of this beam test dataset for the purpose of developing the method was not possible. Instead, electron-induced showers in that calorimeter prototype were simulated as the benchmark dataset using GEANT4 v10.2 and a CMS-specific tune of the FTFP_BERT physics list. Realistic conditions in the beam tests are included into the simulation and a simplified noise rejection is applied like it would be done in a realistic dataset:

- *Electrons inducing showers traverse the upstream material in the beam line and impinge perpendicularly onto the calorimeter.*
- *Impact positions are extrapolated from straight line tracks computed from four position measurements as they would be measured by four DWCs in the beam line with 200 μm resolution each.*
- *The beam profile is modelled with a rectangular geometry and covers an active area of 6×5 cm^2. Its energies are smeared with a 1% uncertainty.*
- *Calorimeter pixels with energy depositions below 2 MIP are removed from each shower to reject noise contributions.*

The primary benchmark dataset, used for both training and evaluation of the WGAN, consists of $5 \times 100,000$ simulated electromagnetic showers induced by 20, 32, 50, 80 and 90 GeV electrons. Another dataset is given by simulated 70 GeV electrons. With the latter, the WGAN's interpolation capacities to energies for which it has not been trained are investigated. All samples simulated with GEANT4 are also referred to as "real" in the following, whereas the WGAN-generated samples are also denoted as "fake".

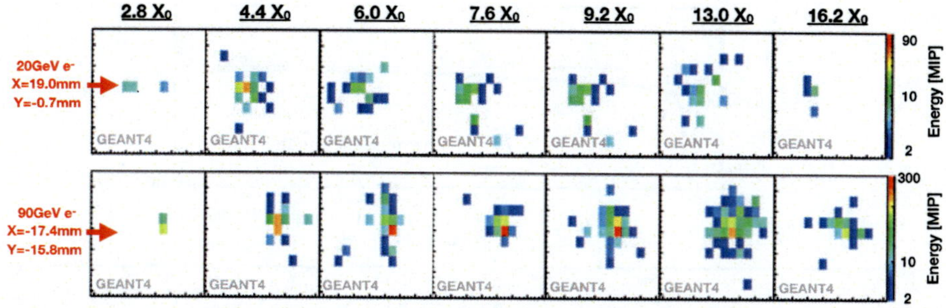

Fig. 11.3 (Top) Energy depositions of an electromagnetic shower induced by a 20 GeV electron and (bottom) induced by a 90 GeV electron with different impact positions (X, Y) simulated using GEANT4. The 3D shower images in the benchmark dataset comprise 12 × 15 × 7 pixels

11.3 Network Architecture and Training

The framework for the generation of electromagnetic calorimeter showers consists of four networks: one generator, one critic and two constrainer networks. One of the constrainer networks is used for conditioning the energy E, while the second is used for conditioning the impact position P. The networks, their training and their evaluation are implemented using the TensorFlow [14] framework (v1.5). Exact details of all architectures can be found in the Appendix G in Tables G.6, G.5 and G.7.

The generator consists of two parts: The first part is separated into seven towers, each of which has the same structure, and a joint part which merges the towers. Each of the seven towers is given ten latent variables z and three labels α describing the energy and the impact position of the calorimeter shower as input. After two fully connected layers and a reshape, a block of three 2D transposed convolutions and a single 2D convolution follows. Next, the seven towers are concatenated to a joint part with three 2D convolutional layers. Finally a locally connected convolutional layer completes the generator architecture. Between the convolutional and transposed convolutional layers batch normalisation and leaky ReLUs as activations are used. After the last layer, batch normalisation is not applied and ReLU is used as activation to allow for the generation of sparse calorimeter images. To enlarge the prior for the generation process, a masking layer masks the dead pixels and regions outside the calorimeter by setting the respective values to zero.

The critic network is given as additional input the three labels, which are processed by two fully connected layers and a reshape to obtain a two-dimensional shape. The following architecture of the critic is straightforward and consists of five 2D convolutional layers followed by a fully connected layer and the output layer. As activation leaky ReLU is used to avert sparse gradients. Between the layers, layer normalisation is used instead of batch normalisation as the gradient penalty loss is used.

For both constrainer networks, a very similar architecture of 3D convolutions is used where only the classification layer is varied. For better convergence and

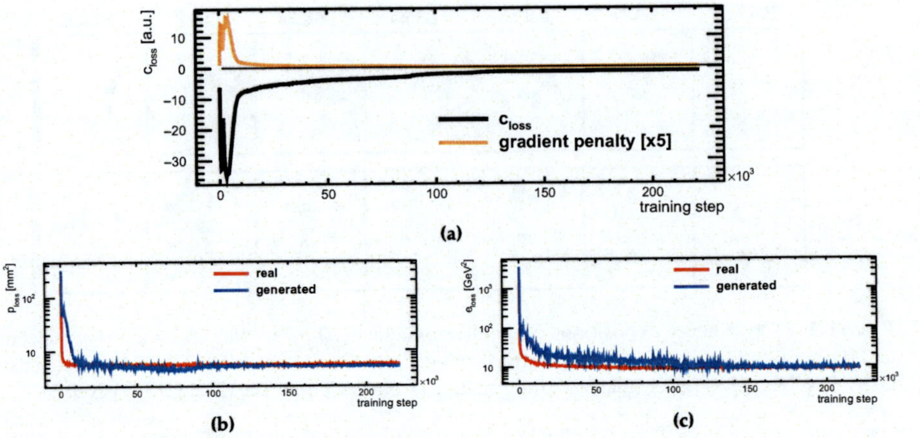

Fig. 11.4 **a** Total critic loss c_{loss} during the training (black) and rescaled gradient penalty (yellow). **b** Loss curves of the constrainer networks during the supervised training for the position regression p_{loss} and **c** the energy reconstruction e_{loss}

regularising effects, batch normalisation between the layers is used. Furthermore, leaky ReLU as nonlinearity are used to ensure sufficient gradients.

During training, the losses of the constrainer networks are scaled with $\kappa_E = \kappa_P = 0.01$. *The gradient penalty scale is set to* $\lambda = 5$. *The constrainer networks and the critic are updated for* $n_{cr} = 9$ *iterations before updating the generator once. A batch size of 256 is used and the framework is trained for 150 epochs on a single NVIDIA GeForce GTX 1080 which takes about 30 h. The ten latent variables are drawn from a uniform distribution* $\mathcal{U}(-1, 1)$. *Furthermore, the Adam optimisers with* $\beta_1 = 0.0,\ \beta_2 = 0.9$ [11] *and different learning rates for the networks are used. The constrainer networks use a small learning rate of* $lr = 5 \cdot 10^{-5}$. *Their training is stopped after 50 epochs. For the generator a learning rate of* $lr = 10^{-3}$ *is used and it drops after 70, 90 and 100 epochs to* $lr = 5 \cdot 10^{-4}$, $lr = 2 \cdot 10^{-4}$ *or rather* $lr = 10^{-4}$. *For the critic an initial learning rate of* $lr = 5 \cdot 10^{-4}$ *is used and it changes to* $lr = 2 \cdot 10^{-4}$, $lr = 10^{-4}$ *and* $lr = 5 \cdot 10^{-5}$ *after 60, 80 and 100 epochs, respectively.*

11.4 Performance Benchmarks

Various benchmarks related to the quality of the generated electromagnetic showers are discussed in the following. It will be demonstrated that generator produces high-quality showers which resemble the benchmark dataset in many aspects while lowering the computing time to simulate a full electromagnetic shower by three orders of magnitude.

The investigation is structured as follows. In Sect. 11.4.1 a visual inspection of WGAN-generated showers is performed. First, two examples are illustrated on which qualitative observations are highlighted. Thereafter, the analysis of pixel occupations reveals that the trained generator considers the radially decreasing occupancy profile. In Sect. 11.4.2 it is then shown that the generated samples reflect characteristics related to the input physics labels. This behaviour was enforced indirectly through the extended generator loss, cf. Eq. 11.5. By contrast, any other physically motivated observable evaluated on the generated showers was not constrained in the training. However, as illustrated in Sect. 11.4.3, many distributions of shower characterising quantities computed on the generated showers match those computed from the real dataset well. Moreover, it is demonstrated that key correlations between calorimeter observables are obtained. For all reported benchmark scenarios, good shower qualities for 70 GeV electron showers are obtained despite the fact that these were not part of the training set. Finally, this section is concluded with a report in Sect. 11.4.4 on the generator's computational time advantage over detailed simulation using GEANT4.

11.4.1 Visual Inspection of Generated Showers

Figure 11.5 shows two exemplary electron-induced showers with 20 and 90 GeV energy labels generated using the WGAN approach. This set of energy depositions is consistent with the physical intuition of how electron-induced cascades in this sampling calorimeter configuration should develop. First, it is noted that both pixel occupancies and pixel intensities scale with the incident electron energy. Second, the positions of the largest energy depositions move according to the input impact position labels. Finally, it is evident that the main activity of generated showers occurs in the central sampling compartments. In particular, the spread and the scale of energy depositions is maximal in intermediate layers, while only a few pixels are active in the first and last layers.

Figure 11.6 shows the average pixel occupancy with energy depositions above the 2 MIP threshold of 90 GeV electron-induced showers simulated using GEANT4 compared to those generated by the WGAN. White spaces correspond to areas of the sensors which are activated above the threshold in less than 1% of the events. The radial development of the pixel occupancy of WGAN-generated showers is similar to GEANT4 while the overall scale appears underestimated.

11.4.2 Label Dependency

Three physics labels, namely the incident's electron energy E and its impact position $P = (P_x, P_y)$, are input to the generator. Ideally, these labels should constrain the shower generation process. As described in Sect. 11.1.2, two constrainer networks

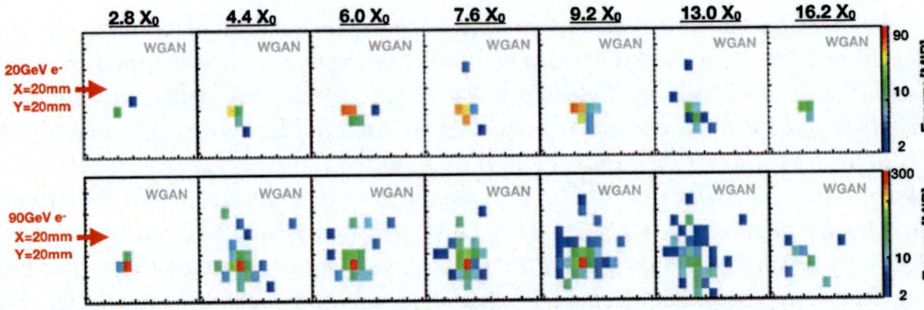

Fig. 11.5 Energy depositions generated with the WGAN for a fixed impact position of an electromagnetic shower (top) for a 20 GeV electron and (bottom) for a 90 GeV electron

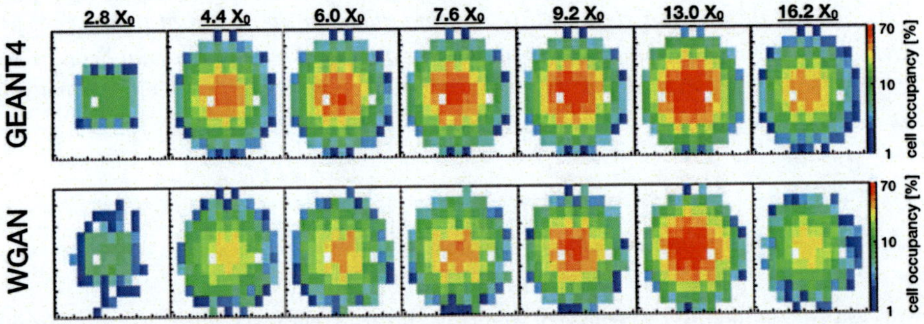

Fig. 11.6 (Top) Cell occupancy for 90 GeV electrons simulated using GEANT4 and (bottom) generated by the WGAN. Dead pixels and areas outside the sensor acceptance are masked in the generator

are trained with real samples for this purpose and then reconstruct these labels based on the full shower information. For the training to be rated as successful, the reconstructed labels of the generated showers should correlate with the imposed physics labels.

Figure 11.7a shows the distribution of reconstructed energies for different energy labels. Their maxima correlate to the true labels. Furthermore, the energy spectra computed on generated and GEANT4-simulated showers exhibit a reasonable agreement. Figure 11.7b, c show the correlation for the position labels. Here, the symbols indicate the mean reconstructed label in bins of the true label. On average, a generated shower with a certain set of labels is reconstructed accordingly. Evidently, shower characteristics which the two constrainer networks are sensitive to are able to condition the generation process. Even the 70 GeV electron cascades, which were not considered in the training, exhibit the same behaviour.

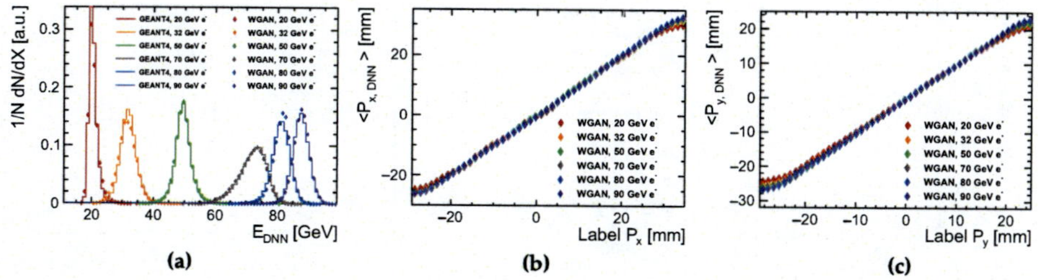

Fig. 11.7 **a** Distribution of energies reconstructed by the constrainer network computed on WGAN-generated and GEANT4-simulated showers for different energy labels. **b** Impact position P_x, and **c** impact position P_y of generated showers reconstructed by the constrainer network as a function of their true labels. Statistical errors on these means are negligible. Note that 70 GeV showers were not part of the benchmark dataset

> ✓ **FSIM.1** The fast simulated showers obey the imposed characteristics that are injected into the generation process, namely their energy and impact location.

11.4.3 Calorimeter Observables

In this section, typical calorimeter observables are presented which are computed both in GEANT4 simulations and generated showers. For better clarity, only the 32 and 90 GeV as well as the additional 70 GeV electron samples are shown. The agreement between real and generated cascades illustrated therein are representative for the entire dataset. (...)

Distributions of Calorimeter Observables

Figure 11.8 shows six sets of representative observables that characterise particle showers in sampling calorimeter configurations. The distributions are normalised to unity.

A reasonable agreement between WGAN-generated and GEANT4-simulated showers is seen in Fig. 11.8a for the total energy deposition summed over all pixels and in Fig. 11.8b for the longitudinal shower depth. Also, the maximum pixel energy for each individual layer exhibits a good match with the GEANT4 simulation (only layers 2 and 4 are shown in Fig. 11.8c, d). Furthermore, the energy-weighted transverse spread in each layer is computed according to Eq. 11.7.

$$\Delta \mathrm{Y}_{\text{layer } l} = \sum_{\text{pixel } i}^{\text{layer } l} \left| y_i - \sum_{\text{pixel } j}^{\text{layer } l} y_j \cdot \frac{E_j}{E_{\text{layer } l}} \right| \cdot \frac{E_i}{E_{\text{layer } l}} \tag{11.7}$$

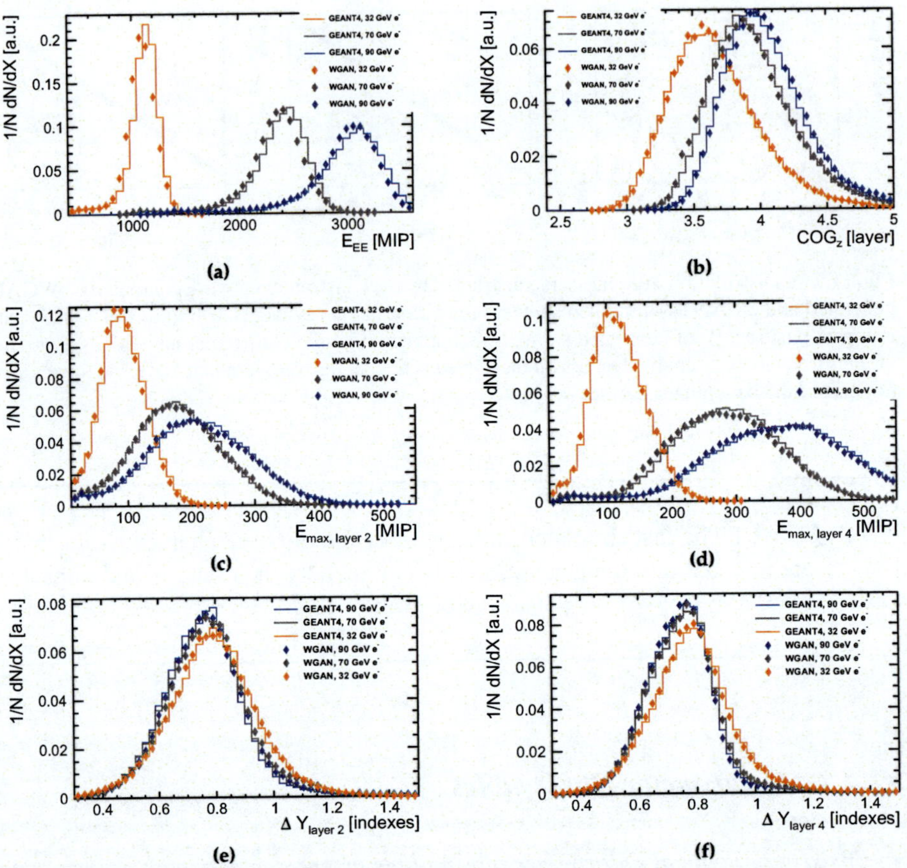

Fig. 11.8 Comparison of calorimeter observables computed in generated showers (symbols) to those computed in fully simulated showers using GEANT4 (histograms). **a** Energy sum of all pixels, **b** energy-weighted shower depth, **c** the maximum pixel energy in layer 2 and **d** layer 4, respectively, and **e** the transverse shower spread along the y-direction in layer 2 and **f** layer 4, respectively. The 70 GeV showers were not part of the benchmark dataset

The distributions of $\Delta Y_{\text{layer 2}}$ *and* $\Delta Y_{\text{layer 4}}$ *are shown in Fig.* 11.8*e, f by way of example. The computation for other layers l and for the x-coordinate is analogous. With the exception of the first layer at 2.8* X_0*, the agreement therein is representative for all other layers and the x-coordinate. Thus the transverse shower shapes are well-modelled by the WGAN.*

In Fig. 11.9*b the energy spectrum of active pixels is investigated. The region of low energy densities,* i.e. *pixels with depositions below* $\sim$ *10 MIP, is underrepresented by the WGAN with respect to GEANT4 causing a mismatch in the number of active pixels (energy* $\geq$*2 MIP) in Fig.* 11.9*a and ultimately also resulting in the underestimate of their energy sum, cf. Fig.* 11.8*a. (...)*

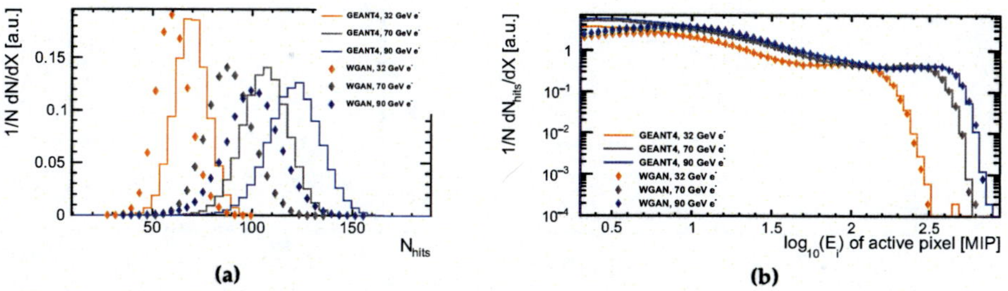

Fig. 11.9 **a** Distributions of the number of pixels with energy depositions above 2 MIPs equivalents for several energies. **b** Single pixel energy spectra show reasonable agreement with the simulation for energy densities above ≈10 MIP per pixel. Note that the 70 GeV showers were not part of the benchmark dataset

Correlations

For energy reconstruction or particle identification in modern particle detection systems, multiple calorimeter observables are needed simultaneously (e.g. see Ref. [15]*). Typical approaches exploit correlations of shower characteristics. Consequently, a crucial quality measure for a simulation tool is the assessment of the pairwise correlations of reconstructed physics observables.*

This part focusses on four examples of such correlations. First, both the summed energy and the maximum pixel energy in a fixed layer is expected to correlate with the sum (maximum) in the previous layer. Figure 11.10*a–d visualise this trend for layers 3 and 4, respectively 4 and 5, by way of example. It must be noted that the generated showers exhibit good agreement with the GEANT4 showers. Second, a greater number of active pixels should correspond to higher values in the sum of energy depositions, which is illustrated in Fig.* 11.10*e. Some discrepancy is to be expected owing to the underestimation of the low energy spectrum by the WGAN. While the positive correlation is also obtained for the WGAN samples, an agreement is not reached.*

Ultimately, the sampling in this specific calorimeter configuration is not uniform as the setup comprises non-equidistant sensitive sampling layers. As a consequence, showers with large energy depositions at a fixed incident electron energy should relate to lower shower depths. Showers whose centre of gravity is located deeper in the calorimeter deposit higher fractions of their energy in the larger amount of passive material between layer 5 and 7, hence resulting in lower values of energy sums in the sensitive layers. Also for this example, a good match between real and WGAN-generated showers is obtained as displayed in Fig. 11.10*f.*

In summary, typical calorimeter observables computed on fast WGAN-generated showers correspond well to those simulated using GEANT4, not only in terms of their spectra but also in their pairwise correlations. Only the number of low-energy depositions is too small. A detailed simulation typically requires extensive tuning of its model parameters using expert knowledge in particle interactions with matter to achieve a similar level of agreement to real data. By contrast, no dedicated knowledge

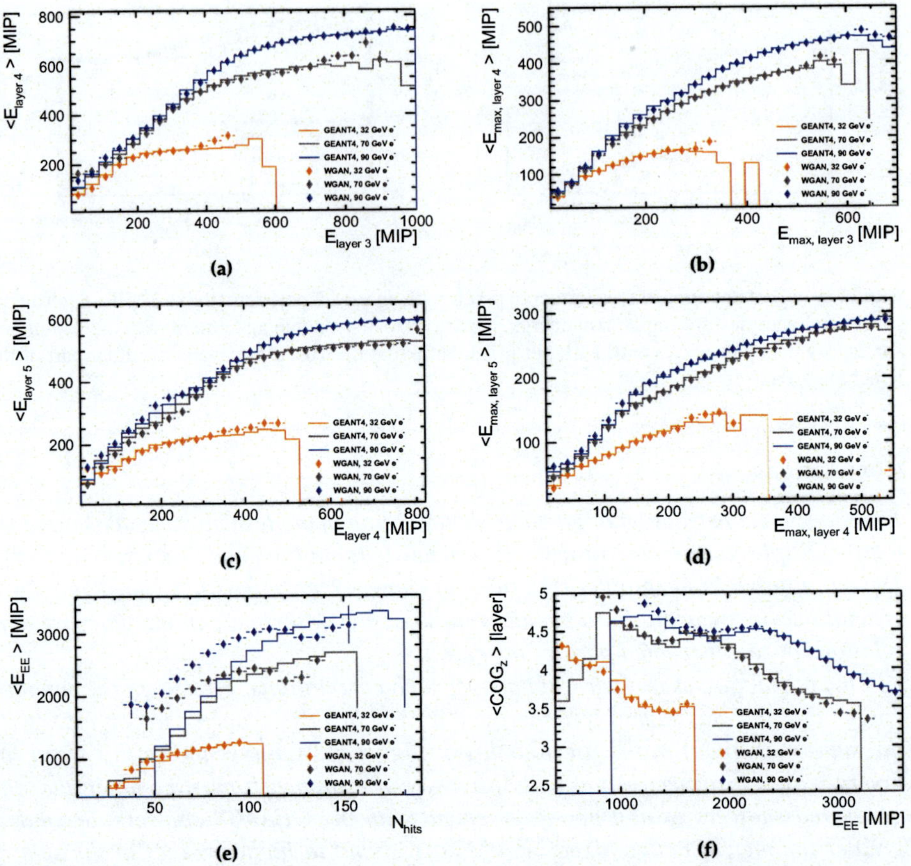

Fig. 11.10 Comparison of calorimeter observable correlations evaluated on generated and showers simulated with GEANT4. **a** Energy sum, respectively **b** maximum pixel energy, in layer 4 versus total energy deposition (maximum pixel energy) in layer 3. **c** Energy sum, respectively **d** maximum pixel energy, in layer 5 versus total energy deposition (maximum pixel energy) in layer 4. **e** Total energy deposition plotted against number of hits. **f** Shower depth versus total energy deposition. Note that the 70 GeV showers were not part of the benchmark dataset

on how particles interact with the material had to be input into the generative model here.

✓ **FSIM.2** Reduced to typical calorimeter observables, the fast simulated shower data are in good agreement to the GEANT4 dataset, both in their distributions and also in their pairwise correlations.

✓ **FSIM.3** The fast simulation approach allows for interpolation to phase spaces of the incidence energy for which it has not been set up explicitly.

Table 11.1 Computational time required for the generation of one 20 GeV, respectively 90 GeV, electron-induced cascade through evaluation of the generator using different hardware setups and enhancement with respect to a full simulation using GEANT4. Using the WGAN approach, a speed-up of more than three orders of magnitude has been achieved with the benchmark dataset

Method	Hardware	20 GeV e^-	xSpeed	90 GeV e^-	xSpeed
GEANT4	standard (2017) CPU	500 ms	–	2000 ms	–
WGAN	Intel© Xeon© CPU E5-1620	52 ms	**x10**	52 ms	**x40**
WGAN	NVIDIA© Quadro© K2000 GPU	3.6 ms	**x140**	3.6 ms	**x560**
WGAN	NVIDIA© GTX™ 1080 GPU	0.3 ms	**x1660**	0.3 ms	**x6660**

11.4.4 Computational Speed-Up

Fast simulation of electromagnetic showers using this generator architecture is up to three orders of magnitude faster than full simulations. By contrast, expert engineered, parameterised fast simulation provides CPU gains of 10–100 with respect to GEANT4 depending on the particle energy, cf. Ref. [16]. *In Table* 11.1, *numbers on the computation time advantage of the generator compared to GEANT4 using three different hardware architectures are provided. As expected, graphics processing units (GPUs) are the preferred hardware unit since the generator's internal set of linear computational operations is more efficiently parallelised and runs faster than in CPUs. Furthermore, it is remarkable that the time for evaluation of the generator is independent of the incident electron energy while it scales significantly in simulations using GEANT4.*

✓ **FSIM.4** Compared to sequential shower simulation with GEANT4, the fast simulation approach is significantly faster in the generation of calorimeter data.

11.5 Further Attempts at Improvement

As demonstrated in the previous section, the generated samples exhibit good agreement to the GEANT4 benchmark dataset. *(...) Only the spectrum at low-energy densities is underestimated by the generator. Hence, the total number of sensor pixels with energies above a threshold energy equivalent to 2 MIP is found to be reduced by about 15%.*

Since the initial publication of this finding, the nominal training of the generator has been re-iterated for a total of ten times. Each time, the internal network parameters were initialised randomly and independently from the previous iteration. Nine of the outcomes resulted in generator performances that are consistent with the reported findings. Only one training led to a non-converging gradient penalty and with it to a failure of the adversarial training.

In the meantime, numerous attempts have also been made aiming at improving the generation of low energy density depositions. The experimental modifications to the framework with respect the nominal strategy are listed below.

1. The architectures of all involved networks were varied both simultaneously and individually. Usage of 3D (transposed-) convolutions both in the critic and the generator, discarding of the towering of layers or of the locally connected layers in the generator are just a few examples. In addition, three other architectures for the auxiliary networks, similar to the three for the electron-pion classification in Sect. 10.4, cf. Appendix G, were tried. None of them nor combinations of them have resulted in any noteworthy improvement in the modelling of low energy depositions.
2. All coupling parameters in the loss function, i.e. λ in Eq. 11.2 and the κ_i's in Eq. 11.5, were enlarged as well as reduced by one order of magnitude yielding consistently worse results than the ones in Sect. 11.4.
3. A third auxiliary, in this case discriminatory, network was introduced that was trained to distinguish real and fake shower images given, among others, sparsity-related quantities. After the critic loss had converged, the generator was trained, besides the normal WGAN procedure, to minimise an additional loss term. This additional loss term is identical to the one from the original GAN approach with this additional network acting as the discriminator. The result was an increase of the Wasserstein distance estimate with increasing duration of that training. Ultimately, this measure did not fix the agreement in the number of hits between the generated and the simulated samples and even worsened the agreement in the other observables.
4. The duration of the training was extended to twice the amount of epochs which showed no effect.
5. As an alternative for regularising the critic network, "spectral normalisation" [17] in the critic was tried. When it was used in conjunction with the gradient penalty, the performance results remained unchanged with respect to the nominal training strategy. By contrast, the spectral normalisation yielded worse generator performances when the gradient penalty was not included in the training simultaneously.
6. In Ref. [18] it is explained that the Wasserstein metric is biased by the sample gradients meaning it is affected by the limited size of batches in the training. The authors of this work suggest that this may have "a serious issue in practice". Circumventing any possibility of such resulting issues, the "Cramér distance as a Solution to Biased Wasserstein Gradients" [18] was experimented with in this

thesis. In all trials, the final performance indicators of the trained generator networks did not improve.

In summary, none of the experiments has yielded any noteworthy improvement in the modelling of low-energy depositions. Currently, their description appears to be a general deficit in the fast calorimeter simulation with WGANs. *(...) Similar mismodelling of sparsity describing quantities has also been reported in the work based on traditional GANs* [8]. *It should be noted that the analysis of such fast simulated showers could always be limited to the well-described range by restricting the analysis to pixels with energies above 10 MIP equivalents, cf. the appendix of Ref.* [1]. *(...) In this specific benchmark calorimeter setup, the rejected part of the spectrum contributes only 10% to the total signal.*

11.6 Application to Real Beam Test Data

With the availability of thousands of recorded positron showers in the *O1E* dataset from October 2018, it is possible to train a generator network directly on realistic calorimeter data. Complementing the studies with the benchmark dataset, the following three aspects can be addressed now:

- Can this fast simulation concept be applied to a fully equipped EE compartment with 28 sensitive layers? This corresponds to a multiplication of the pixelation of the shower images by a factor of four with respect to the benchmark scenario.
- Can it be used also for higher incidence energies? The available energies in the *O1E* dataset increase the covered range in the energy labels from 20–90 GeV to 20–(almost) 300 GeV.
- As the generator is optimised to produce realistic data from the beam tests, do its generated showers exhibit better agreement to the beam test data than the ones produced from full shower simulation with GEANT4?

The considered positron energies[1] in the *O1E* dataset for training and evaluation of the generator are 20, 30, 50, 100, 150, 250 and 300 GeV, whereas the 80 and 200 GeV samples serve for evaluation only. Since the number of recorded showers at a given momentum is not the same, at most 60k showers at each beam momentum are randomly selected for the training. The best knowledge on the actual beam momentum after subtraction of the synchrotron losses are used as energy labels. A selection of events analogous to Sect. 10.1 is applied. In addition, all hits with reconstructed energies below 2 MIP are discarded in order to further reject the contributions from fake hits. Impact position labels are determined by the DWC measurement. Translation alignment corrections of the DWCs to the EE prototype are incorporated hereby.

The generator architecture, cf. Table G.5, remains unchanged overall. However, due to the multiplication of the input layers, the number of 2D transposed convolution-based towers need to be increased accordingly from 7 to 28. It is found practical to

[1] Energy equals momentum, cf. Eq. 10.7.

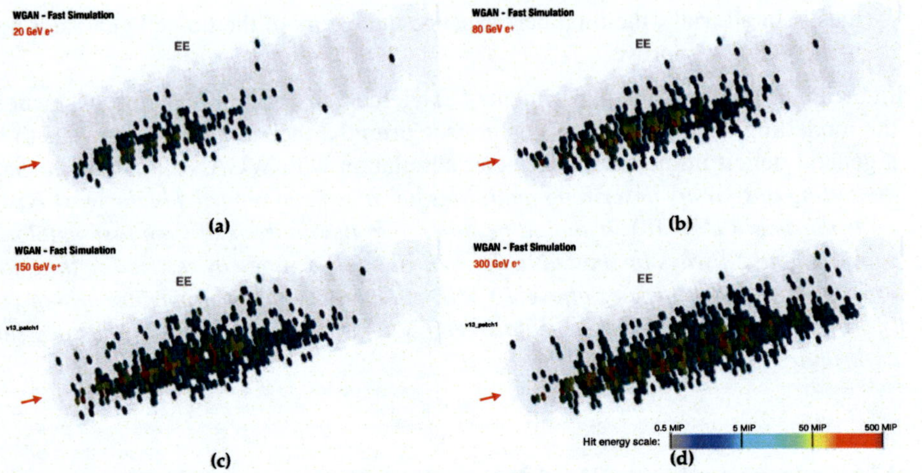

Fig. 11.11 Event displays of fast simulated positron-induced electromagnetic showers inside the 28-layer EE prototype calorimeter compartment. The generator network was trained with showers from the *O1E* dataset

counteract the associated increase of free parameters in the network by halving the number of maps in the 2D transposed convolutions and by setting the number of maps to 32 for each 2D convolution after the concatenation operation.

All networks are again randomly initialised and the network training proceeds from scratch, in particular without any pre-training. All training and loss parameters are set identically to the values in Sect. 11.3. The critic loss as well as the auxiliary losses converge successfully, i.e. analogous to the demonstration in Fig. 11.4. 150 epochs of the training with batch sizes of 256 (1349 minimisation steps per epoch) last approximately two days on the aforementioned GPU hardware. The network training was stopped just after 500 epochs to guarantee full convergence of this framework.

Figure 11.11 shows randomly chosen event displays of WGAN-generated showers whose coordinates are transformed back into the hexagonal grid. These images look reasonable in the sense that the hit energy's, the cell occupancy's as well as the shower depth's scaling with the incident energy are apparent. In this application, the shower images are generated even up to 20,000 times faster with the generator than with full simulation using GEANT4, cf. Table 11.2.

The core findings regarding the performance assessment are illustrated representatively for 150 GeV positrons in Fig. 11.12. Therein, reconstructed observables are shown for the beam test data, as well as for the WGAN-generated and GEANT4-simulated samples. It is indicated that the generated showers, in fact, exhibit better agreement to the beam test data than GEANT4. In this connection, the description of the longitudinal shower profiles at layers 7 and 10, that are known to have slightly less efficiency than the other layers in the data, is remarkable, cf. Fig. 11.12a.

Table 11.2 Computational time required for the generation of positron-induced cascades in the EE compartment of the October 2018 configuration 1, cf. Fig. 6.16a, through evaluation of the generator compared to full simulation using GEANT4. Here, the generator structure has been slightly modified with respect to the nominal one described in Table G.5

	GEANT4 std. (2017) CPU	WGAN Intel© Xeon© CPU E5-1620	WGAN NVIDIA© GTX™ 1080 GPU
20 GeV e^+	550 ms [x1]	10 ms **[x55]**	0.4 ms **[x1375]**
80 GeV e^+	2200 ms [x1]	10 ms **[x220]**	0.4 ms **[x5500]**
150 GeV e^+	4000 ms [x1]	10 ms **[x400]**	0.4 ms **[x10000]**
300 GeV e^+	8000 ms [x1]	10 ms **[x800]**	0.4 ms **[x20000]**

Moreover, higher values in the transverse shower shape characteristic ΔY_{EE}, cf. Eq. 11.8, are observed in the beam test which is not reproduced with GEANT4.

$$\Delta Y_{EE} = \sum_{\text{pixel } i}^{EE} \left| y_i - \sum_{\text{pixel } j}^{EE} y_j \cdot \frac{E_j}{E_{EE}} \right| \cdot \frac{E_i}{E_{EE}} \tag{11.8}$$

By contrast, values in that domain are obtained with the WGAN, as it is demonstrated in Fig. 11.12b.

Nevertheless, the previously discussed failure in the modelling of low energy densities remains. This issue is illustrated both in the longitudinal profile of reconstructed hits ($E_i > 2$ MIP) in Fig. 11.12c and in the underrepresentation of entries in the hit energy spectrum at low energies in Fig. 11.12d. In addition, the aforementioned (in Sect. 9.2.3) discontinuity in the recorded hit energy spectrum at the transition between the low gain and the TOT, i.e. $\log_{10}(E_i/\text{MIP}) \approx 2.4$, does not appear to be learnt by the generator.

Despite this assessment not being as detailed as the documentation of the performance benchmark studies in Sect. 11.4, it may be concluded that this fast simulation concept is certainly not limited to the benchmark scenario only. Instead, it is proven to be also suitable for shower simulation in a fully equipped HGCAL prototype compartment and for a wider range of incident energies than what was published earlier. However, the typical deficit in the generation of low energy depositions remains. When trained on realistic calorimeter data, in this case from *O1E* dataset, the overall agreement of the fast simulated showers to the actual data is evidently better than with GEANT4. These results are encouraging and motivate further development of such deep learning-based methods for the fast simulation of HGCAL data in the future.

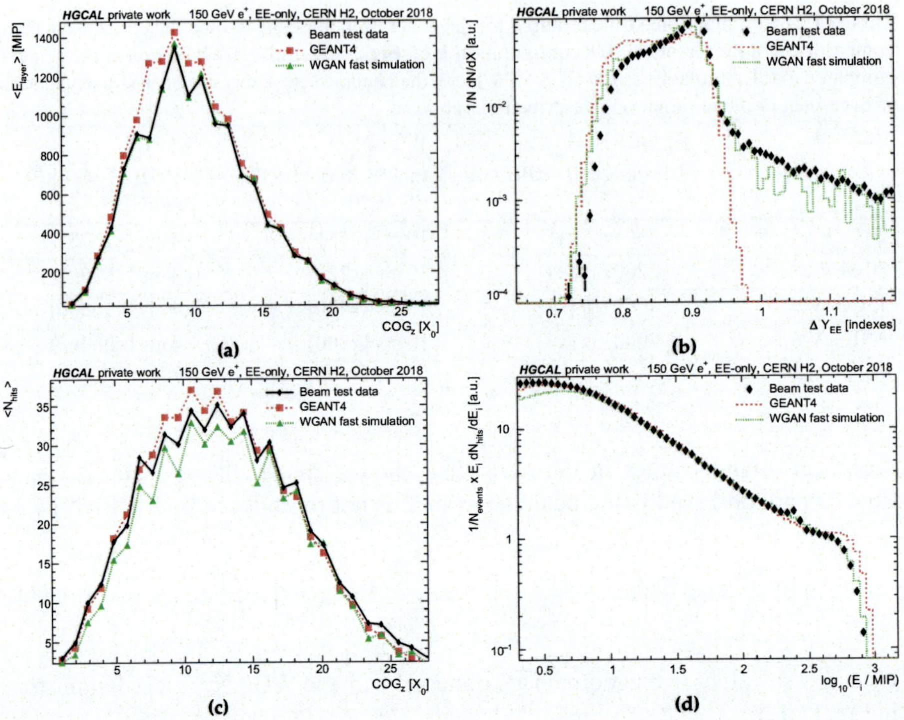

Fig. 11.12 Comparison of calorimetric observables due to 150 GeV positrons between the recorded beam test data (black), full GEANT4 simulation with the FTFP_BERT_EMM physics list (red) and fast simulated samples generated by the WGAN (blue). **a Longitudinal energy profile, b** transverse shower spread along the y-axis, **c longitudinal hit profile and d** energy spectrum of active pixels. Note that the GEANT4 samples in this study incorporate an energy scaling that is different to the one in Eq. 9.5. This inconsistency does not alter the interpretation of these graphics as discussed in the text

✓ **FSIM.5** The fast simulation approach is not only applicable to one particular calorimeter configuration or energy range but is also applicable to others.

✓ **FSIM.6** The fast simulation approach is applicable to realistic beam test data.

References

1. Erdmann M, Glombitza J, Quast T (2019) Precise simulation of electromagnetic calorimeter showers using a wasserstein generative adversarial network. Comput Softw Big Sci 3. https://doi.org/10.1007/s41781-018-0019-7
2. **Erdmann M, et al. (2018) Generating and refining particle detector simulations using the** wasserstein distance in adversarial networks. Comput Softw Big Sci 2. https://doi.org/10.1007/s41781-018-0008-x
3. Bocalli T, Costanzo D (2018) HL-LHC computational challenge for ATLAS and CMS experiments. https://ep-news.web.cern.ch/content/hl-lhc-computational-challenge-atlas-and-cms-experiments. Accessed 17 Sep 2019
4. Goodfellow IJ, et al. (2014) Generative adversarial networks. arXiv:1406.2661 [stat.ML]
5. Shrivastava A, et al. (2016) Learning from simulated and unsupervised images through adversarial training. arXiv:1612.07828 [cs.CV]
6. **Hooberman B, et al. (2017) Calorimetry with deep learning: particle classification, energy** regression, and simulation for high-energy physics. https://dl4physicalsciences.github.io/files/nips_dlps_2017_15.pdf. Accessed 05 Nov 2019
7. Paganini M, de Oliveira L, Nachman B (2018) Accelerating science with generative adversarial networks: an application to 3D particle showers in multilayer calorimeters. Phys Rev Lett 120:042003. https://doi.org/10.1103/PhysRevLett.120.042003
8. Paganini M, de Oliveira L, Nachman B (2018) CaloGAN: simulating 3D high energy particle showers in multilayer electromagnetic calorimeters with generative adversarial networks. Phy Rev D 97:014021. https://doi.org/10.1103/PhysRevD.97.014021
9. Goodfellow I, Bengio Y, Courville A (2016) Deep learning. MIT Press, Cambridge. http://www.deeplearningbook.org
10. Arjovsky M, Chintala S, Bottou L (2017) Wasserstein GAN. arXiv:1701.07875 [stat.ML]
11. Gulrajani I, et al. (2017) Improved training of wasserstein GANs. arXiv:1704.00028 [cs.LG]
12. **Odena A, Olah C, Shlens J (2016) Conditional image synthesis with auxiliary classifier GANs.** arXiv:1610.09585 [stat.ML]
13. Quast T (2018) Construction and beam-tests of silicon-tungsten prototype modules for the CMS high granularity calorimeter for HL-LHC. JINST 13:C02044. http://dx.doi.org/10.1088/1748-0221/13/02/c02044
14. Abadi M, et al. (2016) TensorFlow: large-scale machine learning on heterogeneous distributed systems. arXiv:1603.04467 [cs.DC]
15. **Colas J, et al. (2005) Position resolution and particle identification with the ATLAS EM** calorimeter. Nucl Instrum Meth A 550:96–115. https://doi.org/10.1016/j.nima.2005.05.041
16. **Yung Jun S (2011) Gflash as a parameterized calorimeter simulation for the CMS experiment.** J Phys: Confer Ser 293:012023. http://dx.doi.org/10.1088/1742-6596/293/1/012023
17. Miyato T, et al. (2018) Spectral normalization for generative adversarial networks. arXiv:1802.05957 [cs.LG]
18. Bellemare MG, et al. (2017) The cramer distance as a solution to biased wasserstein gradients. arXiv:1705.10743 [cs.LG]

Chapter 12
Summary, Outlook and Conclusion

Proton-proton collisions at unprecedented centre-of-mass energies and rates are provided by the Large Hadron Collider (*LHC*) at CERN in Switzerland. The collisions are the basis for precise testing of the Standard Model of Particle Physics, currently the best theory to describe matter and the fundamental forces on microscopic scales, and enable searches for physics beyond. During its third long shutdown from 2024–2026, the LHC will be upgraded for running at five times its original design instantaneous luminosity. This so-called High-Luminosity era (*HL-LHC*) of the LHC will last until the end of the 2030s providing around $3000\,\mathrm{fb}^{-1}$ of integrated luminosity to the experiments. This pushes statistical uncertainties of the measurements to a minimum and enhances the sensitivity to rare, new phenomena. However, this increase in instantaneous luminosity is connected to increased pile-up and radiation levels in the detectors. Therefore, various upgrades to the CMS experiment are necessary to fully profit from the future collision datasets.

One of the upgrades in preparation is the replacement of the CMS electromagnetic and hadronic calorimeter endcaps with a highly granular calorimeter, commonly referred to as *HGCAL*. In essence, HGCAL is a sampling calorimeter. Silicon sensors are used as active material in the inner detector regions, whereas the outer, less occupied regions are instrumented with more cost-effective scintillators coupled to silicon photomultipliers as sensitive material. The characteristic of granularity stems from the small size of the foreseen six million hexagonal silicon-readout channels in total. The involved readout electronics need to cover a large dynamic range and should comprise timing capabilities with resolutions down to 50 ps. Ultimately, HGCAL will offer robustness and good performance through the full HL-LHC operational lifetime.

The specifications of the HGCAL design are based on simulation studies, for which experimental validation is essential. For this purpose, prototypes of the silicon-based modules were produced and used for the assembly of HGCAL prototypes. Since 2017, the prototype modules were equipped with the SKIROC2-CMS readout chip

T. Quast, *Beam Test Calorimeter Prototypes for the CMS Calorimeter Endcap Upgrade*, Springer Theses, https://doi.org/10.1007/978-3-030-90202-5_12

that offers prototype functionality of the ultimate HGCAL ASIC. In 2018, a series of tests of those prototypes was conducted under realistic conditions with particle beams. The programme started with the test of single modules at the DESY (Hamburg) beam test facility and ended with the test of the so far largest HGCAL calorimeter prototype equipped with more than 90 prototype 6″ modules of each 128 channels at the CERN SPS. The goals of the tests were the characterisation of prototype modules including their individual components, the overall assessment of their functionality in beam conditions, and the validation of the calorimetric performance by comparison to detailed GEANT4 simulation.

Important contributions to these HGCAL beam test efforts were made as part of the work for this thesis. Besides that, the possibility of time-efficient calorimeter simulation using state-of-the-art machine learning techniques was investigated in this work. The hereby applied strategy has been defined in Chap. 5. In the following, the chapters, in which the original contributions of this thesis to the HGCAL project have been discussed, are summarised first before an outlook and the conclusions of this work are finally presented.

Summary

Experimental Infrastructure

The infrastructure used in the various beam tests and for the electrical characterisation of bare prototype silicon sensors has been described in Chap. 6. Therein, special emphasis was given to the contributions made as part of this work: the operation and readout of the involved beam characterising detectors, the developments of the data acquisition system, and the offline synchronisation of trigger numbers.

Data Reconstruction Algorithms

Due to the new prototype electronics with respect to the initial 2016 HGCAL prototypes, the reconstruction of the recorded data in the beam tests had to be largely implemented from scratch. This is true both for the data from the HGCAL prototype modules equipped with the SKIROC2-CMS, but also for the data from the beam characterising detectors. The reconstruction of the latter was fully developed and the one for the former was co-developed as part of this thesis. Chapter 7 has provided a detailed documentation of the corresponding algorithms and procedures. Furthermore, all reconstruction steps were steered within a comprehensive reconstruction workflow. This workflow was realised as part of this work, and by now it has been established in the CMS HGCAL System Test group. At the end of the data processing flow, the reconstructed test beam data are provided to the group as ntuples facilitating collaborative analysis.

Silicon Sensor and Module Qualification

In Chap. 8, it has been demonstrated that the per-cell leakage currents of the bare prototype silicon sensors are at the order of a few nanoampere above full depletion, and that the pad capacitances decrease with the applied bias voltage until full depletion is reached, as expected. Electrical breakthrough behaviour was only observed for two out of more than 2500 examined sensor pads. Using a beam of electrons of a few GeV/c in conjunction with the DATURA beam telescope, the thickness of the module PCB could be measured to be $1.066 \pm 0.002^{+0.009}_{-0.011}\%$ of a radiation length. From GEANT4 simulations, the impact on the prototype's energy response to electromagnetic showers due to the thickness of the SKIROC2-CMS has been estimated to be below 1%, justifying the modelling of the PCB as homogeneous in the subsequent studies. Pedestals as well as common mode noise of the assembled modules were computed and subtracted from the raw signals. The experienced intrinsic noise level is acceptable, as it is well below the signal induced by minimum ionising particles (*MIPs*). MIP signals are confined to the exposed calorimeter cells only and the efficiency to detect those signals was determined to be close to 100% for most of the module area.

In Situ Calibration of Prototype Modules

Subsequently, it has been shown in Chap. 9 that the prototype modules can be calibrated in situ using beam test data. Typical for sampling calorimeters in particle physics, the energy scale was calibrated using minimum ionising particles both from a beam of muons and from a parasitic, i.e. undefined, beam. Reduction of the noise contributions to the measured MIP signal spectra, reconstructed both in high and low gain, have been achieved by deploying the HGCAL calorimeter prototype as a tracking device. The variation of the MIP calibration constants for channels on the same chip is within 3%. Moreover, they scale with the area of the silicon pads and with the thickness of the depletion zone, as expected. The signal-to-noise ratio amounts to ~7–8 (~5) in high gain and ~4–5 (~3) in low gain for modules with 300 μm (200 μm) thick sensors. The intercalibration of the gains was validated by comparison of the reconstructed and the simulated hit energy spectra. Furthermore, the prototype's timing information could be calibrated using an external time reference detector. By this means, an ad hoc timing resolution of around 60 ps could be achieved for single readout channels. This finding is the first experimental confirmation of the HGCAL's foreseen timing capabilities in beam conditions using the prototype readout electronics.

Performance Validation of the Silicon-Based Calorimeter Prototype

The validation of the prototype's calorimetric performance with electromagnetic and hadronic showers has been discussed in Chap. 10. The studies on electromagnetic showers have revealed that the reconstructed shower shapes are generally in agreement to the simulation. Similarly, the electromagnetic energy resolution is well reproduced by the GEANT4 simulation and reaches ~1.5% for positron energies of 290 GeV. The impact position of perpendicularly impinging, highly energetic, electromagnetically induced showers could be reconstructed with precisions well below 1 mm close to the shower maximum. The direction of the shower axis could be resolved with precisions down to ~5 mrad at the highest accessible test beam energies. Again, those measurements are in reasonable agreement to the simulation when accounting for the finite resolution of the delay wire chambers used for upstream particle tracking. Most of the results on the calorimetric performance are reproduced even when averaged calibration factors were used or when the high gain information in the hit energy reconstruction was ignored. These addressed variations in the hit energy reconstruction are closer to the final HGCAL design and operation.

For hadronic showers, the inferred primary interaction depth follows an exponential decay. The measured resolution for hadronic showers amounts to ~12% at the highest accessible pion energy of 300 GeV at the beam tests. The measured hadronic energy resolution is impaired by transverse and longitudinal shower leakage owing to the limited size of the prototype's hadronic compartment. The hadronic energy resolution was found to be consistent with the expectation from simulation. In addition, it is exemplified that the calorimetric performance for hadronic showers profits from the granularity, e.g. by incorporating software compensation approaches into the energy reconstruction. Granularity is also beneficial for the separation of electrons and pions in the electromagnetic prototype compartment using deep learning-based classifiers. Nevertheless, it was noticed that reliable inferences with the applied classification method relies on a realistic simulation of the calorimeter data.

Fast Generative Modelling of Electromagnetic Calorimeter Showers

Precise simulation of calorimeter data will remain essential as part of the production of Monte Carlo samples in many particle physics analyses at LHC. Unfortunately, sequential shower simulations with GEANT4 may not be feasible on large scales in the future due to the increasing computing demand for the data reconstruction in the High Luminosity era. A new fast simulation approach based on state-of-the-art machine learning techniques has been investigated in Chap. 11. There, empirical evidence was presented that the Wasserstein Generative Adversarial Network (*WGAN*) concept is well applicable for the adversarial training of a generator neural network that is able to produce shower images as realistic calorimeter data. The hereby trained generator was used successfully to generate images of electromagnetic showers in two different HGCAL topologies. The generated calorimeter data are largely in good agreement to their GEANT4-simulated counterpart. Only the description of low energy densities remains unsatisfactory. When performing the adversarial training directly on beam test data, it was observed that the fast simulated WGAN data are partially in better agreement to the beam test data than those from GEANT4. At the same time, this WGAN-based simulation is 3-4 orders of magnitude faster than sequential shower simulation.

Outlook

After submission of this thesis and despite Covid-19-related delays, the CMS HGCAL upgrade project is proceeding at full speed for it to be ready for operation by 2027. Presently, the prototyping is converging towards a final system, and final design specifications are being made. The mass production of the sensors, electronics, mechanics, etc. is being prepared and will commence in 2022. After its restart in the second half of 2021, tests of HGCAL modules with final silicon sensors, readout electronics, and PCBs are foreseen at the CERN SPS. The infrastructure related to the beam characterising detectors prepared as part of this thesis will in principle be re-usable then. All other aspects, such as the data quality monitoring or the data reconstruction, require revisions to be compatible with the new elements on the final HGCAL modules. The procedures and concepts in this thesis may provide helpful guidelines in this regard.

Owing to its speed in generating shower data, the WGAN-based fast simulation as used for the first time in this thesis is overall a promising tool for the future. Graph-based neural networks [1, 2] may be a helpful ingredient for the generalisation of this approach to mixed sensor pad geometries, like they are included in the actual HGCAL design. Recently, other fast simulation applications deploying Bounded-Information-Bottleneck autoencoders [3] or utilising normalizing flows [4] have been demonstrated to outperform the WGAN-based approach, especially in the modelling of low energy deposits. Independent of the underlying methodology, it would be interesting to apply and to evaluate those novel deep learning-based fast simulation alorithms also on more complex shower signatures, such as hadronic showers (cf. Ref. [4]) or even jets, and with pile-up.

Conclusion

In conclusion, the developments presented in this work have supported the successful beam test campaigns with up to more than 90 HGCAL prototype modules in 2018. This was a crucial project milestone towards the realisation of this calorimeter. The obtained results from the analysis of the beam test data in this thesis are in agreement with the full functionality of the silicon-based HGCAL design. Moreover, this thesis has provided a proof of principle that generative modelling based on deep neural networks in conjunction with the Wasserstein distance in the adversarial training is a suitable approach for the fast and precise simulation of HGCAL data.

References

1. Qasim SR, et al. (2019) Learning representations of irregular particle-detector geometry with distance-weighted graph networks. Eur Phys J C 79:608. https://doi.org/10.1140/epjc/s10052-019-7113-9
2. Zhou J, et al. (2018) Graph neural networks: a review of methods and applications. arXiv:1812.08434 [cs.LG]
3. Buhmann E, et al. (2021) Getting high: high fidelity simulation of high granularity calorimeters with high speed. Comput Softw Big Sci 5. http://dx.doi.org/10.1007/s41781-021-00056-0
4. Krause C, Shih D (2021) CaloFlow: fast and accurate generation of calorimeter showers with normalizing flows. arXiv:2106.05285 [physics.ins-det]

Appendix A
CAEN Module Settings for the Readout of the Beam Characterising Detectors at the H2 Beam Line

See Tables A.1 and A.2.

Table A.1 Settings of the CAEN v1290N TDC used for the readout of the DWCs and the XCETs during the HGCAL prototype beam tests at the CERN SPS H2 beam line in October 2018. For more details on the possible options, see Ref. [1]

Quantity	Value	Description
triggerTimeSubtraction	True	Subtracts the trigger timestamp from each measurement
triggerMatchMode	True	Hits are recorded when matched to a trigger, i.e. those that are detected within the trigger time window
edgeDetectionMode	2	Hits are measured corresponding to the leading or rising edge of the signal
timeResolution	3	Index indicating the desired TDC binning. 3 corresponds to a binning of 25 ps
windowWidth	64	Length of the trigger acquisition window in units of 25 ns (clock period). 64 corresponds to an acquisition window of 1.6 μs
windowOffset	−10	Definition of the offset of the acquisition window with respect to the trigger in units of 25 ns
maxHitsPerEvent	9	Maximum number of hits in a readout window. Readouts with more hits are attributed with an error code

T. Quast, *Beam Test Calorimeter Prototypes for the CMS Calorimeter Endcap Upgrade*, Springer Theses, https://doi.org/10.1007/978-3-030-90202-5

Table A.2 Settings of the CAEN v1742 digitiser used for the readout of the MCPs during the HGCAL prototype beam tests at the CERN SPS H2 beam line in October 2018. For more details on the possible options, see Ref. [2]

Quantity	Value	Description
RECORD_LENGTH	1024	Maximum number of waveform samples
DRS4_FREQUENCY	0	Index for the signal sampling frequency. 0 corresponds to the maximum frequency of 5 GHz
EXTERNAL_TRIGGER	ACQUISITION_ONLY	Acquisition is triggered by an external trigger signal
TRIGGER_EDGE	FALLING	Trigger is generated by the falling (otherwise leading) edge of the external trigger signal
FAST_TRIGGER	DISABLED	Disable the local triggering using the input waveform directly
POST_TRIGGER	15	Post trigger size in percent of the whole acquisition window

Appendix B
Threshold Pressure and Efficiency for Threshold Cherenkov Detectors

The optical properties of the Cherenkov detector (XCET) gas determine the XCET's particle discrimination properties and they are related to the applied gas pressure. The threshold pressure at which a particle with mass m and momentum p emits Cherenkov lights as well the efficiency to detect this light are derived in this section.

Preparatory Considerations
For a particle travelling at a speed $v = \beta \cdot c$ in a gas with refractive index n, Cherenkov light can only be emitted if the condition in Eq. B.1 is met.

$$\beta > \frac{1}{n} \tag{B.1}$$

Using the Lorentz-Lorenz formula [3, 4] and assuming the validity of the ideal gas equation, the refractive index can be related to the gas pressure P and to its temperature T according to Eq. B.2, where R is the universal gas constant and A is a gas-specific constant.

$$\frac{n^2-1}{n^2+2} = \frac{A \cdot P}{R \cdot T} \tag{B.2}$$

For values $n \approx 1$, Eq. B.2 can be rewritten as Eq. B.3.

$$n \approx \sqrt{1 + \frac{3 \cdot A \cdot P}{R \cdot T}} \tag{B.3}$$

Given the refractive index n_0 at some reference pressure P_0 and reference temperature T_0, Eq. B.3 is equivalent to Eq. B.4.

$$n = \sqrt{1 + \left(\frac{P}{P_0}\right) \cdot \left(\frac{T_0}{T}\right) \cdot \left(n_0^2 - 1\right)} \tag{B.4}$$

T. Quast, *Beam Test Calorimeter Prototypes for the CMS Calorimeter Endcap Upgrade*, Springer Theses, https://doi.org/10.1007/978-3-030-90202-5

Temperature variations (on the Kelvin scale) in the 2018 HGCAL beam tests amounted to $\mathcal{O}(5\%)$ and are negligible compared to the variations of the gas pressure. Hence, $T \approx T_0$ and Eq. B.4 simplifies to Eq. B.5.

$$n^2 = 1 + \frac{P}{P_0} \cdot \left(n_0^2 - 1\right) \tag{B.5}$$

Threshold Pressure

The threshold pressure P_T is defined as the point above which Cherenkov light is emitted when a particle of mass m and momentum p traverses the gas. It is beneficial to write the kinematical quantity β as stated in Eq. B.6.

$$\frac{1}{\beta^2} = 1 + \left(\frac{m \cdot c}{p}\right)^2 \tag{B.6}$$

Insertion of Equations B.6 and B.5 into B.1 results in the desired expression for the threshold pressure as defined by Eq. B.7.

$$P > P_T = P_0 \cdot \left(\frac{m \cdot c}{p}\right)^2 \cdot \frac{1}{n_0^2 - 1} \tag{B.7}$$

The theoretical pressure threshold curve of Helium, the chosen gas for the October 2018 HGCAL beam test, for selected particle types is shown in Fig. B.1a.

Detection Efficiency

Using the Frank-Tamm formula [5], the average number of photoelectrons N_e from Cherenkov light in the photodetection device can be approximated according to Eq. B.8.

$$N_e = A' \cdot L \cdot \left(1 - \frac{1}{\beta^2 \cdot n^2}\right) \tag{B.8}$$

In this equation, L is the length of the Cherenkov detector tube and A' an empiric quality factor of the detector that is subject to calibration. Application of the relationship of Eq. B.7 one can rewrite Eq. B.6 as Eq. B.9.

$$\frac{1}{\beta^2} = 1 + \frac{P_T}{P_0} \cdot \left(n_0^2 - 1\right) \tag{B.9}$$

For reference refractive indexes close to 1, especially $n_0^2 - 1 \approx 0$, the pressure-dependent refractive index may be rewritten as Eq. B.10.

$$\frac{1}{n^2} = \frac{1}{1 + \frac{P}{P_0} \cdot \left(n_0^2 - 1\right)} \approx 1 - \frac{P}{P_0} \cdot \left(n_0^2 - 1\right) \tag{B.10}$$

The combination of Eqs. B.9 and B.10 yields Eq. B.11.

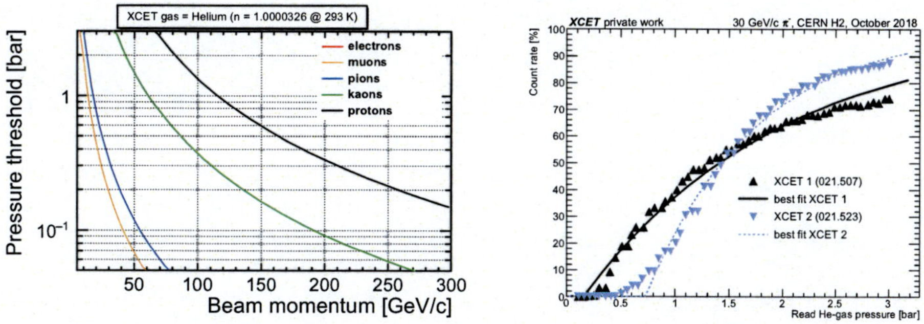

(a) (b)

Fig. B.1 **a** Theoretical pressure threshold for Helium ($n_0 = 1.0000326$) for different beam particles and as a function of the beam momentum. **b** Measured count rate of the two XCETs operated during the HGCAL beam test at the H2 beam line at CERN in October 2018 for 30 GeV/c charged pions as a function of the Helium gas pressure. The count rate corresponds to the XCET's detection efficiency as long as particle contamination in the beam can be neglected. It is found that neither of the two curves are described accurately by the theoretically motivated parameterisation in Eq. B.13 (best fit)

$$\frac{1}{\beta^2 \cdot n^2} \approx 1 + \frac{1}{P_0} \cdot \left(n_0^2 - 1\right) \cdot (P_T - P) + \mathcal{O}\left((n_0^2 - 1)^2\right) \tag{B.11}$$

Finally, the average number of expected photoelectrons is given by Eq. B.12.

$$N_e = A' \cdot L \cdot \frac{n_0^2 - 1}{P_0} \cdot (P - P_T) \tag{B.12}$$

For a sufficiently large number of photoelectrons, a Poisson probability distribution in their detection can be assumed. In that case, the probability W to record no photoelectron from N_e in total corresponds to $W(0) = \exp(-N_e)$. Hence, the efficiency to detect at least one photoelectron is equal to $\epsilon := 1 - W(0)$ as written in Eq. B.13.

$$\epsilon = 1 - \exp\left[-A' \cdot L \cdot \frac{n_0^2 - 1}{P_0} \cdot (P - P_T)\right] \tag{B.13}$$

According to the H2 beam line physicists, typical quality factors for the used XCET in the October 2018 beam test amount to $A' \approx 50\,\text{cm}^{-1}$. Given that the length of the used XCET tubes was less than 2 m, not more than one photoelectron could be expected on average. Hence, the validity of the efficiency formula B.13 does not necessarily hold. The measured XCET efficiency curves in the October 2018 beam test are shown in Fig. B.1a. It is found empirically, that neither of them can be modelled accurately by Eq. B.13.

For integration of XCETs into the future HGCAL beam tests with $\mathcal{O}$ (100 GeV/c) hadrons at the H2 beam line, it is advised to deploy tubes that are sufficiently long to yield more than one photoelectron on average.

Appendix C
Parameters for the Ntlet-Tracking Algorithm

See Table C.1.

Table C.1 Parameter settings of the Ntlet-Tracking algorithm for the reconstruction the DATURA beam telescope data and for the alignment of the different telescope setups. The definition of all parameters is provided in Sect. 7.3

Parameter	Prealign. 1	Prealign. 2	Prealign. 3	Prealign. 4	Align.	Reco.
N_1	3	3	3	3	3	3
N_2	6	6	6	6	6	6
z_{ref} [mm]	370.0	370.0	370.0	370.0	370.0	370.0
χ_1^2/ndf	2000.0	2000.0	2000.0	20.0	20.0	20.0
χ_2^2/ndf	2000.0	2000.0	2000.0	20.0	20.0	20.0
$N^{\mathrm{max}}_{\mathrm{tracks,\ 1}}$	5	5	5	5	5	5
$N^{\mathrm{max}}_{\mathrm{tracks,\ 2}}$	5	5	5	5	5	5
$\Delta D_{z_{\mathrm{ref}}}$ [mm]	4.0	1.5	0.3	0.3	0.1	0.1
Δk [mrad]	10.0	10.0	10.0	10.0	10.0	10.0*

*This parameter is set to 50.0 for the PCB tomography study in Sect. 8.2

T. Quast, *Beam Test Calorimeter Prototypes for the CMS Calorimeter Endcap Upgrade*, Springer Theses, https://doi.org/10.1007/978-3-030-90202-5

Appendix D
Modelling of Multiple Scattering in Layers with Finite Thickness using General Broken Lines

The General Broken Lines (*GBL*) method [6] does not incorporate the stochastic displacements Δx of an incident particle trajectory traversing a material with some finite thickness Δs. In this section, a method of including this effect into the GBL formalism is derived. The general idea hereby is to introduce additional thin scatterer planes into the GBL track fit to mimic the displacement effects. The situation is sketched in Fig. D.1.

The following argumentation must be invariant under the direction of the particle incidence. For this reason, the displacement a of the hypothetical scatterer planes with respect to the material boundaries must be identical on both sides. Note the geometric relations in Eq. D.1 which hold in the limit of small angles θ_1 and θ_2.

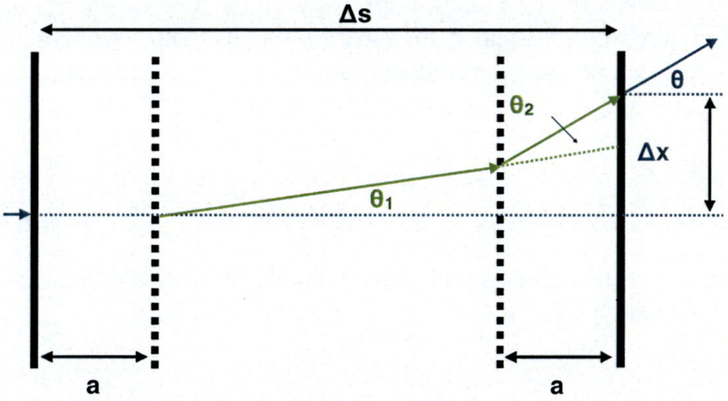

Fig. D.1 Principle idea to describe the multiple scattering of particles traversing material with finite thickness Δs using the GBL formalism: Inclusion of two hypothetical (thin) scatterer planes. (black, dotted lines) The aforementioned hypothetical scatterer planes, displaced by some distance a from the boundaries of the finite thickness scatterer with thickness Δs. (blue, solid lines) The physically observable particle trajectory. (green, solid lines) GBL-based model trajectory with kinks at the two hypothetical scatterer planes

T. Quast, *Beam Test Calorimeter Prototypes for the CMS Calorimeter Endcap Upgrade*, Springer Theses, https://doi.org/10.1007/978-3-030-90202-5

$$\begin{aligned} \theta &= \theta_1 + \theta_2 \\ \Delta x &= (\Delta s - a) \cdot \theta_1 + a \cdot \theta_2 = (\Delta s - a) \cdot \theta_1 + a \cdot \theta_2 \end{aligned} \tag{D.1}$$

Elementary algebraic transformation of Eq. D.1 yields Eq. D.2.

$$\begin{aligned} \theta_1 &= \frac{-1}{\Delta s - 2a} \cdot (a \cdot \theta - \Delta x) \\ \theta_2 &= \frac{-1}{\Delta s - 2a} \cdot (\Delta x + (a - \Delta s) \cdot \theta) \end{aligned} \tag{D.2}$$

Equation D.2 can be rewritten in matrix notation like it is denoted in Eq. D.3.

$$\begin{pmatrix} \theta_1 \\ \theta_2 \end{pmatrix} = \frac{1}{2a - \Delta s} \cdot \begin{pmatrix} a & -1 \\ a - \Delta s & 1 \end{pmatrix} \cdot \begin{pmatrix} \theta \\ \Delta x \end{pmatrix} \tag{D.3}$$

According to the theory of multiple scattering [7], the physically observable quantities θ and Δx are correlated random numbers with zero mean and computable variance according to Eq. D.4.

$$\begin{aligned} \sigma_\theta^2 &:= \mathrm{Var}(\theta) \\ \sigma_{\Delta x}^2 &= \mathrm{Var}(\Delta x) = \frac{1}{3} (\Delta s)^2 \cdot \mathrm{Var}(\theta) \\ \mathrm{cov}(\theta, \Delta x) &= \frac{1}{2} \Delta s \cdot \mathrm{Var}(\theta) \end{aligned} \tag{D.4}$$

Note that the transformation relationship between $(\theta, \Delta x) \leftrightarrow (\theta_1, \theta_2)$ in Eq. D.3 is linear. Therefore, the 2×2-matrix therein corresponds to the Jacobian of this transformation and can be used for computation of the covariance of $\sigma_{\theta_{1(2)}}$ as stated by Eq. D.5.

$$\left(\frac{\sigma_\theta^2}{2a - \Delta s}\right)^2 \cdot \begin{pmatrix} a & -1 \\ a - \Delta s & 1 \end{pmatrix} \begin{pmatrix} 1 & \frac{1}{2}\Delta s \\ \frac{1}{2}\Delta s & \frac{1}{3}(\Delta s)^2 \end{pmatrix} \begin{pmatrix} a & a - \Delta s \\ -1 & 1 \end{pmatrix} \equiv \begin{pmatrix} \sigma_{\theta_1}^2 & 0 \\ 0 & \sigma_{\theta_2}^2 \end{pmatrix} \tag{D.5}$$

Multiplication of the products and comparison of the entries in Eq. D.5, yields the expressions noted in Eq. D.6.

$$\begin{aligned} \sigma_{\theta_1}^2 = \sigma_{\theta_2}^2 &= \frac{\sigma_\theta^2}{(\Delta s - 2a)^2} \cdot \left(a^2 - a \cdot \Delta s + \frac{(\Delta s)^2}{3}\right) \\ \mathrm{cov}(\theta_1, \theta_2) &= a^2 - a \cdot \Delta s + \frac{(\Delta s)^2}{6} \end{aligned} \tag{D.6}$$

The GBL formalism does not allow for correlation between the scattering at the scattering planes. Consequently, $\mathrm{cov}(\theta_1, \theta_2) \equiv 0$ and a, as the parameter in question, should be chosen to fulfill Eq. D.7.

$$\Rightarrow a = \Delta s \cdot \left(\frac{1}{2} \pm \sqrt{\frac{1}{12}} \right) \tag{D.7}$$

The ‘+’ solution therein corresponds to the interchange of the left and right hypothetical scatterer planes. Using Eq. D.7 in Eq. D.6, one finds Eq. D.8 for the kink angles at those planes.

$$\Rightarrow \sigma_{\theta_1}^2 = \sigma_{\theta_2}^2 = \frac{\sigma_\theta^2}{(\Delta s - 2a)^2} \cdot \frac{(\Delta s)^2}{6} = \frac{\sigma_\theta^2}{\left(2 \cdot \Delta s \sqrt{\frac{1}{12}}\right)^2} \cdot \frac{(\Delta s)^2}{6} = \frac{1}{2} \cdot \sigma_\theta^2 \tag{D.8}$$

Conclusion

Modelling the particle trajectory including the displacement due to multiple scattering in a material with finite thickness Δs is possible using the GBL formalism. For this purpose, two additional hypothetical scatterer planes need to be placed at a distance of $a = 12^{-0.5} \cdot \Delta s$ with respect to the middle of the considered material boundaries. The allowed scattering at those planes must be chosen such that the introduced angular variation at each of them amounts to $2^{-0.5}$-times the value associated to the full material thickness. Only then, the angles at both helper planes are uncorrelated, as it is necessary for using the GBL formalism, and the full trajectory with displacement Δx is modelled correctly.

Appendix E
Transverse Beam Profiles in the October 2018 Beam Test

See Figs. E.1 and E.2.

T. Quast, *Beam Test Calorimeter Prototypes for the CMS Calorimeter Endcap Upgrade*, Springer Theses, https://doi.org/10.1007/978-3-030-90202-5

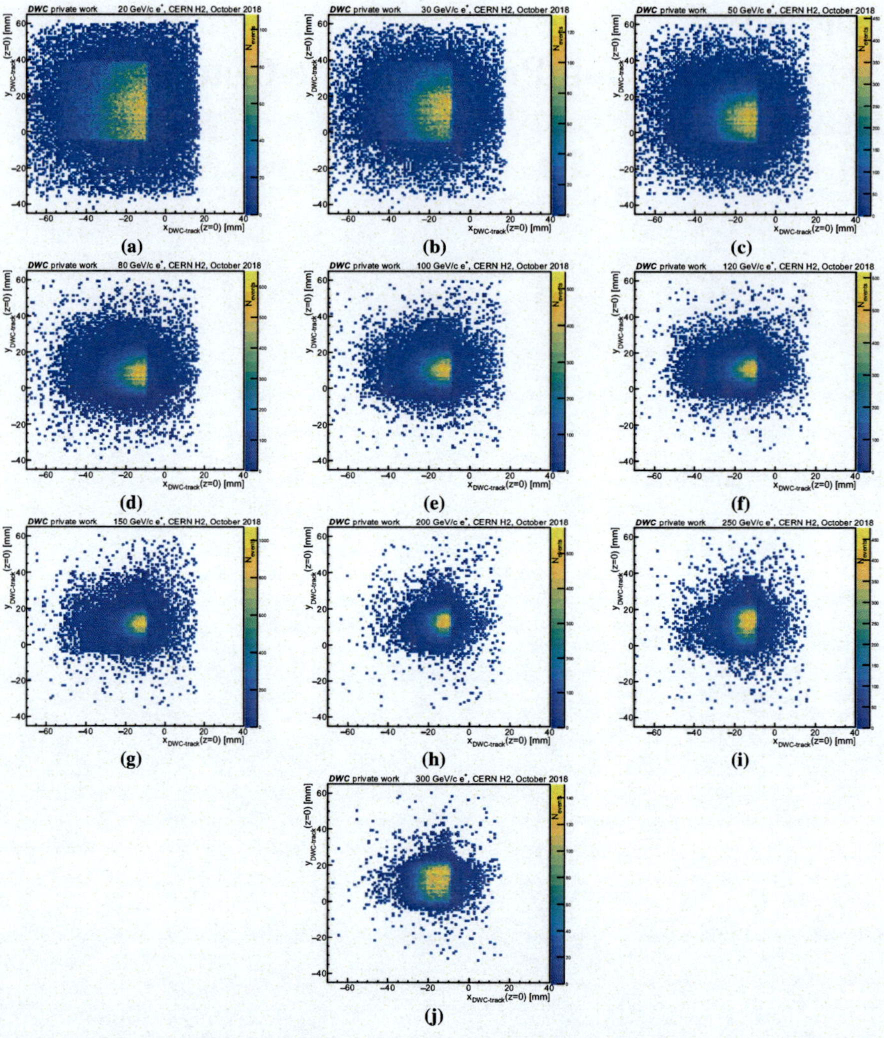

Fig. E.1 Transverse beam profiles of incident positrons measured with the delay wire chamber tracking during the October 2018 HGCAL beam test at the H2 beam line at CERN

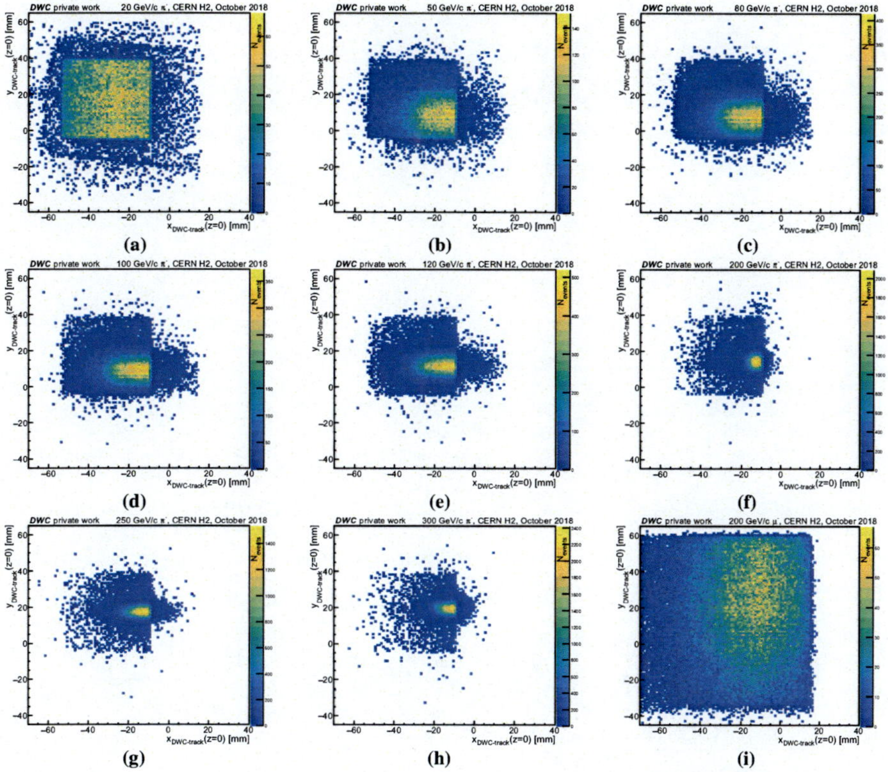

Fig. E.2 Transverse beam profiles of incident charged pions and muons measured with the delay wire chamber tracking during the October 2018 HGCAL beam test at the H2 beam line at CERN

Appendix F
Location of Calibrated Cells in the October 2018 Configuration 1

See Figs. F.1, F.2 and F.3.

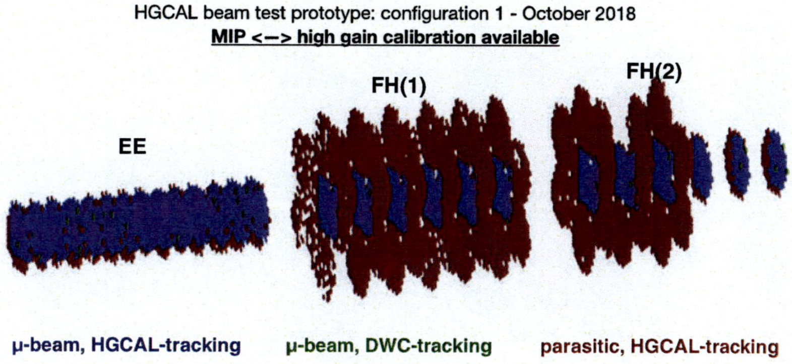

Fig. F.1 Location of cells for which the MIP-high gain calibration constants could be derived in the October 2018 HGCAL beam test. "Parasitic" refers to the *O3* dataset

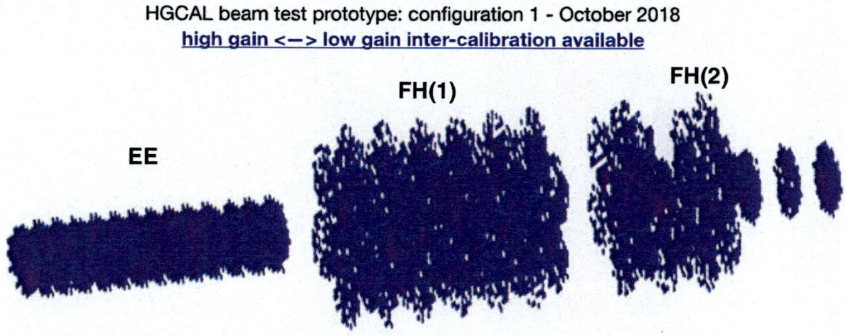

Fig. F.2 Location of cells whose high versus low gain could be intercalibrated in the October 2018 HGCAL beam test

T. Quast, *Beam Test Calorimeter Prototypes for the CMS Calorimeter Endcap Upgrade*, Springer Theses, https://doi.org/10.1007/978-3-030-90202-5

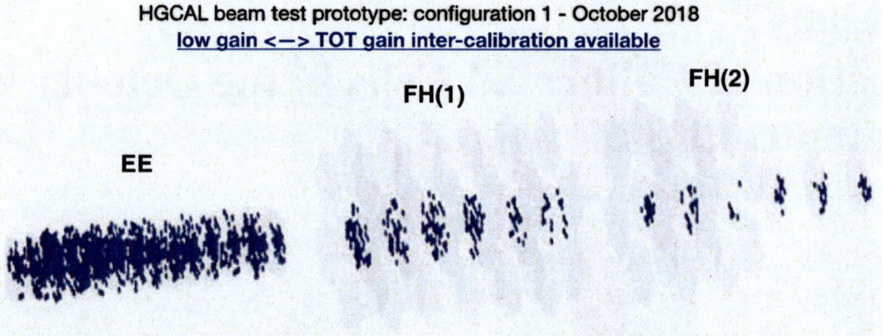

Fig. F.3 Location of cells whose low gain versus TOT could be intercalibrated in the October 2018 HGCAL beam test

Appendix G
Applied DNN Architectures

See Fig. G.1 and Tables G.1, G.2, G.3, G.4, G.5, G.6 and G.7.

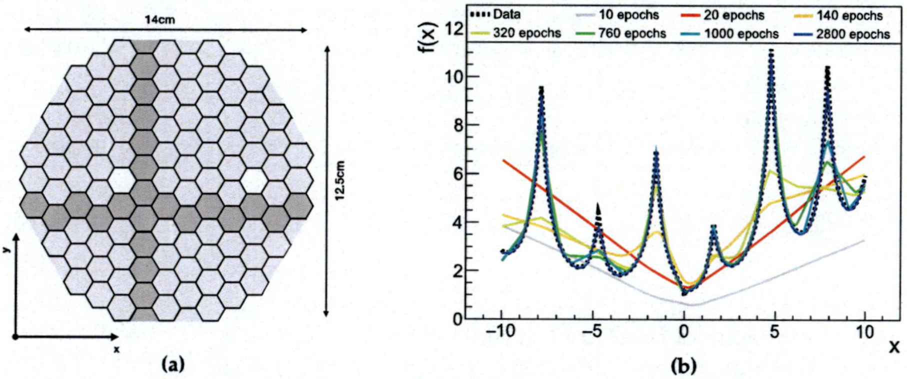

Fig. G.1 **a** Full wafer pixelation of 6″ HGCAL prototype sensors. Shaded pixels are of constant x and y pixel coordinates. The two white pixels in the middle row correspond to pads which are discarded in the transformation that processes the data for input to the deep neural networks deployed in this thesis. **b** Any function can be approximated by a neural network. In this example, a feedforward neural network is trained to predict the points from a complicated function. The agreement improves the longer the network optimisation proceeds

T. Quast, *Beam Test Calorimeter Prototypes for the CMS Calorimeter Endcap Upgrade*, Springer Theses, https://doi.org/10.1007/978-3-030-90202-5

DNN Architectures Used in Sect. 10.3.4

Table G.1 Network ("2D-conv. DNN") for the energy reconstruction of pion-induced showers. The EE and FH compartments are first treated independently and are combined at the end

Operation	Kernel	Features	Stride	Padding	Normalisation	Activation
Input from EE compartment 12 × 15 × 28						
Convolution 2D	12 × 15	32 maps	1×1	same	×	leaky ReLU
Convolution 2D	3 × 3	16 maps	1×1	valid	×	leaky ReLU
Convolution 2D	3 × 4	8 maps	1×1	valid	×	leaky ReLU
Convolution 2D	3 × 3	4 maps	1×1	valid	×	leaky ReLU
Convolution 2D	3 × 3	2 maps	1×1	valid	×	leaky ReLU
Reshape to 4 × 6 × 2						
Input from FH compartment 33 × 45 × 12						
Convolution 2D	33 × 45	32 maps	1×1	same	×	leaky ReLU
Convolution 2D	6 × 8	16 maps	1×1	valid	×	leaky ReLU
Convolution 2D	6 × 8	8 maps	1×1	valid	×	leaky ReLU
Convolution 2D	6 × 8	4 maps	1×1	valid	×	leaky ReLU
Convolution 2D	6 × 8	2 maps	1×1	valid	×	leaky ReLU
Reshape to 13 × 6 × 17						
Concatenation of EE and FH 490 × 1						
Linear	N/A	10 nodes	N/A	N/A	×	ReLU
Linear	N/A	1	N/A	N/A	×	×
Energy regressor output 1						

DNN Architectures Used in Sect. 10.4

Table G.2 Network architecture DNN_1, based on layer-wise evaluation, for the shower discrimination between electrons and pions within the EE compartment

Operation	Kernel	Features	Stride	Padding	Normalisation	Activation
Input from EE compartment 12 × 15 × 28						
Tower ×28 Convolution 2D	12 × 15	16 maps	1×1	valid	×	tanh
Concatenation 28 × 16						
Linear	N/A	1000 nodes	N/A	N/A	dropout	tanh
Linear	N/A	1	N/A	N/A	×	sigmoid
$e - \pi$ discriminator output 1						

Table G.3 Network architecture DNN_2, based on 2D convolutions, for the shower discrimination between electrons and pions within the EE compartment

Operation	Kernel	Features	Stride	Padding	Normalisation	Activation
Input from EE compartment $12 \times 15 \times 28$						
Convolution 2D	12×15	32 maps	1×1	same	$\times$	tanh
Convolution 2D	3×3	16 maps	1×1	valid	$\times$	tanh
Convolution 2D	3×4	8 maps	1×1	valid	$\times$	tanh
Convolution 2D	3×3	4 maps	1×1	valid	$\times$	tanh
Convolution 2D	3×3	2 maps	1×1	valid	$\times$	tanh
Reshape to $4 \times 6 \times 2$						
Linear	N/A	1000 nodes	N/A	N/A	dropout	tanh
Linear	N/A	1	N/A	N/A	$\times$	sigmoid
$e - \pi$ discriminator output 1						

Table G.4 Network architecture DNN_3, based on 3D convolutions, for the shower discrimination between electrons and pions within the EE compartment

Operation	Kernel	Features	Stride	Padding	Normalisation	Activation
Input from EE compartment $12 \times 15 \times 28 \times 1$						
Convolution 3D	$4 \times 6 \times 4$	32 maps	$1 \times 1 \times 1$	same	$\times$	tanh
Convolution 3D	$3 \times 6 \times 3$	16 maps	$2 \times 2 \times 2$	same	$\times$	tanh
Convolution 3D	$3 \times 6 \times 3$	8 maps	$2 \times 2 \times 2$	same	$\times$	tanh
Convolution 3D	$3 \times 6 \times 3$	4 maps	$1 \times 1 \times 2$	same	$\times$	tanh
Convolution 3D	$3 \times 6 \times 3$	2 maps	$1 \times 1 \times 2$	same	$\times$	tanh
Reshape to $4 \times 6 \times 2$						
Linear	N/A	1000 nodes	N/A	N/A	dropout	tanh
Linear	N/A	1	N/A	N/A	$\times$	sigmoid
$e - \pi$ discriminator output 1						

DNN Architectures Used in Chap. 11

Table G.5 Generator network as used to generate electromagnetic calorimeter showers from the benchmark dataset. The table is copied from Ref. [8]

Operation	Kernel	Features	Stride	Padding	Normalisation	Activation
Generator input 10 + 3						
Tower ×7						
Linear	N/A	10 nodes	N/A	N/A	×	leaky ReLU
Linear	N/A	192 nodes	N/A	N/A	×	leaky ReLU
Reshape to 3 × 4 × 16						
Transposed convolution 2D	3×3	16 maps	2×2	same	batch norm	leaky ReLU
Transposed convolution 2D	3×3	32 maps	2×2	same	batch norm	leaky ReLU
Transposed convolution 2D	3×3	64 maps	2×2	same	batch norm	leaky ReLU
Convolution 2D	5×9	1 maps	1×1	same	batch norm	leaky ReLU
Concatenation of 7 towers						
Convolution 2D	3×3	64 maps	1×1	same	batch norm	leaky ReLU
Convolution 2D	5×6	128 maps	1×1	valid	batch norm	leaky ReLU
Convolution 2D	3×3	14 maps	1×1	same	batch norm	leaky ReLU
Locally connected 2D	3×3	7 maps	1×1	same	×	ReLU
Generator output 12 × 15 × 7						

Table G.6 Critic network as used in the adversarial framework for the training with the benchmark dataset. The table is copied from Ref. [8]

Operation	Kernel	Features	Padding	Normalisation	Activation
Critic input 12 × 15 × 7 + 3					
Linear	N/A	10 nodes	N/A	×	leaky ReLU
Linear	N/A	180 nodes	N/A	×	leaky ReLU
Reshape to 12 × 15 × 1					
Concatenation with input to 12 × 15 × 8					
Convolution 2D	5×5	256 maps	same	×	leaky ReLU
Convolution 2D	3×3	128 maps	same	layer norm	leaky ReLU
Convolution 2D	3×3	64 maps	same	layer norm	leaky ReLU
Convolution 2D	3×3	32 maps	same	layer norm	leaky ReLU
Convolution 2D	3×3	16 maps	same	layer norm	leaky ReLU
Linear	N/A	10 nodes	N/A	×	leaky ReLU
Linear	N/A	1 node	N/A	×	×
Critic output 1					

Table G.7 Constrainer network as used in the framework for energy (position) regression with the benchmark dataset. The table is copied from Ref. [8]

Operation	Kernel	Features	Padding	Normalisation	Activation
Constrainer input 12 × 15 × 7 × 1					
Convolution 3D	3×3×3	1 map	same	batch norm	leaky ReLU
Convolution 3D	3×3×2	16 maps	same	batch norm	leaky ReLU
Convolution 3D	3×3×2	16 maps	same	batch norm	leaky ReLU
Convolution 3D	3×3×2	32 maps	same	batch norm	leaky ReLU
Convolution 3D	3×3×2	32 maps	same	batch norm	leaky ReLU
Convolution 3D	3×5×2	64 maps	same	×	leaky ReLU
Linear	N/A	1(2) node(s)	N/A	×	×
Constrainer output 1(2)					

References

1. CAEN (2010) Technical information manual, MOD. V1290 A/N, revision n. 11
2. CAEN (2011) Technical information manual, MOD. V1742, revision n. 0
3. Lorentz HA (1880) Ueber die Beziehung zwischen de r Fortpflanzungsgeschwindigkeit des Lichtes und der Körperdichte. Annalen der Physik 245:641–665. https://doi.org/10.1002/andp.18802450406
4. Lorenz L (1880) Ueber die Refractionsconstante. Annalen der Physik 247:70–103. https://doi.org/10.1002/andp.18802470905
5. Frank I, Tamm I (1991) Coherent visible radiation of fast electrons passing through matter. Springer, Berlin, pp 29–35. https://doi.org/10.1007/978-3-642-74626-0_2
6. Kleinwort C (2012) General broken lines as advanced track fitting method. Nucl Instrum Meth A 673:107–110. https://doi.org/10.1016/j.nima.2012.01.024
7. Patrignani C, et al (Particle Data Group) The review of particle physics. Chin Phys C 40:100001. https://doi.org/10.1088/1674-1137/40/10/100001
8. Erdmann M, Glombitza J, Quast T (2019) Precise simulation of electromagnetic calorimeter showers using a wasserstein generative adversarial network. Comput Softw Big Sci 3. https://doi.org/10.1007/s41781-018-0019-7